Signals and Communication Technology

Volume 182

More information about this series at http://www.springer.com/series/4748

Jari Nurmi · Elena Simona Lohan
Stephan Sand · Heikki Hurskainen
Editors

GALILEO Positioning Technology

Springer

Editors
Jari Nurmi
Elena Simona Lohan
Heikki Hurskainen
Department of Electronics and Communications Engineering
Tampere University of Technology
Tampere
Finland

Stephan Sand
German Aerospace Center DLR
Koeln
Germany

ISSN 1860-4862 ISSN 1860-4870 (electronic)
ISBN 978-94-007-1829-6 ISBN 978-94-007-1830-2 (eBook)
DOI 10.1007/978-94-007-1830-2

Library of Congress Control Number: 2014949337

Springer Dordrecht Heidelberg New York London

Printed on acid-free paper

Springer is part of Springer Science+Business Media (www.springer.com)

Contents

Part I Introductory Notions

1 **Introduction** 3
Jari Nurmi, Elena Simona Lohan, Stephan Sand and Heikki Hurskainen

2 **Overview of Galileo System** 9
Javier Pérez Bartolomé, Xavier Maufroid, Ignacio Fernández Hernández, José A. López Salcedo and Gonzalo Seco Granados

3 **Galileo Signals** 35
Elena Simona Lohan, Heikki Hurskainen and Jari Nurmi

Part II Receiver, the Core Galileo Positioning Technology

4 **Advanced RF Front-End Design Issues** 59
Ernesto Perez and Marco Detratti

5 **Advanced Acquisition and Tracking Algorithms** 85
Elena Simona Lohan

6 **Baseband Hardware Implementations for Galileo Receiver** 121
Heikki Hurskainen and Jari Nurmi

7 **PVT Computation Issues in Mixed Galileo/GPS Reception** 139
Tommi Paakki, Francescantonio Della Rosa and Jari Nurmi

8 **Hybridization with Localization Information from Wireless Communications Systems** 169
Stephan Sand

9 Putting It All Together: TUTGNSS Receiver Platform 209
Jussi Raasakka and Heikki Hurskainen

10 Fully Software Implementations for Galileo Receiver 229
F. Dovis, M. Fantino, M. Pini and A. Molino

Part III Advanced Galileo Positioning Technologies

11 Galileo Receiver Research in Europe 249
Ignacio Fernández Hernández

12 Experimental and Professional Galileo Receivers 273
Andrew Simsky and Jean-Marie Sleewaegen

13 Software Simulators and Multi-Frequency Test Scenarios for GALILEO 289
Sarang Thombre and Jari Nurmi

14 An Introduction to EGNOS: The European Geostationary Navigation Overlay System 323
Javier Ventura-Traveset, Carlos López de Echazarreta, Juan-Pedro Lam and Didier Flament

15 Issues in Implementing the Galileo Safety-of-Life Service 359
Boubeker Belabbas

16 GIMS: Galileo Interference Management System 373
Philippe Morgand and Anne Ferreol

17 Conclusions and Outlook 401
Jari Nurmi, Elena Simona Lohan, Stephan Sand and Heikki Hurskainen

Erratum to: GIMS: Galileo Interference Management System E1
Philippe Morgand, Eugenio Rossini and Anne Ferreol

Abbreviations

3GPP	3rd Generation Partnership Project
ACF	Auto-Correlation Function
ADAS	Advanced Driving Assistance Systems
ADC	Analog to Digital Converter
AltBOC	Alternate Binary Offset Carrier
AM	Amplitude Modulation
AOA	Angle of Arrival
AP	Access Point
API	Application Programming Interface
APME	A-Posteriori Multipath Estimation
APNT	Alternate Position, Navigation and Time
ARAIM	Advanced Receiver Autonomous Integrity Monitoring
ART-X	Advanced Receiver Terminal Extension to Technology and Market Evolution
ASQF	Application Specific Qualification Facility
ATENA	Advanced Techniques for Personal Navigation (Tekes project)
ATENEA	Advanced Techniques for Navigation Receivers and Applications
AU	Acquisition Unit
AV	Allan Variance
AWGN	Additive White Gaussian Noise
BCCH	Broadcast Control Channel
BCU	Baseband Converter Unit
BER	Bit Error Rate
BOC	Binary Offset Carrier
BPSK	Binary Phase Shift Keying
BS	Base Station
BSIC	Base Station Identity Code
BW	Bandwidth
C/A	Coarse Acquisition
CBOC	Composite Binary Offset Carrier
CCDF	Complementary Cumulative Distribution Function
CCF	Central Control Facility
CDDS	Commercial Data Distribution Service

CDMA	Code Division Multiple Access
CE	Constant Envelope
CEP	Circular Error Probability
CIR	Channel Impulse Response
CNES	French space agency
CNR	Carrier to Noise Ratio
CosBOC	Cosine Binary Offset Carrier
CPF	Central Processing Facility
CPICH	Common Pilot Channel
CPU	Central Processing Unit
CRC	Cyclic Redundancy Check
CRLB	Cramér–Rao Lower Bound
CRMM	Complexity Reduced Multipath Mitigation
CW	Continuous Wave
DAT	Differential Augmentation Terminals
DGNSS	Differential Global Navigation Satellite System
DLL	Delay Locked Loop
DOA	Direction of Arrival
DOP	Dilution of Precision
DS-CDMA	Direct Sequence Code Division Multiple Access
DSP	Digital Signal Processor
DS-SS	Direct Sequence-Spread Spectrum
DP	Dot Product
EC	European Commission
ECEF	Earth-Centered Earth-fixed
EDAS	EGNOS Data Access Service
EEIG	European Economic Interest Grouping
EGEP	ESA GNSS Evolutions Programme
EGNOS	European Geostationary Navigation Overlay Service
ELS	Early Late Slope
EM	Expectation Maximization
EML	Early Minus Late
EMS	EGNOS Message Server
ESA	European Space Agency
ESPRIT	Estimation of Signal Parameters via Rotational Invariance Techniques
ESSP	European Satellite Services Provider
EU	Europe
EUROCAE	European Organisation for Civil Aviation Equipment
EWAN	EGNOS Wide Area Network
FAA	Federal Aviation Administration
FAU	Fast Acquisition Unit
FCH	Frequency CHannels
FDIM	Frequency Domain Interference Mitigation

FDMA	Frequency Division Multiple Access
FDIM	Frequency Domain Interference Mitigation
FDOA	Frequency Difference of Arrival
FE	Front End
FEC	Forward Error Correction
FFT	Fast Fourier Transform
FIMLA	Fast Iterative Maximum Likelihood Algorithm
FLL	Frequency Locked Loop
FM	Frequency Modulated
FP	Framework Programme
FPGA	Field Programmable Gate Array
FPGA	Field Programmable Gate Array
GAGAN	GPS Aided Geo Augmented Navigation (India)
GAMMA-A	Assisted Galileo/GPS/EGNOS Mass Market Receiver
GARDA	Galileo user Receiver Development Activity
GATE	GAlileo TErrestrial test facility
GBAS	Ground Based Augmentation System
GDDN	Galileo Data Dissemination Network
GDOP	Geometrical Dilution of Precision
GEO	Geostationary Orbit
GETR	Galileo Experimental Test Receiver
GGTO	Galileo/GPS Time Offset
GIM	Global Lonospheric Map
GIMS	Galileo Interference Management System
GIVD	Grid-point-Ionospheric Vertical Delay
GIVE	Grid-point-Ionospheric Vertical Error Bounding
GJU	Galileo Joint Undertaking
GLONASS	Russian GNSS system
GMS	Global System for Mobile Communication
GMU	Galileo Memory Unit
GNSS	Global Navigation Satellite System
GPS	Global Positioning System
GR	Galileo Receiver
GRAMMAR	Galileo Ready Advanced Mass MArket Receiver (EU FP7 project)
GREAT	Galileo Receivers for Mass Market (EU FP6 project)
GSA	European GNSS Agency
GSS	Galileo Sensor Station
GTK	Generalized Teager-Kaiser
GTRF	Galileo, Galileo Terrestrial Reference Frame
HAL	Horizontal Alert Limit
HDF	Hybrid Data Fusion
HPL	Horizontal Protection Limit
HRC	High Resolution Correlator
I	In phase
I&D	Integrate and Dump

ICAO	International Civil Aviation Organization
ICC	Interference Control Centre
ICI	Intercell Interference
IF	Intermediate Frequency
IFFT	Inverse Fast Fourier Transform
IID	Independently and Identically Distributed
IODE	Issue of Data Ephemeris
IONEX	Ionospheric Exchange
IOS	Interference Observation Site
IRI	International Reference Lonosphere Model
ISM	Integrity Support Message
ISR	Interrupt Service Routine
ITRF	International Terrestrial Reference Frame
KF	Kalman Filter
LAA	Local Augmentation Area
LAAS	Local Augmentation Area System
LBS	Location Based Service
LEN	LENgth (of the observed time interval)
LFSR	Linear Feedback Shift register
LIM	Local Integrity Monitoring
LOS	Line of Sight
LS	Least Square
LTE	Long Term Evolution
LUT	Look Up Table
M&C	Monitoring and Control
MAC	Medium Access Control
MAP	Maximum A Posteriori estimate
MBOC	Multiplexed Binary Offset Carrier
MCC	Mission Control Center
MCRV	Modified Correlator Reference Waveforms
MCU	Master Control Unit
MEDLL	Multipath-Estimating Delay Locked Loop
MEE	Multipath Error Envelope
MET	Multipath Elimination Technique
MGD	Multiple Gate Delay
MIMO	Multiple-Input Multiple-Output
ML	Maximum Likelihood
MMSE	Minimum Mean Square Error
MMT	Multipath Mitigation Technology
Mod	Modulation
MOPS	Minimum Operational Performance Standards
MOT	Mobile Positioning (Tekes project)
MSAS	Multi-transport Satellite based Augmentation System
MUSIC	Multiple Signal Classification
MUX	Multiplexing

NCE	Non-Constant Envelope
NCO	Numerically Controlled Oscillator
NEML	Narrow Early Minus Late
NFC	Near Field Communication
NLOS	Non Line of Sight
NMEA	National Maritime Electronics Association
NOB	Number of Bursts
NSE	Navigation System Error
OCXO	Oven Compensated Oscillator
ODTS	Orbit Determination and Time Synchronization
OEM	Original Equipment Manufacturer
OFDM	Orthogonal Frequency Division Multiplexing
OS	Open Service
P(Y)	Precise (Encrypted) GPS code
PA	Power Amplifier
PAC	Pulse Aperture Correlator
PACF	Performance Assessment and Check-out Platform
PD	Probability of Detection
PDF	Probability Distribution Function
PFA	Probability of False Alarm
PIT	Predetection Integration Time
PLL	Phase Locked Loop
PM	Phase Modulated
PMU	Phase Measurement Unit
PND	Personal Navigation Device
POCS	Projection Onto Convex Sets
PPP	Precise Point Positioning
PPS	Pulse Per Second
PRECISIO	Professional Receivers via Software Radio
PRF	Pulse Repetition Frequency
PRN	Pseudorandom Noise
PRS	Public Regulated Service
PSD	Power Spectral Density
PW	Pulse Width
PVT	Position Velocity and Time
Q	Quadrature phased
RAAN	Right Ascension of Ascending Node
RAE	Running Average Error
RAMS	Reliability Availability, Maintainability and Safety
RF	Radio Frequency
RHCP	Right-Hand Circularly Polarised
RIMS	Ranging and Integrity Monitoring Stations
RINEX	Receiver Independent Exchange
ROM	Read-Only Memory
RSS	Received Signal Strength

RSSML	Reduce Search Space Maximum Likelihood
RTCA	Radio Technical Commission for Aeronautics
RTCM	Radio Technical Commission for Maritime services
RTK	Real-Time Kinematic
RTOS	Real-Time Operating System
RUT	Receiver Under Test
SAGE	Space-Alternating Generalized Expectation maximization
SARP	Standards and Recommended Practices
SBAS	Satellite Based Augmentation System
SBME	Slope Based Multipath Estimator
SCM	Sidelobe Cancellation Method
SCPC	Sub-Carrier Phase Cancellation
SCU	Sample Counter Unit
SDCM	System of Differential Correction and Monitoring (Russia)
SDD	Service Definition Document
SDIM	Spatial Domain Interference Mitigation
SDR	Software Defined Radio
SESAR	Single European Sky Air Traffic Management Research
SH	Spherical Harmonics
SIL	Sub/Integration Length
SinBOC	Sine Binary Offset Carrier
SISA	Signal-in-Space Accuracy
SISE	Signal in Space Error
SIS-ICD	Signal in Space—Interface Control Document
SISMA	Signal-in-Space Monitoring Accuracy
SLAM	Simultaneous Localization And Mapping
SNAS	Satellite Navigation Augmentation System (China)
SNR	Signal to Noise Ratio
SoL	Safety of Life
SPI	Serial Peripheral Interface
SS	Spread Spectrum
SWIRLS	Galileo Professional Receiver Development
TCAR	Triple Carrier Ambiguity Resolution
TCXO	Temperature Compensated Oscillator
TDOA	Time Difference of Arrival
TEC	Total Electron Content
Tekes	Finnish Funding Agency for Technology and Innovation
TK	Teager-Kaiser
TMBOC	Time Multiplexed BOC
TOA	Time of Arrival
TOW	Time of week
TROPEX	Tropospheric Exchange
TRP	Technology Research Programme
TRU	Tracking Result Unit
TTA	Time to Alert

TTFF	Time To First Fix
TU	Tracking Unit
TUR	Test User Receiver
UAL	Unsuppressed Adjacent Lobes
UDRE	User Differential Range Error
UERE	User Equivalent Range Error
UERRE	User Equivalent Range Rate Error
UMTS	Universal Mobile Telecommunication System
URA	User Range Accuracy
WAAS	Wide Area Augmentation System
WAD	Wide Are Differential
VAL	Vertical Alert Limit
WBMOD	Wide Band model
WCDMA	Wideband CDMA
WGS-84	World Geodetic System 84
VHDL	VHSIC Hardware Description Language
WLAN	Wireless Local Area Network
WN	Week Number
VPE	Vertical Positioning Error
VPL	Vertical Protection Limit
XOR	Exclusive OR

Part I
Introductory Notions

Chapter 1
Introduction

Jari Nurmi, Elena Simona Lohan, Stephan Sand and Heikki Hurskainen

We are living in an era of significant and fast changes in the field of Global Navigation Satellite Systems (GNSS). The US Navstar Global Positioning System (GPS) is being modernized, Europe is building its own, civilian GNSS, named Galileo, China is launching Compass/Beidou 2 GNSS with both regional and global components and Russia has finalized the deployment of its renewed "Globalnaya navigatsionnaya sputnikovaya sistema" or GLONASS, by launching the 24th satellite of the constellation in 2011. While about a decade ago, the world satellite navigation was mainly on the shoulders of Navstar GPS, the situation has now changed and it is forecast that a decade from now, the world will benefit of more than 100 navigation satellites in the sky. In addition, regional Satellite-Based Augmentation Systems (SBAS) are being built in order to support and enhance the satellite positioning given by the GNSS. In Europe, such an SBAS is the European Geostationary Navigation Overlay Service (EGNOS), operated by the European Space Agency.

Interference issues in GNSS are also becoming more and more challenging, not only because of the increased constellation of the satellites and inter-system interferences, but also because of unintentional interferences from adjacent frequency bands (Boulton et. al. 2011) and from intentional interferences such as jamming (narrowband interfering signal), spoofing (false signals), or meaconing (delaying and rebroadcasting) (Bull 2010; Motella et al. 2008).

J. Nurmi · H. Hurskainen
Department of Electronics and Communications Engineering,
Tampere University of Technology, Tampere, Finland

E.S. Lohan (✉)
Department of Electronics and Communications Engineering,
Tampere University of Technology, Tampere, Finland
e-mail: elena-simona.lohan@tut.fi

S. Sand
German Aerospace Center DLR, Oberpfaffenhofen , Germany

J. Nurmi et al. (eds.), *GALILEO Positioning Technology*,
Signals and Communication Technology 182, DOI 10.1007/978-94-007-1830-2_1

Table 1.1 Galileo milestones

Year	Milestones
1998	EU decides to develop its own satellite navigation system, Galileo
2000	Definition phase of Galileo starts
2002	Statement on GPS-Galileo cooperation; establishment of Galileo joint undertaking (GJU) for the management and concession of Galileo
2003	Joint statement between European commission and US regarding GPS-Galileo
2004	EU and US signed GPS-Galileo agreement and an official European union regulatory authority, the European GNSS supervisory authority (GSA) is established
2005	First Galileo test satellite launched on orbit (GIOVE-A)
2006	GSA replaces GJU completely; First Galileo standardization documents for open space (OS) signal are made available
2007	MBOC modulation adopted for common GPS-Galileo signal for civilian use
2008	Second Galileo test satellite (Giove-B) launched on orbit; GIOVE-A + B signal in space (SIS) interface control documentation (ICD) published (GALILEO SIS-ICD 2008)
2009	EGNOS OS is declared operational
2010	Issue 1.1 of Galileo OS SIS ICD released (GALILEO SIS-ICD 2010)
2011	First two IOV Galileo satellites launched (in Oct); European parliament approved detailed rules governing the future "Public Regulated Service" of Galileo. A drawing competition for children is launched for naming the Galileo satellites (Galileo competition 2012)
2012	3rd and 4th Galileo IOV satellites are launched (Nov) and all three Galileo bands are being broadcast from Galileo IOV-3 (Dec); also EGNOS satellite was successfully launched (Jul) and Galileo Pathfinder GIOVE-A Retires (Jul)

Compared with its 'older' siblings, GPS and GLONASS, that have initially emerged in the late 1970s (then GLONASS was not maintained and it had to be renewed in the last decade), the Galileo system is a relatively new system, and its basic concepts were started to be built around the end of the 1990s. In 2003, an official agreement between the European Union and the European Space Agency was drawn regarding the first stage of Galileo. This marked the completion of the first definition stage of Galileo, followed by an US-EU agreement in 2004 regarding common civil signals. In there, the baseline modulation for Galileo and modernized GPS signals has been agreed upon as being the sine Binary Offset Carrier modulation, but was later (in 2007) changed into the Multiplexed Binary Offset Carrier modulation. The first Galileo standardization documents became available in May 2006, and the updates are still in progress, as of 2013. The first Galileo test satellites were launched in 2005 and 2008, respectively, followed in 2011 by the launch of the first two In-Orbit Validation (IOV) Galileo satellites. Table 1.1 shows the main milestones in Galileo history. The name of the European GNSS has been chosen to honor the Italian astronomer, scientist, mathematician and philosopher Galileo Galilei (Kreye et al. 2006).

Motivation

As the Galileo system is currently built, a book dedicated solely to the Galileo system is still missing. It is our target to fill this gap with the present book, born as a joint effort of several European scientists participating within the European-funded project "Galileo Ready Advanced Mass MArket Receiver" (GRAMMAR, 2009—2011) and affiliated institutions.

Currently, several GNSS related books dedicate one or few Chapters to Galileo system description. For example, in (Hofmann-Wellenhof et al. 2008) there is a 33-page Chapter dedicating with Galileo, touching mainly the Galileo architecture (space and segments) and Galileo signals. An additional sub-section deals with a comparison between GPS, GLONASS and Galileo.

In (Strang and Borre 2012), there is a whole chapter dedicated to Galileo modulations, but the main focus on the book is on GPS. (Borre et al. 2006) deals more extensively with the baseband receiver blocks in GPS and Galileo single frequency receivers (focus on L1 band of GPS and E1 band of Galileo).

In (Xu 2007; Samama 2008; Gleason and Gebre-Egziabher 2010) brief discussions about developments of Galileo can be found.

Our book covers multi-band Galileo receivers (especially E1-E5 bands of Galileo) and addresses all receiver building blocks starting from the antenna and front end, going through details of the baseband receiver processing blocks and up to the navigation processing (including the Galileo message structure and Position Velocity, Time (PVT) computation). Moreover, hybridization solutions with communications systems for a better localization are discussed and an open-source GNSS receiver platform developed at Tampere University of Technology (TUT) is addressed in detail and it will be later made available at TUT webpages. Both hardware and software solutions are discussed for the Galileo receivers. A chapter dedicated to EGNOS is also included in the book, motivated by the fact that EGNOS is to be integrated within Galileo and it is an integral part of the European satellite radio-navigation policy (Delfour et al. 2012).

The Emerging GNSS Landscape

Currently there are four Global Navigation Satellite Systems: Galileo (Europe), GPS (US), GLONASS (Russia) and Compass or Beidou 2 (China). Two of them are fully operational (GPS and GLONASS) and the other two are currently being deployed. Each of them, when reaching the full operational capability, is able to operate in a stand-alone mode and to offer global coverage, and several agreements have been reached regarding the inter-compatibility between these systems, in such a way that multi-system receivers will be able to offer an improved navigation solution for the end user. The next sub-sections briefly summarize the constellations and status of deployment of the four GNSSs.

Galileo

Galileo system, an initiative of European Union, has currently four satellites in the sky from the final orbit constellation of 27 satellites (plus 3 spare). The full operation capability of Galileo is currently predicted for the end of 2019. Galileo satellites are placed in 3 orbital planes, at 23,222 km altitude, which means that they are executing 17 orbits in 10 days (i.e., an orbital period of 14 h and 5 min). It is to be noted that the Galileo altitude is 3,022 km higher than the GPS altitude. The orbital inclination is 56 degrees, 1 degree higher than in the GPS case. The first two IOV Galileo satellites, Thijs and Natalia, were launched in October 2011, and the third and fourth IOV Galileo satellites, David and Sif, were launched in October 2012.

GPS

The Navstar GPS system is the oldest among all its GNSS siblings and it has been up and running almost for 2 decades. Its current constellation consists of 31 satellites (as of June 2013) from four generations of satellites, namely blocks IIA (8 operational satellites), IIR (12 operational satellites), IIR-M (8 operational satellites), and IIF (3 operational satellites). None of the original Block I and Block II satellites is currently operational anymore. The first Block IIA satellite was launched in November 1990, while the first satellite in Block IIF was launched on May, 2010. The last GPS satellite (the 32nd one), still under commissioning phase, was launched in May 2013. GPS satellites are placed into 6 nearly circular orbital planes (meaning an orbital period of 11 h and 58 min per satellites).

Glonass

GLONASS system is also fully operational since November 2011. It consists of 24 satellites, placed in 3 orbital planes with 64.8 degrees inclination, at an altitude of 19,100 km above the Earth. The satellite orbital period is 11 h and 15 min. By distinction with all the other three GNSS, GLONASS is the only one using the Frequency Division Multiple Access (FDMA) technique to make the separation between satellites. The other three GNSS rely on Code Division Multiple Access (CDMA) technique, as it will be described later in this book. A CDMA component is also forecast for GLONASS in the years to come. Currently, all GNSS manufacturers selling GNSS receivers on Russian territory support both GPS and GLONASS at the receiver side.

Compass/Beidou-2

The Compass or Beidou-2 GNSS is being developed by China and has a regional and a global component. The regional component, initially named Beidou-1 (the name allegedly referring to the brightest star in 'Ursa Major' constellation) has been deployed in 2000 and was formed only of Geostationary Earth Orbit (GEO) satellites (Boon et al. 2007). It is to be noted that the global coverage can be achieved through Medium Earth Orbit (MEO) satellites, such as those used in Galileo, GLONASS, GPS and partly in Compass/Beidou-2. Later on, Beidou-1 was renamed to Compass or Beidou-2 and a global component has been added. The final Compass/Beidou-2 constellation will consist of 24 MEO satellites, 5 GEO satellites and 5 Inclined Geosynchronous Earth Orbit (IGSO) satellites. The MEO satellites are placed at around 21500 km above the Earth, in 3 orbital planes. The first full Interface Control Document about Compass/Beidou-2 was published in Dec 2012.

Multi-Frequency Multi-Systems Receivers of Tomorrow

With the advent of multiple satellite constellations, coming from the four existing GNSS, and supporting SBAS, the GNSS receivers of tomorrow not only have to support the multi-frequency multi-system operations, but they also have to deal with interference challenges and the challenges related to low-cost implementations for mass market purpose. Some of our chapters in the book address the issue of dual-frequency dual system (Galileo-GPS) receivers.

Moreover, the smartphones are increasingly using assisted GNSS (van Diggelen 2009) and hybrid localization techniques, thus when we talk about multi-system, we refer not only to multiple GNSS systems, but also to systems belonging to very different categories, such as cellular Wideband CDMA (WCDMA) (Muratore 2001), cellular Long Term Evolution (LTE) (Dahlman et al. 2007) and GNSS.

Book Structure

The book structure evolved from the core research interests of editors; receiver architectures, multipath mitigation solutions and combined navigation and telecommunication technologies. Having their knowledge available the editors asked recognized contributors in field to enhance the message about Galileo receiver technologies and also to give valuable insight of other Galileo related programs.

We divided the book into three parts: Part 1 "Introductory notions" deals with an overview of Galileo system and with the Galileo signal and frequencies. Part 2 "Receiver, the core Galileo technology" discusses the three main building blocks of a Galileo receiver: front-end, baseband processor and navigation unit, it presents both the hardware and software solutions for Galileo receivers, and it addresses the hybridization with communication systems, that is likely to be a must in the future cellular devices. The TUTGNSS open-source Galileo-GPS receiver is also presented in this part. Part 3 "Advanced Galileo positioning technologies" starts with an overview of Galileo receiver-related research in Europe and then goes through various Galileo-related issues, such as simulators and multi-frequency test scenarios for Galileo, an EGNOS overview as a complementary European navigation program to Galileo, issues in Safety Of Life (SoL) implementation in Galileo (a service that currently has been discontinued from Galileo), and Galileo Interference management system (GIMS).

References

Boon F, Wilms F, Sleewaegen JM, de Wilde W (2007) More compass points: tracking China's MEO satellites on a hardware receiver. Inside GNSS, Jul/Aug 2007. http://www.insidegnss.com/node/157

Borre K, Akos DM, Bertelsen N, Rinder P, Jensen SH (2006) A software-defined GPS and Galileo receiver: a single-frequency approach. Birkhäuser, Bosten

Bouloton P, Borsato R, Butler B, Judge K (2011) GPS interference testing: lab, live and lightsquared, Inside GNSS, Jul/Aug 2011. http://www.insidegnss.com/node/2674

Bull T (2010) A new high performance way of detecting and mitigating the jamming meaconing and spoofing of commercial GNSS signals. In: 2010 5th ESA workshop on satellite navigation technologies and European workshop on GNSS signals and signal processing (NAVITEC), pp 1, 5, 8—10 Dec 2010, doi:10.1109/NAVITEC.2010.5708050

Dahlman E, Parkvall S, Sköld J, Beming P (2007) 3G evolution: HSPA and LTE for mobile broadband. Elsevier, Amsterdam

Delfour H, Gicquel JA, Larhantec P, Joly D, Lekaim D, Jeannot M, Melinotte JM (2012) Support platform for EGNOS evolution and demonstration (SPEED): operational SBAS test bed as real as EGNOS. 6th ESA workshop on satellite navigation technologies and European workshop on GNSS signals and signal processing, (NAVITEC), pp 1—8, Dec 2012, doi:10.1109/NAVITEC.2012.6423046

European GNSS (2010) (GALILEO) Open service—signal in space interface control document, Feb 2010

Galileo GNSS Through Children's Eyes (2012) Inside GNSS, industry view column. http://www.insidegnss.com/node/2906

Galileo Open Service (2008) Signal in space interface control document (OS SIS ICD). European Space Agency/European GNSS Supervisory Authority (2008)

Gleason S, Gebre-Egziabher D (2010) GNSS applications and methods. Artech House, Norwood

Hofmann-Wellenhof B, Lichtenegger H, Wasle E (2008) GNSS—global navigation satellite systems GPS, GLONASS, Galileo, and more. Springer, New York

Kreye C, Niedermeier H, Heyen R, Stelkens-Kobsch T, Boedecker G (2006) Galileo and the earth's gravity field: using GNSS for airborne gravimetry—an overview. Inside GNSS, Nov/Dec 2006. http://www.insidegnss.com/node/689

Motella B, Savasta S, Dargaria D, Dovis F (2008) An interference impact assessment model for GNSS signals. In: Proceedings of the ION GNSS conference. Savannah, Georgia

Muratore F (2001) UMTS-mobile communications for the future. Wiley, Chichester

Samama N (2008) Global positioning-theory and performance. Wiley, New York

Strang G, Borre K (2012) Algorithms for global positioning. Wellesley-Cambridge Press, Wellesley

van Diggelen F (2009) A-GPS: Assisted GPS, GNSS and SBAS. Artech House, Boston

Xu G (2007) GPS-theory, algorithms and applications. Springer, Berlin

Chapter 2
Overview of Galileo System

Javier Pérez Bartolomé, Xavier Maufroid, Ignacio Fernández Hernández, José A. López Salcedo and Gonzalo Seco Granados

Abstract This chapter provides an introduction to the Galileo program and architecture. It starts by presenting the program context, rationale and history, including the early definition phases and test beds and the GIOVE experimental satellites. It then presents an overview of the Galileo services. Later, an architectural overview is provided, including the Galileo segments: the Space Segment, the Ground Mission Segment, and the Ground Control Segment. The chapter also provides a description of Galileo's contribution to the Search And Rescue services through COSPAS/SARSAT, and finalizes with an overview of the user segment and highlighting interoperability and compatibility issues with other GNSS.

Galileo Program Context, History and Implementation Phases

Institutional Context of the Galileo Program

As the reader can imagine, the process of developing Galileo has not been easy. In exchange, the Galileo program is, apart from a satellite navigation system whose signals will be received worldwide, the evidence that different sovereign nations can collaborate for the benefit of the EU and the world. Galileo has pioneered the cooperation in EU space industries to develop a highly complex operational and globally distributed infrastructure.

The context of the European Union under which Galileo has been conceived and developed is a very important aspect that has driven the program significantly. Saying that Galileo has been developed by the twenty-seven Member States that

J.P. Bartolomé (✉) · X. Maufroid · I.F. Hernández
European Commission, DG Enterprise and Industry, Brussels, Belgium
e-mail: Javier.PEREZ-BARTOLOME@ec.europa.eu

J.A. López Salcedo · G.S. Granados
Universitat Autònoma de Barcelona, Barcelona, Spain

J. Nurmi et al. (eds.), *GALILEO Positioning Technology*,
Signals and Communication Technology 182, DOI 10.1007/978-94-007-1830-2_2

form the European Union would already be a simplification of the facts. Galileo is therefore different to any other system like GPS, GLONASS and Beidou, which are developed under a single nation (the United States of America, the Russian Federation and the People's Republic of China, respectively). Perhaps this does not represent a difference from the technical point of view or from a usage perspective, but it represents a big challenge for the program. The program management had to reconcile the views of many sovereign states, liaise with independent intergovernmental bodies, and deal with intricate decision processes.

The European Union is the main sponsor and the owner of Galileo, according to the current regulation (Europeon Commission 2008). Whereas the program was financed jointly with ESA in the first stages, its full deployment has been financed with European Union budget funds. The European Commission, as the executive arm of the European Union, is the program manager. As any other undertakings of the European Union, the EU Member States are the ultimate stakeholders of the program and play a major role in the decision-making process. The European Commission reports to them regularly on the program developments through a dedicated forum called the GNSS Programmes Committee. Many European Member States also support the Galileo program through dedicated activities performed by their national space agencies, as the French Centre National d'Etudes Spatiales (CNES), the German Deutsches Zentrum für Luft-und Raumfahrt (DLR), the Italian Agenzia Spaziale Italiana (ASI) or the recently created National Space Agency in the UK, just to name some. Other national ministries, as the Transport or Defence ones, are also involved. More details about the EU functioning and its Member States can be found in Europeon Commission (2007) .

The European Space Agency has the leading role in the technical direction of the program. After pioneering research in satellite navigation in Europe in the late 1990s and successfully demonstrating the program's technical feasibility over the first decade of the 21st century, ESA is currently responsible for the development, deployment, integration and operational validation of the Galileo system infrastructure.

As the European Commission's main activities are related to policymaking and EU budget administration, as part of its normal functioning it is assisted by Community Agencies which deal with more specific tasks. In case of the GNSS programs, the agency in charge of assisting the EC is the GSA (former GNSS Supervisory Authority, now European GNSS Agency). The role of the GSA and the EC have slightly evolved over the years, and at the time of writing the GSA is foreseen to be in charge of the Galileo service provision and exploitation, market development and security.

The Early Stages

In the early nineties, the European Union started to consider the development of its own satellite navigation system, first through the deployment of a regional infrastructure, called GNSS-1 at the time, and later by developing its own global system,

GNSS-2. A formal agreement was concluded on 18 June 1996 between the European Community, Eurocontrol and ESA for the development of GNSS-1, which would become European Geostationary Navigation Overlay Service (EGNOS), an SBAS or satellite-based augmentation system aimed at augmenting GPS to improve air navigation operations. GNSS-2 would later become Galileo. In 1995, at the time the GNSS-2 program was being outlined, GPS had just declared its Full Operational Capability, and the U.S. Government had already committed to provide the GPS signals to the civil user community (Kaplan and Hegarty 2006; Pace et al. 1995), although the selective availability, a functionality to intentionally degrade GPS position accuracy for unauthorized users, was still on. In this context, the European Commission's Communication "Towards a Trans-European Positioning and Navigation Network—including a European strategy for GNSS" (Kaplan and Hegarty 2006) issued in 1998 formally opened a serious debate for Europe to develop its own GNSS.

In 1999, the European Commission, with the support of the European Space Agency (ESA), prepared the Communication "Galileo—Involving Europe in a New Generation of Satellite Navigation services" (Europeon Commission 1999) that was seminal to the GNSS program development in Europe. Three reasons for the EU to develop its own system were stated in this Communication:

- To increase control on satellite-based safety-critical navigation systems.
- To ensure a positioning service for European users in the long term, not subject to the risk of potential U.S. policy changes affecting GPS.
- To support EU industry competitiveness in the global market of satellite navigation and grant access to the system's technological developments.

The Communication presented the results of consultations with worldwide stakeholders to define how the EU satellite navigation system would look alike. Influenced by ESA GNSS-2 Comparative System Studies (ESA 1998) and EC GNSS-2 Forum (Fairbanks 1999), it proposed to develop a system very similar to GPS or GLONASS to minimize technical risk and provide the highest value for money. This included a signal structure compatible and interoperable with GPS as much as possible.

Once the political drive to build Galileo was clear, and its basic principles outlined, the EC and ESA embarked in the first studies to define the Galileo mission and system requirements that would ultimately determine the Galileo system and receiver technologies. The EC formed the Galileo Task Force (GTF) and launched the GALA project to define the future Galileo service levels and receiver functional concepts, and ESA launched the GalileoSat program to support the definition of ground and space infrastructure and study the signal design and transmission performance (Schweikert et al. 2000; De Gaudenzi et al. 2000).

In parallel, a considerable effort in international cooperation and research was carried out in the early 2000s to agree on the spectrum allocation, carrier frequencies selection, signal design, code selection and timing and geodetic references for Galileo, in order to make Galileo and GPS, the only operational system at the time, as compatible and interoperable as possible. In the frame of the ONU International Telecommunications Union (ITU), these efforts eventually led to the allocation of the Galileo frequency plan in June 2000 by the World Radio Communications Conference held at Istanbul, granting protection until June 2006. Some years later, in 2004, the EU and the U.S. signed a Cooperation Agreement "On The Promotion, Provision And Use Of Galileo And GPS Satellite-Based Navigation Systems And Related Applications" (United States of America and European Community 2004) that set the framework to achieve full interoperability and radio frequency compatibility between both systems.

Galileo Early Technology Demonstrator

Several years of development and qualification of critical technologies have been necessary for the deployment of an operational system like Galileo. This is particularly true for the satellite on-board clocks. In the late 1990s, the European Space Agency started the development of the Rubidium Atomic Frequency Standard (RAFS) and the Passive Hydrogen Maser (PHMs) that would be integrated in the satellites and, in early 2000s, these two technologies were environmentally qualified on-ground.

Later on, ESA launched in 2002 the GSTB-V1 (Galileo System Test Bed 1) program aimed at developing an experimental ground mission segment for the validation of Galileo navigation and integrity determination algorithms and products based on raw GPS measurements collected by a global network of sensor stations.

In 2003, the European Space Agency began the development of two test satellites, GIOVE-A and GIOVE-B (Galileo In-Orbit Validation Element), as part of the GSTB-V2 (Galileo System Test Bed 2) program.

The GIOVE-A satellite was built by SSTL and launched in December 2005. The satellite launch mass was about 600 kg for a total power of 700 W. The satellite was designed to transmit simultaneously 2 out of the 3 frequencies allocated to Galileo and included 2 cold redundant RAFS with a stability of 10 ns per day. A major Galileo program milestone was achieved on the 12th January 2006, when GIOVE-A transmitted for the first time a Galileo-like signal in space (SIS) towards the Earth from orbit, several months in advance of the ITU filing protection expiration date.

The GIOVE-B satellite was built by a consortium led by Astrium GmbH and was launched in April 2008. The satellite launch mass was about 530 kg for a total power of 1100 W. The satellite included 1 PHM and 2 RAFS. It was also the first satellite to transmit the Multiplexed Binary Offset Carrier modulation (MBOC), the latest signal waveform agreed between the European Union and the United States.

Both satellites were initially designed for a lifetime of about 2 years but thanks to very good performances at the end of their lifetime, their missions were extended. GIOVE-A satellite lasted more than 6 years in-orbit before being finally decommissioned on 30 June 2012. GIOVE-B was decommissioned on 23 July 2012 (Fig. 2.1).

Fig. 2.1 GIOVE-A (*left*) and GIOVE-B (*right*), artist's impression, courtesy of ESA

Galileo Implementation Phases

During the early definition phase, a staggered approach was adopted for the development, deployment, integration and validation of the Galileo system infrastructure. Two major implementation phases were considered, namely:

1. The In-Orbit Validation phase (IOV),
   whose main goal was the end-to-end validation of the Galileo service concept based on a mini constellation with four operational Galileo spacecraft and a limited ground system configuration.
2. The Full Operational Capability phase (FOC),
   intended to complete the deployment of the Galileo constellation and ground infrastructure and achieve full operational validation and service performance.

More details on the satellites and ground infrastructure of Galileo IOV and FOC phases are provided in the following sections on the Galileo segments.

Galileo Services

Like other GNSS systems such as GPS or GLONASS, the Galileo navigation concept relies on the measurement of the time of arrival (TOA) of electromagnetic signals transmitted from Medium Earth Orbiting (MEO) satellites. As the signals are synchronously transmitted by the satellites, the minimum number of independent measurements (i.e. transmitters) required to compute a 3D position is four, to account for the three-dimensional position unknowns plus the unknown offset of the receiver clock, which is not supposed to be synchronized with the GNSS and affects all measurements equally.

In addition to the time of arrival of the synchronized signals, the receiver needs to know the position of the transmitters at the exact time the signals were transmitted, in order to compute a position. More details on GNSS navigation principles and the related equations can be found in most satellite navigation references in the literature, such as (Kaplan and Hegarty 2006; Misra and Enge 2011 or Spilker and Parkinson 1996).

In order to maximize the potential user base and the potential benefits that Galileo could offer, Galileo has been developed to provide different services, all based on the TOA positioning method as described above, and some of them with additional features, such as signal encryption, digital signature authentication or search and rescue services. The Galileo services are briefly introduced hereafter:

- **Open Service**: The Open Service (OS) provides positioning and timing information worldwide through ranging signals and data broadcast by the Galileo constellation. The detailed definition of the Galileo OS signals is publicly available and can be found in Europeon Union 2010. The OS will be accessible free of charge by any user equipped with a Galileo compatible navigation receiver. It will be provided in the E1 and E5 bands, and it will be comparable to the service offered by GPS open civil signals L1C/A, L2C or L5.
- **Public Regulated Service**: The Public Regulated Service (PRS) provides positioning and timing information worldwide through ranging signals and PRS data broadcast by the Galileo constellation. The access to the PRS will be restricted to government-authorised users, for sensitive applications. The control access policy is implemented through the encryption of the PRS signals and the management of decryption keys. The PRS will only be accessible through receivers equipped with a PRS security module loaded with a valid PRS decryption key. It will be provided in the E1 and E6 bands and will be similar to the services offered in the L1 and L2 through the P(Y) and L2M signals by GPS.
- **Commercial Service**: The Commercial Service (CS) is intended to provide 'added value' data with respect to the Open Service. At the time this chapter is being written, the Commercial Service is still under definition. However, it is already foreseen that these 'added-value' services are related to high accuracy and authentication (Fernandez et al. 2014). One of the main features the CS will bring, with respect to other GNSS, is the capability to broadcast globally external data in real time. It will be provided in the E6 band.
- **Search and Rescue Service**: The Search And Rescue (SAR) service is intended to support the Cospas-Sarsat Program in search-and-rescue operations. Due to its particularities, more details about this service are given in the following sections.

In addition to the above, the Galileo program foresaw a **Safety-Of-Life** (SOL) **service**, which included the provision of worldwide system integrity. The implementation of this service has been postponed until later phases of the program, and will rely on the reuse of regional solutions and a collaborative approach with other global constellation providers. Another chapter in this book explains in more detail the Galileo Integrity Concept behind the SOL service, as it was initially conceived.

The Galileo signals will be described in Chap. 3 of the book. The Galileo signals include all advanced features that are considered in modern GNSS such as transmission at several frequencies, secondary codes and pilot components, and usage of higher power levels, larger bandwidths and error correction codes. Other features include higher chip rates, longer pseudo-noise codes and Binary Offset Carrier (BOC) modulations and variants thereof. All these aspects and their implications will be reviewed in a subsequent chapter.

Galileo Architecture Overview

The provision of the Galileo signals and services relies on the continuous and coordinated operation of a network of specialized system facilities covering different functional needs. Figure 2.2 presents the most relevant elements of the Galileo system infrastructure in an end-to-end service context.

These facilities can be grouped into three main categories: the Galileo Core Infrastructure, the Galileo Service Facilities and the Galileo Support Facilities.

The **Galileo Core Infrastructure** (CI) comprises a Medium Earth Orbit (MEO) satellite constellation continuously transmitting Galileo Signal-in-Space (SIS), i.e. the Galileo Space Segment, and a global ground system infrastructure providing all the functionality required to sustain the provision of Galileo navigation services in an independent manner.

The Galileo CI ground infrastructure comprises two main subsystems or segments, the Galileo Ground Control Segment (GCS) and the Galileo Ground Mission

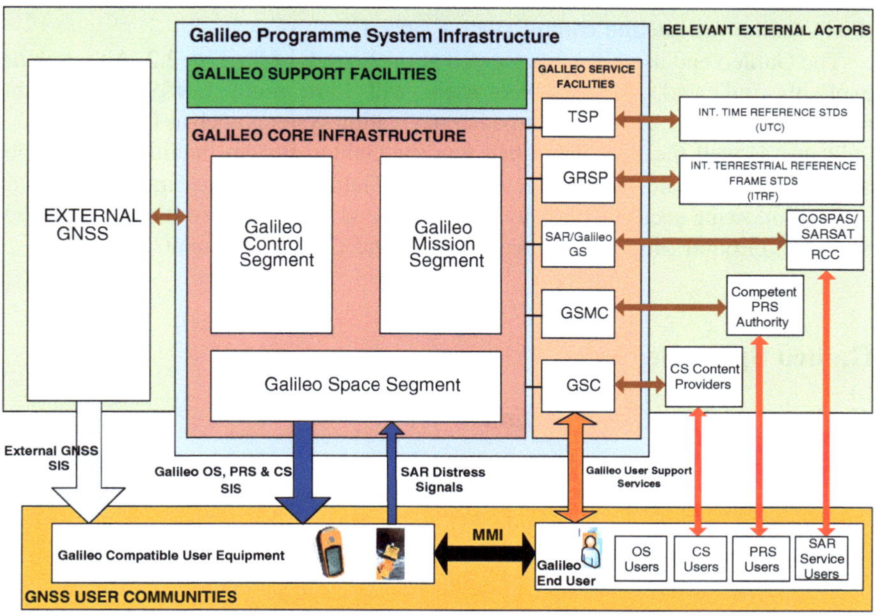

Fig. 2.2 Galileo architectural overview

Segment (GMS). While the GCS provides the Galileo constellation monitoring and control functions, the GMS supports the generation and distribution/uplink of navigation products and other mission data required for the onboard generation of the navigation messages modulated on some of the Galileo SIS components.

The operations of the Galileo CI are centrally managed from two fully redundant Ground Control Centres (GCC) located in Oberpfaffenhofen (Germany) and in Fucino (Italy) respectively.

The principal service offered by Galileo to the end users is the Galileo SIS, which can be processed by Galileo compatible receiver equipment for accurate positioning and time determination in the Galileo terrestrial reference frame (GTRF) and Galileo System Time (GST) scale respectively. Two **Galileo Service Facilities**, the so called Geodetic Reference Service Provider and Time Service Provider, monitor the alignment of GTRF and GST with the international metrological standards (ITRF and UTC) and provide the Galileo CI steering corrections to ensure a very high level of consistency between reference systems.

Further to the GTRF and the TSP, there are other Galileo Service Facilities to provide Galileo-related services to the wide public and to specific user communities. They are the GNSS Service Centre (GSC), the Galileo Security Monitoring Centre (GSMC) and the SAR Galileo Ground Segment.

The **Galileo Support Facilities** is a further category of facilities not directly involved in the routine provision of services but playing an essential role in the deployment, commissioning and maintenance of Galileo. These include among others two external satellite control centres supporting the Launch and Early Operations Phase (LEOP) of each Galileo spacecraft and a ground In Orbit Test (IOT) station for satellite commissioning operations.

The Galileo end users are represented at the bottom of the Fig. 2.2. Although not explicitly indicated, most Galileo users will have multi GNSS interoperable receivers able to track signals from other navigation systems such as GPS. Figure 2.2 indicates as well the main links between the Galileo system facilities, the Galileo service end users and other actors outside the Galileo system perimeter.

The following sections provide a more detailed description of the Galileo Space Segment, Ground Mission Segment and Ground Control Segment.

Galileo Space Segment

Galileo Satellite Constellation

At the time of writing, the Galileo reference constellation or constellation standard foresees 24 nominal orbital positions or operational slots in Medium Earth Orbit (MEO) homogeneously distributed in three orbital planes (i.e. 8 slots equally spaced per plane). At the end of the FOC phase all the orbital positions defined in the constellation standard will be populated with operational Galileo satellites. Besides the core reference constellation, additional satellites will be deployed on each orbital plane in order to ensure the maintenance of the Galileo services upon satellite outages. At present, there are not reference orbital positions defined for these in-orbit spare satellites.

It shall be noted that the Galileo constellation standard has experienced some changes since the Galileo early definition phase. During the constellation design trade-offs analysis, the number of satellites in the core constellation was essentially driven by the Safety of Life (SoL) service of Galileo, initially defined as global integrity service with demanding requirement in terms of time-to-alert, availability and continuity. In order to meet the stringent SoL service requirements, the initial Galileo reference constellation was based on 27 orbital positions (Walker 27/3/1).

Following the program decisions in 2012 to re-profile the SoL service, the impact on the constellation design has been reassessed and several analyses have shown that a reduced constellation with 24 operational satellites deployed in Walker 24/3/1 configuration can meet the Galileo services requirements in terms of accuracy and availability.

The nominal trajectory followed by the operational Galileo satellites is a circular orbit with a radius of approximately 29,600 km (equivalent to 23,229 km altitude over the Earth surface) and an orbital period of approximately14 h. This choice ensures a satellite ground-track repeat cycle every 17 orbits (or 10 days) (Fig. 2.3).

The main orbital parameters of the reference Galileo constellation are summarized below:

- Semimajor axis: 29,600 km
- Eccentricity: 0.0001 (i.e. circular orbit)
- Inclination 56°
- Argument of perigee: ±180° (i.e. not defined for circular orbits)
- RAAN: 0°, 120° and 240°

Fig. 2.3 Galileo reference constellation representation—24 operational satellites (courtesy of EC)

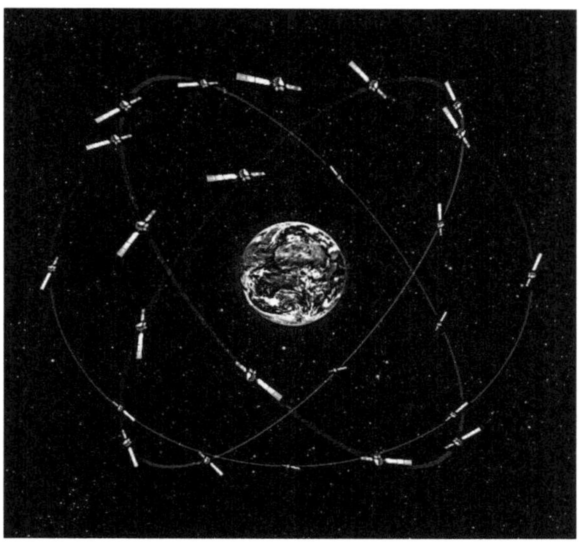

Table 2.1 Galileo satellite position on May 1st, 2013, 00:00:00 UTC

S/C	Position		Semi-major axis	Eccentricity	Inclination	RAAN	Arg. perigee	True anomaly
	Plane	Slot	km		deg	deg	deg	deg
GSAT0101	B	5	29599.8	0.0001	56.0	113.6	0.0	295.9
GSAT0102	B	6	29599.8	0.0001	56.0	113.6	0.0	335.9
GSAT0103	C	4	29599.8	0.0001	56.0	233.6	0.0	269.2
GSAT0104	C	5	29599.8	0.0001	56.0	233.6	0.0	309.2

Source European GNSS Service Centre website—www.gsc-europa.eu

More specifically, the satellites are positioned in a Walker 24/3/1 configuration, which means that satellites in each plane are equally spaced by 45° and satellites in adjacent planes are phased by 15° between each other. The satellite plane inclination for Galileo is 56°.

The Right Ascension of the Ascending Node (RAAN) defines the relative angular phasing between the constellation orbital planes and the vernal equinox. The True Anomaly indicates the angular position of each satellite within a given orbital plane.

The first Galileo operational satellites, resulting from the Galileo IOV phase, were launched respectively on 21st October 2011 (IOV PFM and FM2) and on the 12th October 2012 (FM3 and FM4). Their nominal position can be represented by the following orbital elements (Table 2.1) for the reference time 1 May 2013 at 00:00:00 UTC (note that the difference in the argument of perigee between 180° and 0° is irrelevant as the orbits are almost circular). It must be noted that the orbital positions assigned to these first 4 IOV Galileo satellites correspond to the former Walker 27/3/1 reference constellation geometry.

The coordinate reference frame used is the Celestial Intermediate Reference System CIRS (McCarthy and Petit 2003) (true equator). In order to represent the RAAN precession, the RAAN has to be modified at a rate of −0.027644°/day while the True Anomaly evolves at a rate of around 613.7°/day, equivalent to 1.7 revolutions per day, or 17 in 10 days, as mentioned above.

Galileo Satellites

At the time of writing, the four Galileo IOV operational satellites launched from Kourou (French Guyana) and Soyuz launchers and manufactured by ASTRIUM GmbH (now renamed Airbus Defense and Space) have been deployed in their nominal orbits, as mentioned above. The other 22 additional Galileo operational satellites are being manufactured by OHB-System AG as part of the Galileo FOC phase and will be launched with Soyuz and Ariane 5 launchers from Kourou as well.

Table 2.2 Evolution of galileo satellites from GIOVE-A to Galileo FOC

Parameter	GIOVE-A	GIOVE-B	IOV (PFM-FM4)	FOC (FM5-FM26)
Launch mass	600 kg	530 kg	730 kg	732.8 kg
Total power	700 W	1100 W	1980 W	1900 W
Size	1.3 × 1.8 × 1.65 m	0.95 × 0.95 × 2.4 m	2.7 × 1.2 × 1.1 m	2.5 × 1.2 × 1.1 m
Design lifetime	27 months	27 months	12 years	12 years
Launch dates	28 Dec 2005 (decommissioned	26 Apr 2008 (decommissioned	PFM/FM2: 21 October 2011	FM5/FM6: Q3 2014 (scheduled)
	5 Jun 2012)	23 Jul 2012)	FM3/FM4: 12 October 2012	1 launch every 3/4 months (plan)

The remaining constellation satellites as well as replenishment satellites will be ordered in 2015 and are expected to be available in the 2018–2019 timeframe.

The Galileo IOV satellites perform essentially the same functions as the FOC satellites and their respective designs share a number of common elements dealing with mission critical technologies (e.g. atomic frequency standards, navigation signal generators, or mission data receivers). However, the transition from IOV to FOC has also been used to improve the signal performances in terms of effective radiated power and bandwidth, leading to a change of technology in some units, in particular the high power amplifiers. The satellite description presented in the following sections is based on the Galileo FOC satellites, which will constitute the main part of the Galileo constellation when fully deployed (Table 2.2).

The Galileo satellite in flight configuration and its high-level decomposition block diagram is shown on Fig. 2.4. The Galileo satellites are made of two main components: the platform and the payload, which are further decomposed into modules as shown in the figure. The first four Galileo satellites (called IOV satellites) exhibit some small differences with regards to the Galileo FOC satellites but are very similar in terms of functionalities, overall budget envelope and performances.

The Galileo satellites are designed to be launched in dual launch configuration on Soyuz and in a quadruple launch configuration on Ariane 5. The satellites are directly injected by the launcher into their final orbit and as a result the propulsion capabilities of the satellites can be limited to small out-of-plane orbital corrections, in-plane slot adjustment for constellation spare relocation strategy and graveyarding operations at spacecraft's end of life.

In addition to the main navigation and SAR payloads, the Galileo satellite design includes also the following secondary payloads:

- The Laser Retro-Reflector Array (LRA), a passive instrument that allows Galileo satellites to be tracked by Satellite Laser Ranging (SLR) stations on the ground.
- The Environmental Monitoring Unit (EMU) whose main purpose is to measure the number of heavy ion sand the electric charges in the MEO orbit over the whole 11-year solar cycle.

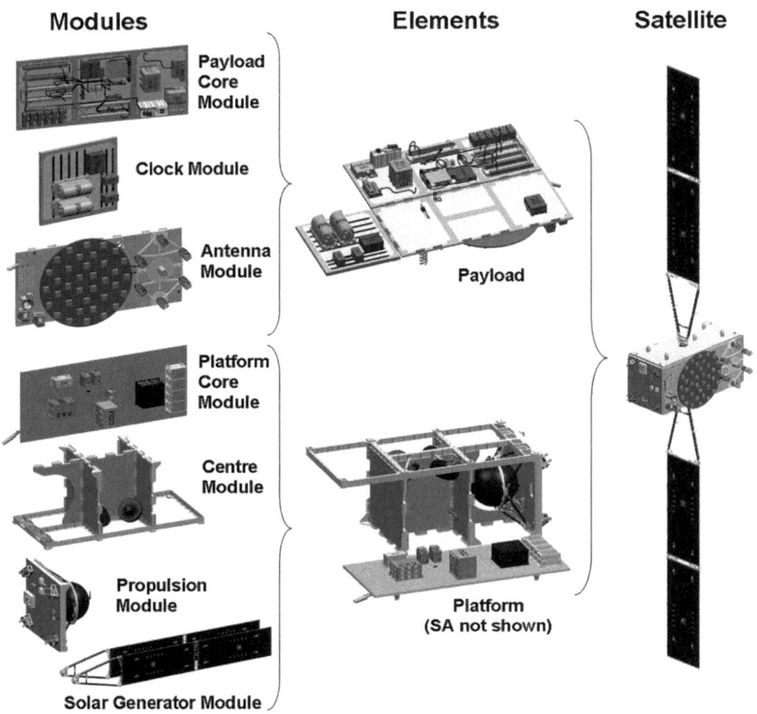

Fig. 2.4 Galileo satellite flight configuration (courtesy of OHB)

Galileo Ground Mission Segment

The main roles of the Galileo Ground Mission Segment (GMS) are the generation of the Galileo C-Band uplink signals including the data required in the Galileo navigation downlink signals and the online monitoring of the downlink navigation signals in closed-loop.

The main GMS functional chains are the *mission data generation chain*, responsible for the generation of the OS and PRS navigation products, and the *mission data distribution chain* responsible for the timely distribution of the navigation products to the Galileo satellites. Moreover, the GMS interfaces with the Galileo Service Facilities briefly introduced previously in this Chapter and it also exchanges mission data required to deliver the Galileo services.

More specifically, the GMS is responsible for the following main functions:

- Generation and distribution of Galileo System Time (GST) to all elements within the Galileo Core Infrastructure perimeter, including the satellites, and support the overall system time steering.
- Generation and distribution to Galileo spacecrafts of mission products for generation of Galileo OS and PRS navigation messages.

- Distribution to Galileo spacecrafts of mission data forwarded by Galileo Service Facilities supporting provision of Galileo CS and SAR services.
- Provision of the physical interface for the exchange of data products with Galileo Service Facilities to support smooth provision of Galileo services.
- Provision of the physical interface for the exchange of data products with the US Naval Observatory (USNO) for coordinating the generation of GPS Galileo interoperability mission products defined in the OS navigation message (i.e. GPS To Galileo Time Offset)
- Galileo mission monitoring and control and archiving

Besides the main service related functions listed above, the GMS supports other internal support functions related to the management and operations of the ground infrastructure. The coordination of the system operations both in ground and in orbit is paramount to ensure the continuity of the Galileo services. In order to support this coordination as efficiently as possible, the GMS design includes physical interfaces with the Galileo Ground Control Segment (GCS) at each Galileo Control Centre.

GMS Architecture

The GMS comprises a worldwide network of facilities that include the following three types of ground elements:

1. Ground Sensor Stations (GSS)
2. Mission Ground Control Centres (GCC)
3. Mission Up-link Local Station (ULS),

A dedicated low latency global telecommunication network ensures the permanent connectivity of the remote sites (i.e. GSSs and ULSs) with the mission ground control center for the routing of mission data and system monitoring and control signals.

Figure 2.5 shows the geographical location of the GMS facilities already deployed or under procurement. It shows also the GCS facilities that are described later in the chapter.

The Ground Sensor Stations

The GSS is an unmanned GMS facility whose main role is the collection of L-Band sensor data from all Galileo satellites in view in all frequencies and forward those data to the GCC for navigation processing and mission monitoring. It is essentially a high-quality Galileo receiver with some additional elements. The collected sensor data includes carrier phase and code phase measurements based on the processing of the pilot components, navigation message symbols demodulated from the data components and some signal quality indicators. The GSS can support SIS measurement rates up to 1 Hz (one measurement every second) in order to ensure continuous SIS monitoring and, in case of anomaly, reduce the notification time to Galileo users.

Fig. 2.5 Galileo GMS and GCS sites

The GSS are equipped with rubidium atomic frequency standards and high performance Galileo receivers, which are the core equipment of the Galileo Receiver Chain (GRC). The GSS design allows hosting two GRC types, the first one supporting the GMS PRS mission data generation chain and the second one supporting the OS and CS mission data generation chains. In addition, each GSS is equipped with Dual-Frequency COTS GPS receivers to support site synchronization with Galileo System Time and realization of the Galileo Terrestrial Reference Frame during the early system operations phase until the number of Galileo satellites deployed ensures the system standalone synchronization worldwide.

The Mission Ground Control Centres

The GMS operations are managed from two fully redundant Galileo Ground Control Centres located in Oberpfaffenhofen (Germany) and Fucino (Italy).

The GCCs centralise several GMS key functions critical for the provision of the Galileo services such as GST generation, navigation data generation and distribution, management of the Galileo system interfaces with external entities, and mission monitoring and control. Outside the perimeter of the GMS, the GCCs play also a central role in the monitoring and control operations of the Galileo satellite constellation, as explained later in the section devoted to the Ground Control Segment (GCS).

The Mission Up-Link Local Stations

The continuous routing of Galileo ground mission data from the GCC to each operational spacecraft is required for the onboard generation of Galileo Signal In Space with meaningful service information (i.e. navigation messages). The ULSs are unmanned facilities which realize the physical ground to space interface supporting the mission data distribution functional chain.

Each operational ULS receives mission data from the GCC for uploading to the Galileo constellation according to a satellite contact plan received also from the GCC. In order to perform this function, each ULS can host up to four dish steerable antennas of 3.5-m diameter to upload mission data in the C-Band part of the RF spectrum (around 5 GHz). Each ULS antenna can track a single Galileo spacecraft at a time.

The accuracy of the Galileo Positioning Velocity Timing (PVT) services depends directly on the accuracy of the mission data uploaded to the Galileo satellites. It shall be noted that a part of the mission data are predictions based on dynamic or empirical models and therefore their accuracy degrades rapidly with time ("ageing" effect). This applies in particular to the satellite GST clock offset prediction broadcast model. In order to meet the minimum Galileo PVT accuracy performance requirements continuously worldwide, it is essential that the design of the ULS network allows for refreshing the satellite onboard mission information of the whole constellation at the required latency on a continuous basis. When the Galileo ULS network is fully deployed, the maximum time between navigation data uploads to any Galileo spacecraft should not exceed 100 min under nominal operational conditions.

Galileo Ground Control Segment

The Galileo GCS is responsible for the management of the Galileo constellation during the normal operation of the system. For achieving this goal, the GCS can exchange monitoring and control signals with individual Galileo spacecraft at scheduled contacts. Further to the monitoring and control data, the GCS can also upload Galileo mission data through the telecommand uplink channel to ensure the continuity of Galileo navigation services under degraded GMS operation modes.

GCS Architecture

The GCS elements are located worldwide and comprise two main facility types:

1. Constellation Galileo Control Centres (GCC)
2. Telemetry, Tracking and Commanding (TT&C) Stations

The Constellation Ground Control Centres

The GCCs host all centralised functions of the GCS, including spacecraft Constellation Monitoring & Control, spacecraft Constellation Planning, Flight Dynamics and Operations Preparation. The GCC implements as well the GCS network interface with external Galileo entities involved in the system operations such as the External Satellite Control Centers (ESCCs) and the In Orbit Test (IOT) station located at Redu, Belgium.

The Telemetry, Tracking and Commanding Facility

The GCS architecture comprises a worldwide network of Telemetry, Tracking and Control ground stations or facilities connected to the Galileo Control Centres. At the time of writing the Telemetry, Tracking and Commanding Facility (TTCF) network includes stations in 5 locations. Additional stations might be added to the TTCF network in the future.

Each TTCF supports telemetry (TM) downlink, telecommand (TC) uplink and can also support collection of satellite tracking data for the management of the constellation. The transmitted TC signal data, and received TM data, together with the TT&C monitoring and control data is exchanged between the TT&C stations and the GCC via a dedicated communication network. Under nominal operations conditions, the TTCFs are autonomous and manned intervention is only required for the purpose of either anomaly investigation or maintenance purposes.

In routine operations, the TT&C stations are utilised to upload TCs to and to receive TM from the Galileo spacecraft through an RF data channel. Besides supporting routine satellite housekeeping tasks, the TTCFs have also the capability to collect satellite tracking data (i.e. two-way ranging measurements) intended for off-line satellite orbit determination under special spacecraft operation scenarios.

The link between the TTCF ground station and the Galileo satellites is established through a 11 m dish antenna. The TT&C facility operates over specifically allocated S-Band RF ranges, between 2 and 2.2 GHz approximately (Fig. 2.6).

Cospas-Sarsat and Galileo

The LEOSAR system developed by the International Cospas-Sarsat Program currently provides accurate and reliable distress alert and location data to help search and rescue (SAR) authorities to assist persons in distress. In 2000, consultations started between the Cospas-Sarsat Program and the European Commission on the feasibility to install 406 MHz SAR instruments on the Medium Orbit navigation satellites systems in order to develop a 406 MHz MEOSAR component to the Cospas-Sarsat system. The main benefits of the MEOSAR system will be the near instantaneous global coverage with accurate independent location capability

Fig. 2.6 TTCF S-band 11 m dish antenna (courtesy of ESA)

(in opposition with the current LEO system which has a higher latency to provide location information). The USA MEOSAR program based on GPSIII is called the Distress Alerting Satellite System, (DASS), the European System based on Galileo is called SAR/Galileo, and the Russian program based on GLONASS is referred to as SAR/GLONASS. This has a direct impact on the probability of survival of the person in distress at sea or on land.

The Galileo Program involvement into Cospas-Sarsat goes beyond the space component of the MEOSAR system. Indeed, the European Union has deployed a significant Ground Segment infrastructure, which provides localization services for distress alerts transmitted by SAR beacons over a wide area comprising continental Europe, and vast oceanic areas around the continent (see Fig. 2.7 below). The SAR/Galileo Ground Segment can receive and process SAR distress signals relayed by any operational Galileo spacecraft or other satellite of the COSPAS/SARSAT MEOSAR constellation and determine thereby the location of the beacon within the coverage area.

The ground segment of the Search and Rescue Service of Galileo consists of 3 receiving ground stations, called Medium Earth Orbit Local User Terminal (MEOLUTs), which receive the distress signals relayed by the Galileo Search and Rescue repeater in the 1544 MHz band. Each MEOLUT includes a minimum of 4 antennas tracking different Galileo satellites.

Receiving the signal relayed from four different satellites makes it possible to determine the distressed beacon position by triangulation using Time of Arrival (TOA) and Frequency of Arrival (FOA) techniques. The MEOLUT then decodes the distress signal message, determines the beacon localization and provides this information to the Cospas-Sarsat Mission Control Center (MCCs).

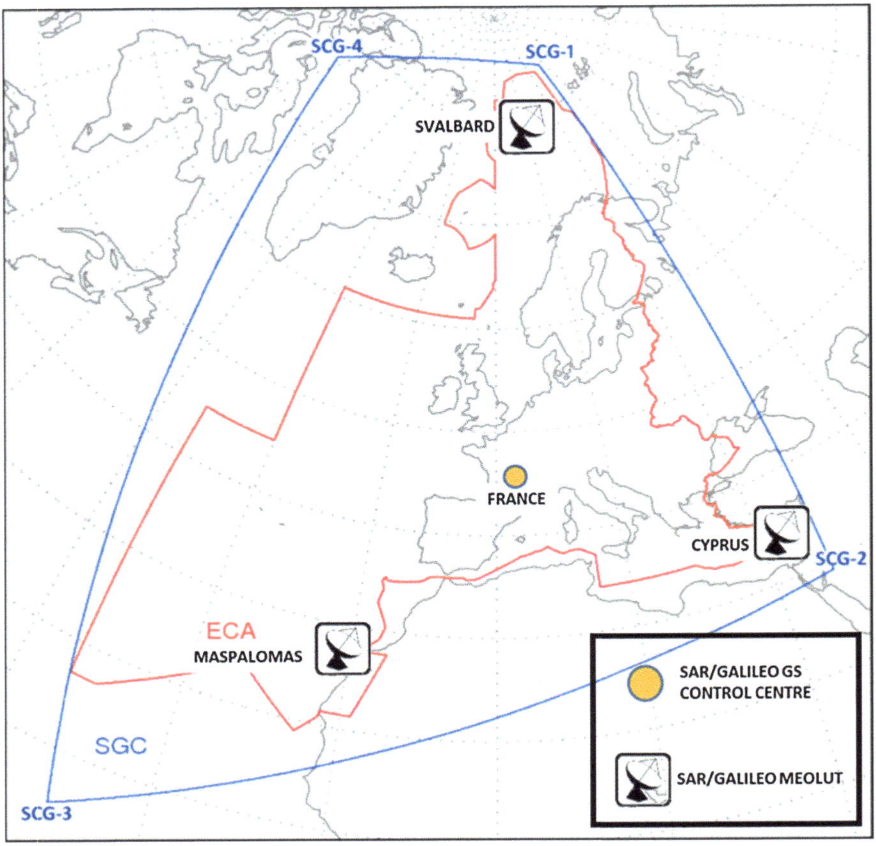

Fig. 2.7 SAR/Galileo European coverage area and ground facilities

The 3 European MEOLUTs are located in Svalbard (Norway), Makarios (Cyprus) and Maspalomas (Spain) and provide the SAR/Galileo service over the European Coverage Area (ECA) as shown in Fig. 2.7. Each MEOLUT is connected to a central facility, the MEOLUT Tracking Coordination Facility (MTCF) located at the SAR/Galileo control centre in Toulouse, (France) and which optimizes the MEOLUT tracking plan of the 3 European MEOLUT in order to achieve the best location accuracy and availability over the European Coverage Area.

As a component of the Cospas-Sarsat MEOSAR system (Cospas-Sarsat 2012), the SAR/Galileo ground segment is also capable of receiving the distress signal relayed by the MEOSAR payloads embarked on the Glonass and GPS satellites (SAR/Glonass and GPS/DASS payloads).

The performances achieved by the SAR/Galileo Service when the full Galileo constellation is operational are indicated in Table 2.3

The SAR/Galileo ground segment also includes the Return Link Service Provider (RLSP), which is responsible for providing Return Link Acknowledgment

Table 2.3 SAR/Galileo service performance recorded at ground segment

Performance parameter	Value
Detection probability	99.5 %
Localization probability	98.0 %
Localization accuracy	<5 km (2σ) within 10 min
Worst case service availability	95.0%

Messages to the COSPAS-SARSAT distress beacons equipped with a Galileo receiver. The Return Link Messages are embedded within the navigation message of the E1 signal.

Galileo User Segment

Definition

The Galileo user segment (GUS) is the third and largest segment that forms the Galileo system. Contrary to the other two segments, namely, the Galileo space segment and the Galileo ground segment, the GUS is certainly the one with the closest connection to the end user. In particular, the GUS is targeted at providing the means for allowing end users to fully exploit all the capabilities of signals being transmitted by the Galileo satellites. It is therefore, the segment encompassing user applications and services, as well as user receiver technologies.

Galileo Receivers, the Key to Galileo Success

The Galileo user receiver is the key technological equipment for translating the Galileo signals received from space into practical applications and services. The goal of Galileo receivers, as it happens with many other GNSS receivers, is to provide the end user with an accurate estimate of its current position and time. This is typically achieved in three steps:

1. The signals from space impinging on the user's receiver antenna are amplified and filtered by the receiver RF front-end in order to compensate for the huge propagation losses undergone in their way from the satellites to the Earth. This signal conditioning is typically followed by an analog-to-digital conversion, which allows processing the received signal in a reliable manner by means of digital signal processing (DSP) techniques.
2. The user receiver generates a local replica of the Galileo signal of interest, which is gradually aligned to the actual signal received from space. As a result of this alignment process, the receiver is able to coarsely determine the actual time-delay and carrier parameters of the received signal (i.e. this is the so-called

"acquisition" stage). Later on, these parameters are further refined and monitored as a function of time (i.e. the so-called "tracking" stage).

3. Once the receiver is precisely synchronized to the actual received signal, and more than four satellites are in view, the refined and accurate signal parameters are used to solve the navigation equations, which lead to the user position, velocity and time information (i.e. the so-called "navigation" or "PVT" solution).

The above three steps are rather standard in most GNSS receivers, but for the particular case of a Galileo receiver, the key point is being able to fully exploit the specific features that Galileo signals bring to end users:

From a physical-layer perspective, this involves receivers taking advantage of the Binary-Offset Carrier (BOC) modulation format adopted in Galileo signals. Due to this format, which can be understood as a one-bit quantization of a sinusoidal signal, and for a given total bandwidth, it is possible to achieve a larger mean square bandwidth (also known as Gabor bandwidth) than with a traditional binary phase shift keying (BPSK) modulated signal. This involves that Galileo receivers may provide an improved accuracy in the determination of the user position, particularly when compared to traditional mass-market receivers that make use of the traditional BPSK-modulated GPS C/A signal.

From a frequency planning perspective, this involves paving the way for the development of multi-frequency receivers, which take advantage of processing different frequency bands with the goal of providing additional diversity and compensating ionospheric effects. These bands correspond to the E1 band compatible with current GPS mass-market receivers and centered at 1.575,42 MHz; the E6 band, centered at 1.278,75 MHz; and the E5 band, which is further split into two adjacent bands, namely the E5a and E5b bands, with center frequencies at 1.176.45 and 1.207,14 MHz, respectively. The availability of these frequency bands is certainly a key feature to be exploited by advanced Galileo receivers. Moreover, the coexistence with GPS within the E1 band opens the door for the use of the so-called multi-constellation GNSS receivers, capable of processing both GPS and Galileo signals with the same front-end, and thus solving the problems of limited visibility of satellites that are suffered in certain working scenarios (e.g. in urban canyons).

From a data perspective, Galileo signals allow the emergence of a myriad of new applications and services due to the definition of different types of data channels, such as the case of open and publicly available data channels, authenticated data channel and even possibly encrypted data channels.

Paving the Way Forward for Galileo User Receivers

The development and widespread deployment of Galileo receivers is certainly the necessary piece to guarantee the success of Galileo. In that sense, the validation of the above-mentioned Galileo features, and their proof-of-concept, has been undertaken in the framework of several R + D initiatives carried out in the recent years.

Most of these efforts have been supported by public funding, either through the 6th and 7th Framework Programs funded by the European Commission, through research projects funded by the European Space Agency (ESA), or through different initiatives funded by national authorities. The goal of all these efforts has been to foster the development of Galileo user receiver technologies and to achieve the deepest and broadest development of Galileo applications. This is particularly important when taking into account the wide range of fields and applications where the potential benefits of Galileo have already been recognized. They span across several domains such as location-based services (LBS) powered by smartphones or portable devices, road and maritime transportation, aviation, precision agriculture, safety or environment protection, just to mention a few. In general terms, GNSS positioning and timing information is expected to have a critical impact in economic terms of more than 6 % of the whole GDP of the European Union (European Commission 2010).

From this perspective, it becomes clear that the expected economical impact of Galileo in a commercially competitive environment is certainly one of the main differences with respect to previous GNSS systems. GPS and GLONASS were mainly developed for military applications with military funding, without a predetermined mass-market orientation, although in the end GPS in particular has had a huge commercial impact. This is inline with the fact that barely 10 % of the US total public spending in R + D has a market or application-driven orientation, whereas for the EU, this amount exceeds 40 % (Fernández-Hernández 2011). This example shows the actual importance of the GUS, and the determined efforts that are being taken in order to make it successful. With Galileo receivers at the forefront of the GUS, most of these efforts have concentrated on the development of Galileo user receivers and technologies. This is the case of the three calls (in 2002, 2003 and 2005) launched by the former Galileo Joint Undertaking (GJU) under the Galileo work program, where nearly 90MEur were dedicated to promote the development of Galileo receiver technologies, applications as well as the standardization of different aspects of Galileo and its mission implementation.

One of the key initiatives being funded under the first GJU call was the "Galileo receiver development activities" (GARDA) project, whose goal was to develop a Galileo receiver analysis and design application (GRANADA) in order to serve as a test-bench for integration and evaluation of technologies, on one side, and as a software receiver tool for Galileo application developers, on the other side (Marradi et al. 2006). The validation of this software tool was carried out by the development of a Galileo simulator according to the Galileo signal-in-space (SIS) interface control document (ICD) available at that time. For the subsequent calls, GJU funding on the GUS targeted the introduction of mass-market Galileo receivers, such as in the GAMMA project (Abwerzger et al. 2006) (which was followed by the GAMMA-A project), where low-cost and low-power receiver solutions were developed for specific market applications. Two main applications were identified: location-based services (LBS) and the automotive area, with the latter involving route guidance, fleet management or road tolling. Complementarily, the GRAIL project focused on the development of Galileo receivers for rail applications, where the emphasis was placed on safety-critical applications related to the European Train Control System (ETCS) (Albanese et al. 2006). Finally, the HIMALAYA activity focused on the design and development of a "ready-to-market" single chip GNSS mass-market receiver for GPS, EGNOS and GALILEO signals. Many other projects have been funded in the last years, and being exhaustive exceeds the goals of this introductory chapter.

Security, environment protection and SoL applications are also key research lines were efforts are being devoted for the development of advanced Galileo receivers. Integrity (not necessarily provided by the system, but by the receiver) is currently emerging as a critical requirement for high-end Galileo receivers, due to the increasing concerns on potential threats to GNSS users caused by the presence of interference, spoofing and other disturbing effects (e.g. multipath, non-line-of-sight). The study of this problem is indeed the goal of the iGNSSrx project funded by the European Commission and initiated in 2012, whose objective is to improve positioning services to terrestrial users by developing user receiver technologies for integrity monitoring, at signal and observable level.

Finally, the integration of Galileo receivers with external technologies is also one of the main topics that are expected to pave the way for the success of Galileo receivers. Such integration can be used to further improve the overall performance of Galileo, particularly in the presence of harsh working scenarios with signal blockages and limited sky visibility. For instance, the work carried out within the IADIRA project has focused on the tight integration of Galileo with external information coming from inertial sensors (INS) (Silva et al. 2006). A step forward was addressed in the DINGPOS project funded by ESA, where apart from the use of INS, wireless communication technologies such as WiFi were also considered for developing a fully integrated platform capable to provide seamless location in indoor environments (López-Salcedo et al. 2008). Indeed, the advent of new applications and services are gradually pushing the use of GNSS in working conditions different from the ones for which the system was designed to operate (Seco-Granados et al. 2012). This is the case of LBS in handheld devices (e.g. location-aware marketing), social networking applications, indoor navigation, etc. In that sense, the integration of the GUS with communication technologies (e.g. WLAN or cellular networks such as Long Term Evolution—LTE Del Peral-Rosado et al. 2012) is certainly one of the most promising future research lines for which significant efforts will be devoted in the coming years.

Galileo Interoperability and Compatibility

While the independence of systems may be a way to promote competition among systems and could virtually provide greater reliability and integrity, the advantages of interoperability and compatibility are widely recognized by all actors. An early (February 1999) Communication of the European Commission (EC) stated: "Galileo must be an open, global system, fully compatible to GPS, but independent of it…" Later on, the EU-US Agreement on the Promotion, Provision and Use of Galileo and GPS Satellite-Based Navigation Systems and Related Applications signed June 26, 2004, in Dublin, Ireland, set up the models and methodology for the radio frequency compatibility of satellite navigation systems, in particular between GPS and Galileo. Further, Global and regional system providers at the third meeting of the Providers' Forum of the International Committee on Global Navigation Satellite Systems (ICG) in 2008 agreed that at a minimum, all global navigation satellite systems (GNSS) signals and services must be compatible. To the maximum extent possible, open signals and services should also be interoperable, in order to maximize benefit to all GNSS users. For many applications, common carrier frequencies are essential to interoperability and commonality of other signal characteristics is desirable. In some cases, carrier frequency diversity may be preferable to improve performance. The definitions of *compatibility* and *interoperability* according to the ICG are as follows (International Committee on Global Navigation Satellite Systems 2008).

Interoperability refers to the ability of global and regional navigation satellite systems and augmentations and the services they provide to be used together to provide better capabilities at the user level than would be achieved by relying solely on the open signals of one system:

i. Interoperability allows navigation with signals from different systems with minimal additional receiver cost or complexity;
ii. Multiple constellations broadcasting interoperable open signals will result in improved observed geometry, increasing end-user accuracy everywhere and improving service availability in environments where satellite visibility is often obscured;
iii. Geodetic reference frames realization and system time steerage standards should adhere to existing international standards to the maximum extent practical;
iv. Any additional solutions to improve interoperability should be encouraged.

Compatibility refers to the ability of global and regional navigation satellite systems and augmentations to be used separately or together without causing unacceptable interference and/or other harm to an individual system and/or service:

i. The International Telecommunication Union provides a framework for discussions on radiofrequency compatibility. Radiofrequency compatibility should involve thorough consideration of detailed technical factors, including effects on receiver noise floor and cross-correlation between interfering and desired signals;
ii. Compatibility should also respect spectral separation between each system's authorized service signals and other systems' signals. Recognizing that some signal overlap may be unavoidable, discussions among providers concerned will establish the framework for determining a mutually acceptable solution;
iii. Any additional solutions to improve compatibility should be encouraged.

In the particular case of Galileo, the application of the interoperability principle with GPS was guided by the following considerations (Hein 2006).

- Signals-in-Space. The driver was the selection of common center frequencies (at least for some signals such as L1/E1 and L5/E5a) since this aspect has a clear impact of the cost of multi-frequency receivers and it is necessary to make combined processing of phase observation possible. The choice of CDMA instead of FDMA was also a definite step in favor of the interoperability between Galileo and GPS. GLONASS, which uses FDMA, can be considered to be "system interoperable" but not "signal interoperable" with Galileo. Although the use of the same modulations, data messages, etc. is not considered as an essential feature to facilitate interoperability because digital receivers can easily encompass a variety of formats, the military GPS-M code and the Galileo Public Regulated Service (PRS) have signal interoperability on L1 band.
- Coordinate Reference Frame. Although the Galileo Terrestrial Reference Frame (GTRF) is different to the GPS coordinate reference frame (WGS84), both of them differ less than very few centimeters with respect to the International Terrestrial Reference Frame (ITRF), hence guaranteeing interoperability for most applications.
- Time Reference. Galileo System Time (GST) as well as GPS Time are different real-time realizations of UTC (Universal Time Coordinated)/TAI (Atomic Time), which is the international civilian time standard. Both times are expected to be within the nanoseconds order of magnitude. The service providers have agreed to broadcast the GPS/Galileo time offset, and alternatively the offset can be determined in a combined receiver with very high accuracy at the very small cost of using one satellite observation.

References

Abwerzger G et al. (2006) GAMMA—status and project details of a GNSS mass market receiver development. In: Proceedings of the 3rd ESA workshop on satellite navigation technologies (NAVITEC), pp 1–8, Dec 2006

Albanese A et al. (2006) The GRAIL project: GNSS-based user terminal introduction in the rail application. In: Proceedings of the 3rd ESA workshop on satellite navigation technologies (NAVITEC), pp 1–8, Dec 2006

Cospas-Sarsat (2012) MEOSAR implementation plan, Issue 1.8, Oct 2012

De-Gaudenzi R et al. (2000) Galileo signal validation development, John Wiley & Sons, Ltd, UK

European Union (2010) *European* GNSS (Galileo) Open service—signal in space interface control document, Issue 1 Revision 1—Galileo OS SIS ICD, Sept 2010

European Commission (1999) Communication from the commission—Galileo—involving Europe in a new generation of satellite navigation services. COM(1999)54. Off J Eur Union 1999

European Commission (2010) Action plan on global navigation satellite system (GNSS) applications. Communication from the commission, COM(2010)308, 14 June 2010

European Commission (2007) How the European union works—your guide to the EU institutions. Directorate-General for Communication, 2007

European Commission (2008) Regulation (EC) No 683/2008 on the further implementation of the European satellite navigation programmes (EGNOS and Galileo). Off J Eur Union 2008
ESA (1998) DaimlerChrysler Aerospace, GNSS-2 Comparative System Study Executive Summary. ESA contract 13217/98/NL/DS, 1998
Fairbanks M (1999) Global navigation satellite systems (GNSS) area report 1998, BX5329/D0133, May 1999
Fernández-Hernández I (2011) GNSS programmes and R&D landscape in the EU, In: ICL-GNSS conference, Tampere, 30 June 2011
Fernández-Hernández I et al. (2014) The Galileo commercial service: current status and prospects. In: Proceedings of the European navigation conference 2014, April 2014
Hein G (2006) GNSS interoperability: achieving a global system of systems or does everything have to be the same? InsideGNSS, Jan/Feb 2006
International committee on global navigation satellite systems (2008) Providers' forum principles of compatibility and interoperability and their further definition, http://www.oosa.unvienna.org/oosa/SAP/gnss/icg/providersforum.html
Kaplan ED, Hegarty CJ (2006) Understanding GPS—principles and applications, 2nd edn. Artech House, Norwood
López-Salcedo JA et al (2008) DINGPOS: a hybrid indoor navigation platform for GPS and Galileo. In: Proceedings of the ION GNSS, pp 1–12, Sept. 2008
Marradi L et al (2006) The GARDA project. building a Galileo receiver. Inside GNSS, pp 40–53, Nov/Dec 2006
McCarthy DD, Petit G (eds) (2003) IERS conventions (2003), IERS Convention Centre, 2003
Misra P, Enge P (2011) Global positioning system: signals, measurements, and performance, revised 2nd edn. Ganga-Jamuna Press, Lincoln
G. Mocci et al. (2001) The GAUSS (Galileo and UMTS synergetic system) user terminal. In: Proceedings of the 1st ESA workshop on satellite navigation technologies (NAVITEC), pp 1–8, Dec 2001
Pace S et al (1995) The global positioning system—assessing national policies, Critical Technologies Institute—RAND
Del Peral-Rosado JA et al (2012) Analysis of the positioning capabilities of 3GPP LTE. In: Proceedings of the ION GNSS, pp 1–10, Sept 2012
Schweikert R et al (2000) On signal structures for GNSS-2. Int J Satell Commun, John Wiley & Sons, Ltd
Seco-Granados G et al (2012) Challenges in indoor global navigation satellite systems. IEEE Signal Process Mag 29(2):108–131
Silva PF et al (2006) IADIRA: Inertial aided deeply integrated receiver architecture. In: Proceedings of the 3rd ESA workshop on satellite navigation technologies (NAVITEC), pp 1–10, Dec 2006
Spilker J, Parkinson B (1996) The global positioning system: theory and applications, vol 1. American Institute of Aeronautics and Astronautics, Washington
United States of America and European Community (2004) Agreement on the promotion, provision and use of Galileo and GPS satellite-based navigation systems and related applications

Chapter 3
Galileo Signals

Elena Simona Lohan, Heikki Hurskainen and Jari Nurmi

Abstract This chapter focuses on the Galileo signal models. We start this chapter with the single-link signal model, explaining the main tasks of GNSS signals. We then describe the main modulation classes used in Galileo, namely Binary Phase Shift Keying (BPSK), Sine and Cosine Binary Offset Carrier Modulations (Sin/CosBOC), Multiplexed Binary Offset Carrier modulations (MBOC), and Alternate Binary Offset Carrier modulations (AltBOC). Both time-domain and frequency-domain characteristics of such signals are discussed.

Basics on Galileo Signals

Single-Link Model

This section will focus on the single-link model, as illustrated in Fig. 3.1. The signal sent from a Galileo transmitter (i.e., a satellite), in one of the Galileo frequency bands (i.e., E1, E5a, E5b, E5 or E6) may suffer various ionospheric (Richert and El-Sheimy 2005) and tropospheric (Collins 1999) delays, and finally reaches the Galileo receiver, through a Line Of Sight (LOS), multipaths or a combination of these waves. The situation when LOS is absent is called Non-Line-Of-Sight (NLOS) situation. The underlying physical phenomena that are creating the so-called multipaths components are well described for example in (Rappaport 1996) and include phenomena such as reflection, refraction, scattering, diffraction, and shadowing.

The main signal tasks are to differentiate between different transmitters (i.e., multiple access), to enable the receiver to compute the pseudorange (i.e., an estimate of the true range or distance between the satellite and the receiver) and to carry

E.S. Lohan (✉) · H. Hurskainen · J. Nurmi
Department of Electronics and Communications Engineering,
Tampere University of Technology, Tampere, Finland
e-mail: elena-simona.lohan@tut.fi

J. Nurmi et al. (eds.), *GALILEO Positioning Technology*,
Signals and Communication Technology 182, DOI 10.1007/978-94-007-1830-2_3

Fig. 3.1 Single-link model: wave propagation from a single satellite to the GALILEO receiver through a wireless channel including multipath

the information needed by the receiver to compute its final position, velocity and time (more about the later in Chap. 7).

In our models, for clarity reasons, we assume that the multiple access interference, i.e., the interference due to the presence of the other signals in the frequency band of interest, is lumped into the noise term and it is not addressed separately. For specific references to the effects of the cross-correlation issues given by the multiple access scheme in GNSS, the reader is directed for example to (Glennon and Dempster 2005; ODriscoll and Murphy 2007; Balaei and Akos 2011). Multiple access interference can be created not only by the presence of other satellites' signals, but also due to terrestrial beacons sharing the same frequency bands as Galileo signals. A side effect of the multiple access interference is the so-called near-far effect, when an interfering signal (terrestrial or from another satellite) reaches the receiver at a higher power than the power of the signal of interest. Specific references dealing with near-far effects in the context of GNSS are for example (Morton et al. 2007; Nunes and Sousa 2012).

As we have mentioned multipaths, an illustration of LOS and NLOS scenarios is shown in Fig. 3.2. The left-hand plot has a direct path between the transmitter and the receiver (LOS component) and it may have also one or several delayed paths (the multipaths). The NLOS scenario refers to the situation when the direct Line of Sight is absent and only the multipath signals reach the receiver. The multipath effects on Galileo signals are discussed later in Chap. 5 in more detail. The reader is

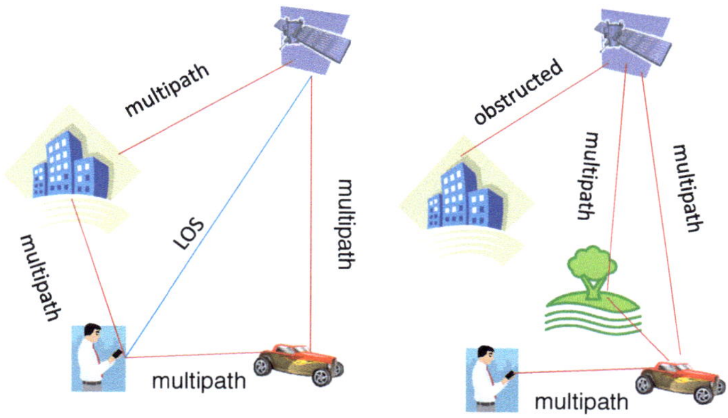

Fig. 3.2 Illustration of LOS (*left*) and NLOS (*right*) situations

directed also to (Seco-Granados et al. 2012) for additional reading on the challenges regarding GNSS signals (including multipaths).

For the final PVT computation, we would need at least four such single-link connections (see Chap. 7), but from the baseband processing point of view, the same operations take place in each of the single-link connections.

Fundamental Signal Components in Satellite Navigation

For each single link, the principle of satellite signal modulation in transmission and demodulation in reception is illustrated in Fig. 3.3. At the satellite side, the low-rate navigation data is modulated via two modulation stages: a spreading modulation and a carrier modulation part. The spreading modulation also spreads the narrow-band signal and transforms it into a Direct Sequence Spread Spectrum Code Division Multiple Access (DS-SS CDMA) signal. This is realized through multiplication with a Pseudo Random Noise (PRN) code. A deeper insight to the principles of code division multiple access modulation (CDMA) in satellite navigation can be found for example in (Braasch and van Dierendonck 1999; Kaplan and Hegarty 2006). The carrier modulation part, that may include also a sub-carrier phase as explained later in this Chapter, is adapting the signal to the transmission medium and moves the baseband signal into the passband, at the desired carrier frequency (e.g., E1, E5, E5a, E5b or E6).

At the receiver side, similar operations in reverse order need to take place: the carrier, the sub-carrier (when present) and the spreading code needs to be removed to enable the data decoding.

The schematic block diagram in Fig. 3.3 assumes that the replica code and received signals are perfectly aligned. In practice, this is achieved through some supplementary acquisition and tracking blocks, which are illustrated in Fig. 3.4.

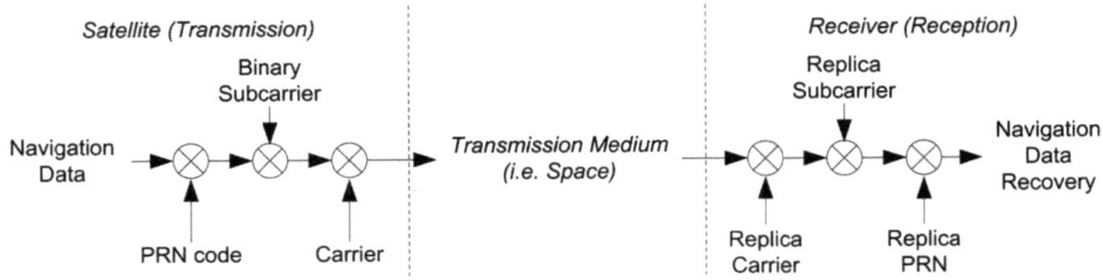

Fig. 3.3 Fundamental signal components in satellite navigation signal transmission and reception

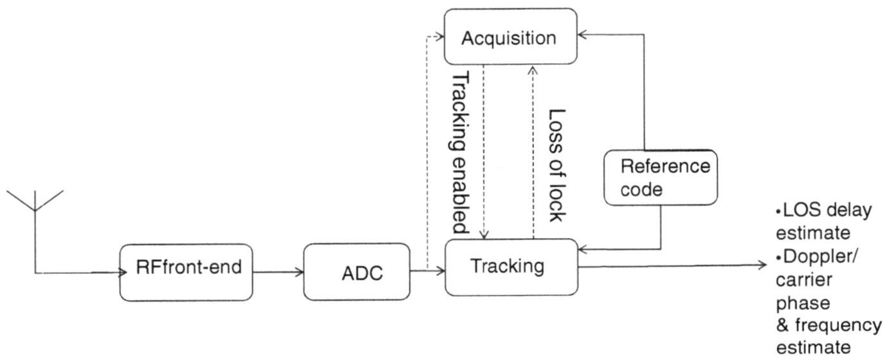

Fig. 3.4 Schematic block diagram of GNSS receiver processing (before the navigation data recovery block)

The received signal is first processed via the RF-front end described in Chap. 4. In the RF front-end, the signal is down-converted to baseband or to a sufficiently small Intermediate Frequency IF, through the Analog-to-Digital Converter (A/D). After the A/D, the acquisition engine is started. In the acquisition engine, described in more detail in the Chap. 5, correlations are performed between the received signal and the reference (or replica) codes (corresponding to each Galileo satellite), at different tentative delays and Doppler frequencies. The acquisition is successful when the satellite number and a coarse estimate of its signal propagation delay and Doppler frequency are identified. After the signal is acquired, the tracking stage is enabled and the receiver operates in the tracking mode. In the tracking stage, finer estimates of the signal LOS delay and carrier frequency and phase are obtained, and these values are continuously tracked in order to compensate for the channel dynamics (Doppler spreads and shifts, multipath delays, etc.). The tracking mode continues until the loss-of-lock occurs, at which point the acquisition is re-started and the signal is again re-acquired. The tracking stage is described in more detail in Chap. 5.

After the tracking stage, the signal is further decoded and the data navigation is extracted. The sub-sequent processing after acquisition and tracking is out the scope of this chapter and will be addressed in Chap. 7.

The Main Differences Between GPS and Galileo Signals at Signal Level

Galileo signals contain some features non-existing in the legacy GPS signals regarding the spreading modulation and carrier modulation stages. The identified differences pertain mainly to the spreading code and sub-carrier, whereas the carrier frequencies are partially shared between these two systems.

Firstly, one of the differences in GPS and Galileo signal components is in the spreading code creation for some of the frequency bands (for example GPS L1 and Galileo E1). In GPS, the so-called Gold codes are used for the code generation, where the PRN codes are generated with dedicated linear feedback shift register (LFSR) structures. These code generators are able to reproduce long streams of code with a little amount of dedicated hardware. The spreading codes are repeated in predefined periods, for example the GPS L1 spreading code repeats itself every 1 ms, and the repetition is synchronized with the transmitted data symbol rate of 50 symbols per second. Thus each data symbol equals one PRN code epoch and when the PRN code is removed and integrated over the epoch time the sign of the result represents the symbol value.

Galileo E5a and E5b signals use the same principles of code generation as GPS signals. Figure 3.5 illustrates the architecture of Galileo E5a/b code generator. The code generating architecture produces 10230 chip-long spreading codes by

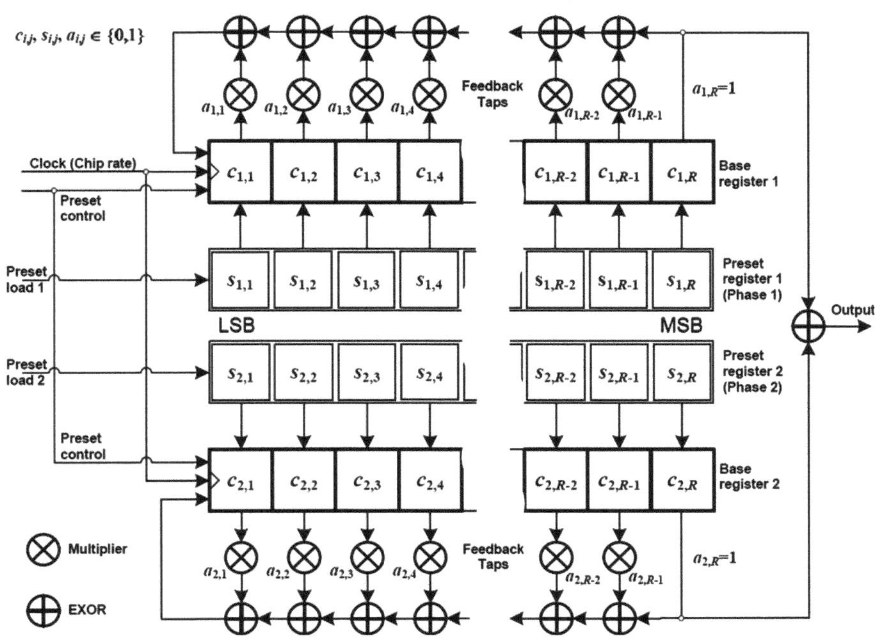

Fig. 3.5 Galileo E5 code generator (Galileo SIS-ICD 2008)

combining the outputs from two 10230 chip-long code streams generated by the base registers 1 and 2. Both registers have tapped feedback and their initial state is driven from the initial state register. Each signal type has a unique feedback tapping selection and for each satellite has a unique preset states are listed in signal specification. Fundamentally this is the same approach as the one used in generation of GPS spreading codes (GPS 2003, 2004). The code generating structures are in general relatively simple; containing only few registers and exclusive OR gates (XOR). Thus, the implementation in hardware is not an issue, even when the initialization values for each register are stored locally in the generators. For example, when implemented in hardware the combined Galileo E5a/b, GPS l1 (C/A) and GPS L5 coders utilize 261 logic cell combinatorial look-up-tables and 120 logic cell registers in Altera's FPGA platform. In a rough estimation, this is about 20 % of the hardware required for a single tracking channel with a TUTGNSS receiver implementation (described in Chap. 9).

Unlike GPS L1 spreading codes, Galileo E1 codes are specified and implemented as memory codes. These memory codes are hand-selected, pre-defined spreading code sequences that are stored to a local memory to be available to channels when needed. The codes are selected so that there appears to be no identifiable means to reproduce them correctly. Such memory codes were obtained starting from known code families (e.g., Gold codes) and using genetic algorithms to obtained optimized chip sequences. The optimization criteria were the conditions of having a good balance of zeros and ones in the codes and of achieving good auto-correlation and cross-correlation properties (Winkel 2006).

The advantages of the memory code approach are claimed to be in increasing receiver tracking performance, since the codes are optimized to minimum cross-correlation and improvement interference mitigation properties. The downsides identified for this approach include the increased hardware cost, increased implementation complexity, increased signal processing resource utilization and additional overhead in power consumption of the receiver due the memory needed (Gerein et al. 2005).

Galileo Interface Control Documentation (ICD) specifies 50 pairs of 4092 chip long codes (GALILEO SIS-ICD 2010). First 32 of both E1-B (with data) and E1-C (data-less or pilot) codes are selected for transmitting in Galileo SVs 1 to 32. Thus, implementing the codes at the receiver side requires 32.7 kilobytes of memory ($2 \times 32 \times 4{,}092$ chips). Practically all 50 codes should be stored in case of future updates of system. This will use 51.15 kilobytes. In a worst case, this should be multiplied with the number of channels, thus the need of memory might increase to over 0.5 (or 0, 8 with 50 codes) megabytes in a 16 channel receiver. Yet, this is just a worst case scenario and alternative approach is discussed in the latter part of Chap. 6.

Another major update in Galileo signals compared to GPS signals is the introduction of the sub-carrier modulation. The Binary Phase Shift Keying modulation (BPSK) was selected for first GPS signals and it has been de facto signal modulation scheme in GNSS ever since. The origins of BPSK modulation are in communications and it basically inverts the phase of sinusoidal carrier whenever there is a bit transition in decoded data. Since navigational signals need not just to carry data but to maintain precise timing as well, a new modulation technique, Binary Offset Carrier (BOC) modulation was introduced around 1999, allegedly at MITRE corporation (Betz 1999, 2001, 2002) to be used in future navigational signals. The next section describes in detail the various modulation classes used in Galileo.

The third significant change in Galileo and also the modernized GPS signals is the introduction of new types of composite signals. In this improvement the traditional data signal, consisting of low rate navigation data, spreading code, possible modulation signal and carrier wave is enhanced with a pilot signal. Pilot signals are transmitted orthogonally with the data signals, they have own sets of spreading codes and they do not contain any navigation data, excluding only a short predefined secondary code.

The real advantages of this new arrangement come from the dispersion of the two main properties of GNSS signal: navigation data and timing information. While the data channel holds the data containing ephemeris and almanac information of the satellite constellation, the pilot channel can be integrated over longer periods without crossing data transitions which can enhance drastically the carrier-to-noise ratio of the received signals.

The new open access Galileo signals are listed in Table 3.1 with the current and future the GPS signals in shared frequencies. From the table it is easy to see that the pilot signals are not only used in future Galileo but also in future GPS signals as well.

Table 3.1 Summary of open Galileo signal components (GALILEO SIS-ICD 2010; GPS 2003, 2004, 2006)

Frequency band	Signal	PRN code size	PRN code rate (MHz)	PRN epoch time (ms)	Modulation
E1/L1 1575.45 MHz	Galileo[a, b] E1-B E1-C	4,092 4,092	1.023 1.023	4 4	CBOC CBOC
	GPS C/A GPS L1C[b]	1,023 10,230	1.023 1.023	1 10	BPSK TMBOC
E5a/L5 1176.45 MHz	Galileo[b] E5a-I E5a-Q	10,230 10,230	10.230 10.230	1 1	BPSK BPSK
	GPS[b] L5-I L5-Q	10,230 10,230	10.230 10.230	1 1	BPSK BPSK
E5b 1207.14	Galileo[b] E5b-I E5b-Q	10,230 10,230	10.230 10.230	1 1	BPSK BPSK

[a] Memory based PRN codes [b] Composite signals (data—pilot)

Galileo Signal Models

The Galileo signals are DS-SS CDMA signals (Simon et al. 1994; Strang and Borre 2012) using various channel modulations, according to the signal type and carrier frequency. A simplified block diagram of the signal transmitted from each satellite is given in Fig. 3.6. On each carrier (i.e., in each frequency band) we can have several signals and each of them has its own PRN code (represented here by the spreading modulation block) and its own channel modulation. As discussed in previous section, some of the signals at each carrier transmit navigation data and some other signals are the so-called **pilot signals**, where no navigation data is transmitted. Sometimes, secondary pseudorandom codes of short duration (e.g., 25–100 bits) are concatenated on pilot signals.

For a unified model, the Data blocks illustrated in Fig. 3.6 are modeling both the navigation data (when present) and the secondary codes data, without any loss of generality. The different signals on one frequency band are multiplexed and sent to the corresponding carrier frequency. Different power amplification can be (and are) used for each frequency band.

The signal at the output of the spreading modulation can be modeled via (Lohan et al. 2006a, b), as:

$$s(t) = \sum_{n=-\infty}^{+\infty} b_n \sum_{k=1}^{+S_F} c_{k,n} \delta(t - nT - kT_c) \tag{3.1}$$

where T_c is the chip rate, b_n is the data bit corresponding to the n-th code epoch, $c_{k,n}$ is a ±1 chip value corresponding to the k-th pseudorandom chip of the n-th data bit,

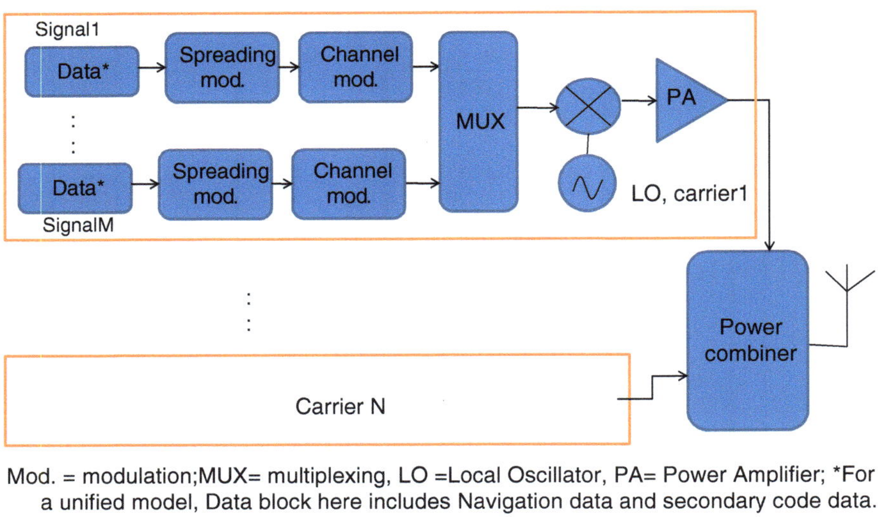

Fig. 3.6 Galileo signal transmitted from each satellite

T is the code epoch interval, and $\delta(t)$ is the Dirac function. A continuous signal model has been adopted here for the sake of simplicity of the notations. It is to be noted however that we usually deal with discrete signal samples and this continuous signal model is only an equivalent model. This is particularly true for GNSS signals where there is no bandwidth limitation at the transmitter (infinite bandwidth assumption).

As seen in Fig. 3.6, the spreading modulation is followed by a channel modulation, which is channel-and frequency band-specific, as shown in Table 3.2 (Galileo SIS-ICD 2010). The N/A values in the table correspond to undisclosed information (e.g., codes used for military applications).

There are basically 5 types of channel modulations employed in Galileo, directly or indirectly, described briefly in the next sub-sections:

- Binary Phase Shift Keying modulation (BPSK),
- Sine Binary Offset Carrier modulations (SinBOC), used only as a building block for the Composite Binary Offset Carrier modulation,
- Cosine Binary Offset Carrier modulation (CosBOC),
- Composite Binary Offset Carrier modulation (CBOC), which is a combination of SinBOC modulations and a particular case of a more general class, named Multiplexed Binary Offset Carrier (MBOC) modulation,
- Alternate Binary Offset Carrier modulation (AltBOC).

The following sub-sections use the equivalent baseband representation of the signals. It is recalled that, if $s_{\text{mod}}(t)$ is the baseband signal (e.g., the output of the channel modulation in Fig. 3.6, then the passband signal $m(t)$ is:

$$m(t) = A_c s_{\text{mod}}(t)\cos(2\pi f_{carrier} t) \tag{3.2}$$

Table 3.2 Galileo signals structure

Signal name	Frequency band edge frequencies [MHz]	Spreading modulation	Modulation type	Chip rate [MHz]	Code epoch length [chips]
Galileo E1-A	1,587–1,591	DS-SS	CosBOC (15,2.5)	2.5575	N/A
Galileo E1-B	1,587–1,591	DS-SS	CBOC(+)	1.023	4,092
Galileo E1-C	1,587–1,591	DS-SS	CBOC(−)	1.023	4,092
Galileo E6-A	1,260–1,278.75	DS-SS	CosBOC (10,5)	5.115	5,115
Galileo E6-B	1,260–1,278.75	DS-SS	BPSK(5)	5.115	5,115
Galileo E6-C	1,260–1,278.75	DS-SS	BPSK(5)	5.115	5,115
Galileo E5a	1,164–1,191.795	DS-SS	BPSK(10)	10.23	10,230
Galileo E5b	1191–795–1,214	DS-SS	BPSK(10)	10.23	10,230
Galileo E5	1,164–1,214	DS-SS	AltBOC (15,10)	10.23	10,230

Where A_c is the sub-sequent power amplification factor, and $f_{carrier}$ is the carrier frequency. The next sub-sections focus on the $s_{\text{mod}}(t)$ signal model, for various modulations.

Bpsk

The BPSK-modulated signal has the form:

$$s_{BPSK}(t) = p_{T_c}(t) \otimes s(t) \tag{3.3}$$

where $p_{T_c}(t)$ is the pulse shape equal to the rectangular waveform of duration T_c as shown in Fig. 3.7, and $\otimes$ is the convolution operation. The above formula simply states that a +1 chip is transmitted as a +1 symbol, and a −1 chip is transmitted as a −1 symbol. The notation BPSK(n) stands for a chip rate of $n \times 1.023$ MHz, n being a positive non-zero number.

In the acquisition and tracking processes, the basic building block is the correlation between the incoming signal and a reference code. For this purpose, a characterization of the correlation properties of the modulation waveform is tremendously important. The Auto-Correlation Function (ACF) $R_{BPSK}(\tau)$ of a BPSK-modulated code is given by:

$$R_{BPSK}(\tau) = \Lambda_{T_c}(\tau) \otimes R_s(\tau) \tag{3.4}$$

where $\Lambda_{T_c}(\tau)$is a triangular shaped function of unit amplitude and width $2T_c$, representing the convolution between the rectangular pulse shown in Fig. 3.7 and itself, and $R_s(\tau)$ is the auto-correlation function of the pseudorandom code, including the data bits. Typically, an approximation is used for $R_{BPSK}(\tau)$, based on the assumption that the PRN code has an infinite length and that the chips and data symbols are independent and identically distributed. Under this assumption, $R_s(\tau) \simeq \delta(\tau)$ and the ACF of a BPSK-modulated chip sequence of chip interval T_c can be approximated via the triangular shape shown in the left plot of Fig. 3.8:

$$R_{BPSK}(\tau) \approx \Lambda_{T_c}(\tau) \tag{3.5}$$

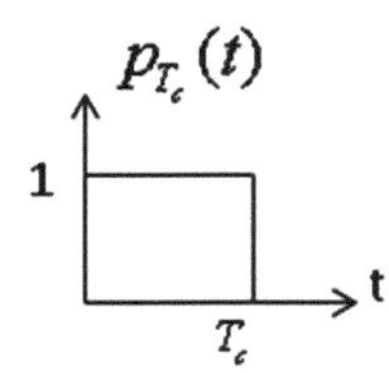

Fig. 3.7 Rectangular pulse shape

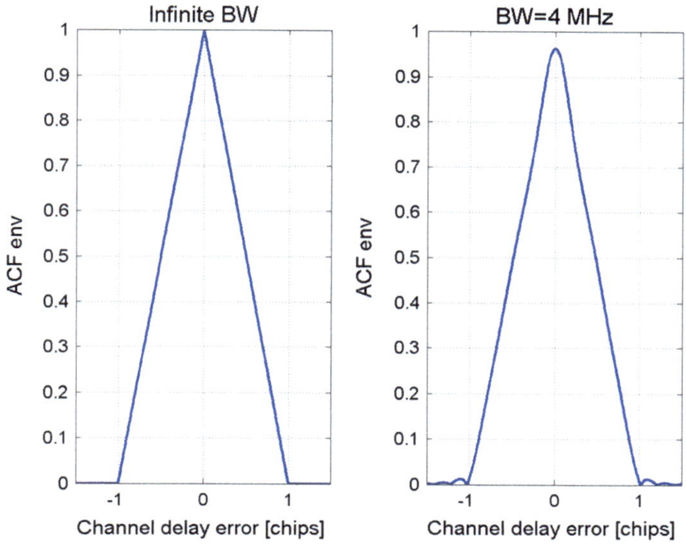

Fig. 3.8 Autocorrelation envelopes for a BPSK(1)-modulated ideal code. *Left* infinite bandwidth, *Right* 4 MHz double-sided bandwidth

We remark that the above formula is valid only for infinite bandwidths. If the bandwidth is limited at the receiver side, then the actual ACF observed at the receiver side is the convolution between the triangular pulse and the limiting bandwidth filter impulse response, as shown in the right plot of Fig. 3.8.

The Power Spectral Density (PSD) $G_{BPSK}(f)$ of the BPSK-modulated signal is the Fourier transform of the ACF and it is given by (Proakis 2008)

$$G_{BPSK}(f) = \frac{1}{T_c}|P_{T_c}(f)|^2 G_b\left(e^{j2\pi f T_c}\right) \tag{3.6}$$

where $G_b\left(e^{j2\pi f T_c}\right)$ is the power spectral density function of the transmitted discrete-time symbol train $\{b_n\}_n$ and $P_{T_c}(f)$ is the Fourier transform of the rectangular pulse shape shown in Fig. 3.7, which satisfies the relationship:

$$\frac{1}{T_c}|P_{T_c}(f)|^2 = T_c \sin c^2(\pi f T_c), \quad \operatorname{sinc}(x) \triangleq \frac{\sin(x)}{x} \tag{3.7}$$

It is to be noticed that for infinite-length ideal codes, $G_b\left(e^{j2\pi f T_c}\right) = 1$, i.e., the PSD is shaped only by the modulation part.

It is to be noted that BPSK is also the modulation of choice in GPS C/A signal (Kaplan and Hegarty 2006).

SinBOC

SinBOC modulation is a square sub-carrier modulation, where the pseudorandom code at chip rate $f_c = \frac{1}{T_c}$ is multiplied by a rectangular sub-carrier of frequency f_{sc}, which splits the spectrum of the signal. The typical notation is SinBOC(m, n), where $m = \frac{f_{sc}}{1.023 \times 10^6}$ and $n = \frac{f_c}{1.023 \times 10^6}$, when f_{sc} and f_c are expressed in Hz. The ratio $N_B = \frac{2m}{n}$ is typically known as BOC modulation order (Betz 1999, 2000; Hein et al. 2004; Lohan et al. 2006a, b) and half of its values shows the number of sub-carrier cycles within one chip interval. According to Betz definition (Betz 1999), the SinBOC waveform is given by:

$$s_{SinBOC(m,n)}(t) = sign\left(\sin\left(\frac{N_B \pi t}{T_c}\right)\right), \quad N_B = \frac{2m}{n} \tag{3.8}$$

Three example of SinBOC time waveforms are shown in Fig. 3.9 for one chip interval. The SinBOC waveforms are shown in red dashed line, and the underlying sinusoid shape is shown in continuous blue line. It is seen that N_B represents the number of sub-intervals (or BOC intervals) per chip.

The motivation of proposing a BOC modulation for Galileo signals was to ensure a maximum spectral separation in L1/E1 band with the existing GPS signals. Indeed, as seen in what follows, BOC is a split-spectrum modulation, that moves away the signal energy from the carrier frequency and has a low spectral correlation coefficient with GPS signals (Betz 1999, 2000).

By using the equivalent representation of SinBOC signals of (Lohan et al. 2006a, b), which allows a straightforward modeling of the baseband signal, the SinBOC-modulated signal can be written as:

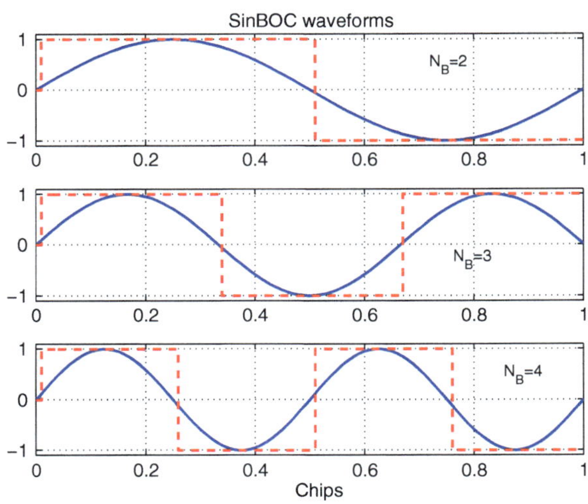

Fig. 3.9 Examples of SinBOC time-domain waveforms. The time axis is given in chips

$$s_{SinBOC(m,n)}(t) = p_{T_B}(t) \otimes \sum_{i=0}^{N_B-1} (-1)^i \delta(t - iT_B) \otimes s(t), \quad T_B = \frac{T_c}{N_B} \tag{3.9}$$

The ACF $R_{SinBOC(m,n)}(\tau)$ of a SinBOC-modulated code is given by (Lohan et al. 2006a, b):

$$R_{SinBOC(m,n)}(\tau) = \sum_{k=0}^{N_B-1} \sum_{i=0}^{N_B-1} (-1)^{i+k} \delta(\tau - iT_B + kT_B) \otimes \Lambda_{T_c}(\tau) \otimes R_s(\tau) \tag{3.10}$$

An example of the ACF envelope is shown in Fig. 3.10 for a SinBOC(6,1)-modulated ideal code.

The PSD $G_{SinBOC(m,n)}(f)$ of the SinBOC-modulated signal is (Lohan et al. 2006a, b):

$$G_{SinBOC(m,n)}(f) = \begin{cases} \frac{1}{T_c}\left(\frac{\sin\left(\pi f \frac{T_c}{N_B}\right)\sin(\pi f T_c)}{\pi f \cos\left(\pi f \frac{T_c}{N_B}\right)}\right)^2 G_b\left(e^{j2\pi f T_c}\right), \mathit{if} \quad N_B = 2\frac{m}{n} \quad \mathit{even} \\ \frac{1}{T_c}\left(\frac{\sin\left(\pi f \frac{T_c}{N_B}\right)\cos(\pi f T_c)}{\pi f \cos\left(\pi f \frac{T_c}{N_B}\right)}\right)^2 G_b\left(e^{j2\pi f T_c}\right), \mathit{if} \quad N_B = 2\frac{m}{n} \quad \mathit{odd} \end{cases} \tag{3.11}$$

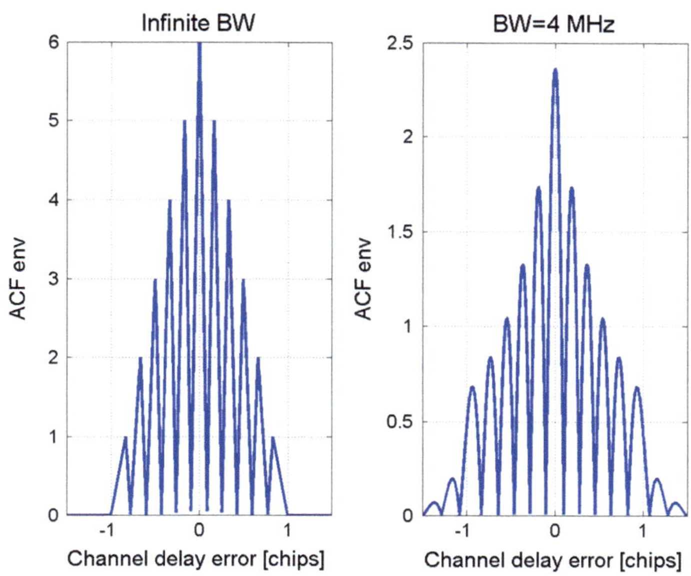

Fig. 3.10 Autocorrelation envelopes for a SinBOC(6,1)-modulated ideal code. *Left* infinite bandwidth, *Right* 4 MHz double-sided bandwidth

CosBOC

Cosine-BOC modulation uses a similar principle with SinBOC modulation, but now the sub-carrier is in cosine phase, instead of sine phase (Betz 2001; Weiler et al. 2008):

$$s_{CosBOC(m,n)}(t) = sign\left(\cos\left(\frac{N_B \pi t}{T_c}\right)\right), \quad N_B = \frac{2m}{n} \tag{3.12}$$

Three example of CosBOC time waveforms over one-chip interval are shown in Fig. 3.11 in dashed lines, together with the underlying cosine waveform, in continuous line.

The CosBOC-modulated signal can be modeled via (Lohan et al. 2006a, b):

$$s_{CosBOC(m,n)}(t) = p_{\frac{T_B}{2}}(t) \otimes \sum_{k=0}^{1} \sum_{i=0}^{N_B-1} (-1)^{i+k} \delta\left(t - iT_B - k\frac{T_B}{2}\right) \otimes s(t), \; T_B = \frac{T_c}{N_B} \tag{3.13}$$

The ACF $R_{CosBOC(m,n)}(\tau)$ of a CosBOC-modulated code is given by (Lohan et al. 2006a, b):

$$R_{CosBOC(m,n)}(\tau) = \sum_{m=0}^{1} \sum_{n=0}^{1} \sum_{k=0}^{N_B-1} \sum_{i=0}^{N_B-1} (-1)^{i+k} \delta\left(\tau - iT_B + kT_B - m\frac{T_B}{2} + n\frac{T_B}{2}\right) \otimes \Lambda_{T_c}(\tau) \otimes R_s(\tau) \tag{3.14}$$

An example of the ACF envelope is shown in Fig. 3.12 for a CosBOC(6, 1)-modulated ideal code.

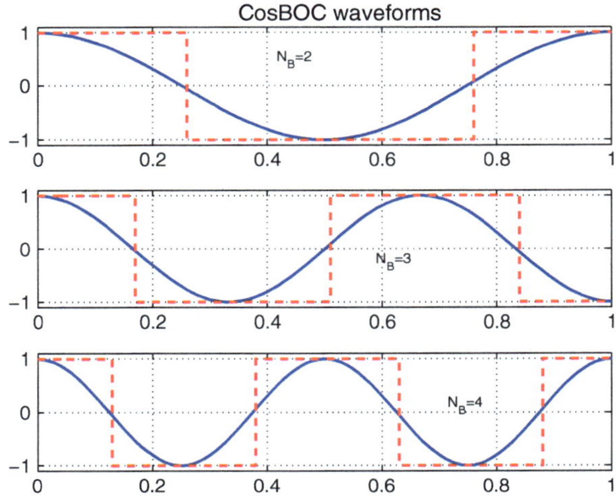

Fig. 3.11 Examples of CosBOC time waveforms. The time axis is given in chips

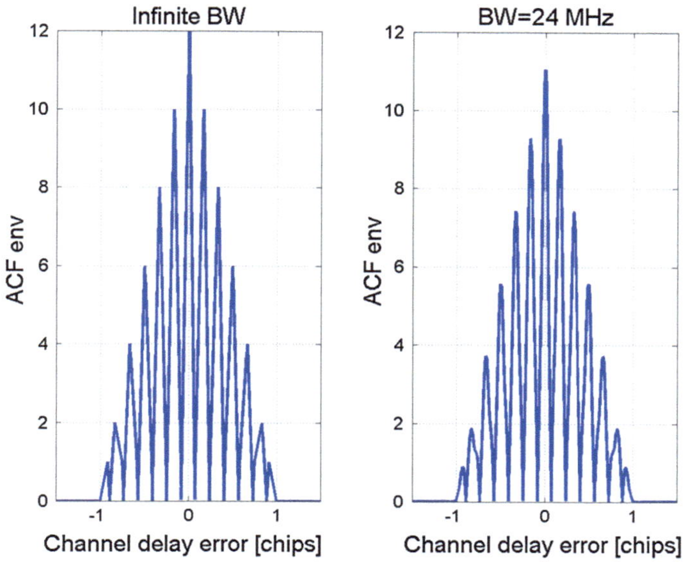

Fig. 3.12 Autocorrelation envelopes for a CosBOC(6,1)-modulated ideal code. *Left* infinite bandwidth, *Right* 24 MHz double-sided bandwidth

The PSD $G_{CosBOC(m,n)}(f)$ of the CosBOC-modulated signal is (Lohan et al. 2006a, b):

$$G_{CosBOC(m,n)}(f) = \begin{cases} \dfrac{1}{T_c}\left(\dfrac{\sin\left(\pi f \frac{T_c}{2N_B}\right)\sin\left(\pi f \frac{T_c}{N_B}\right)\sin(\pi f T_c)}{\pi f \cos\left(\pi f \frac{T_c}{2N_B}\right)\cos\left(\pi f \frac{T_c}{N_B}\right)}\right)^2 G_b(e^{j2\pi f T_c}), & if \; N_B \, even \\ \dfrac{1}{T_c}\left(\dfrac{\sin\left(\pi f \frac{T_c}{2N_B}\right)\sin\left(\pi f \frac{T_c}{N_B}\right)\cos(\pi f T_c)}{\pi f \cos\left(\pi f \frac{T_c}{2N_B}\right)\cos\left(\pi f \frac{T_c}{N_B}\right)}\right)^2 G_b(e^{j2\pi f T_c}), & if \; N_B \, odd \end{cases} \tag{3.15}$$

MBOC

In 2006, an agreement between GPS and Galileo authorities defined the usage of Multiplexed BOC modulation (MBOC) in future navigational signals in E1/L1 frequency band. This was made to ensure the compatibility between GPS and Galileo signals. The MBOC was defined only by its power spectral density (PSD), which left the implementation details open (Galileo SIS-ICD 2008):

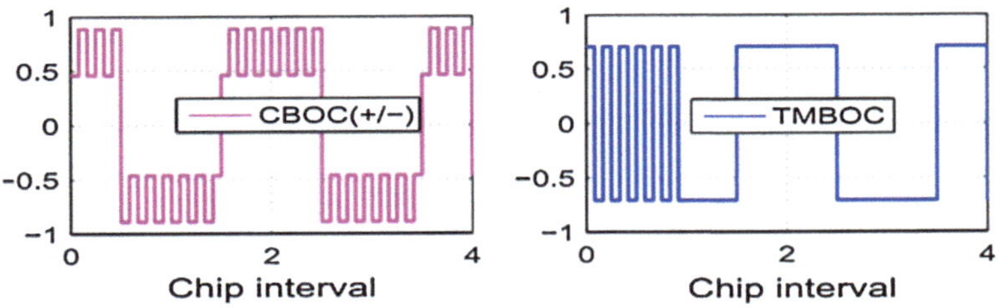

Fig. 3.13 Time domain samples of CBOC (*on left* figure) and TMBOC (*on right*)

$$G_{CBOC(+)} = \frac{10}{11}\text{SinBOC}(1,1) + \frac{1}{11}\text{SinBOC}(6,1) \tag{3.16}$$

For Galileo E1 signals, MBOC is chosen to be implemented as composite BOC (CBOC), described next. In addition to the CBOC(+) and CBOC(−) implementations, other two possible MBOC implementations are the alternating CBOC(±) and the Time Multiplexed BOC (TMBOC), used in modernized GPS signals (GPS 2006). Figure 3.13 illustrates the difference between CBOC and TMBOC implementations in time domain.

CBOC

The Composite Binary Offset Carrier modulation is a particular case of Multiplex BOC modulations (MBOC) and it is a weighted combination of SinBOC(1,1) and SinBOC(6,1) code symbols (Avila-Rodriguez et al. 2006, 2006bis; Dovis et al. 2008; Hein et al. 2005, 2006; Hoult et al. 2008; Macchi et al. 2008). The two variants employed by Galileo are named CBOC(+) and CBOC(−) respectively, and they represent a weighted combination of SinBOC(1,1) and SinBOC(6,1)-modulated chips, as shown below:

$$s_{CBOC(+)}(t) = w_1 s_{SinBOC(1,1)}(t) + w_2 s_{SinBOC(6,1)}(t) \tag{3.17}$$

and

$$s_{CBOC(-)}(t) = w_1 s_{SinBOC(1,1)}(t) - w_2 s_{SinBOC(6,1)}(t) \tag{3.18}$$

Where $w_1 = \sqrt{\frac{10}{11}}, w_2 = \sqrt{\frac{1}{11}}$ are two weighting factors, defined in (Galileo SIS-ICD 2010). An example of CBOC time-domain waveforms, together with the underlying pseudorandom (PRN) code is shown in Fig. 3.14.

The expression of an ACF of a CBOC-modulated signal is given below (Lohan and Renfors 2007) and an example is shown in Fig. 3.15:

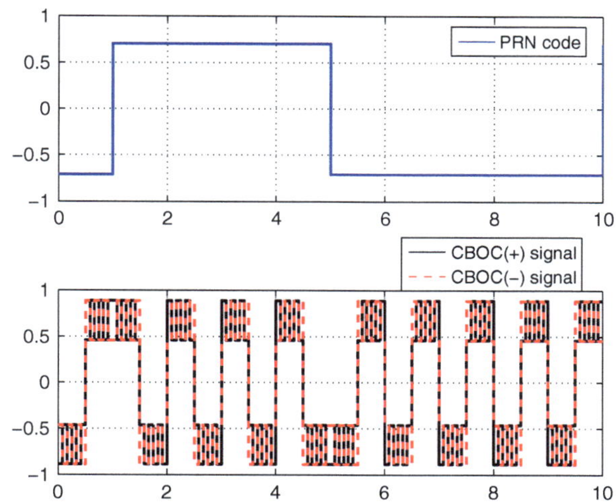

Fig. 3.14 Example of CBOC time-domain waveforms

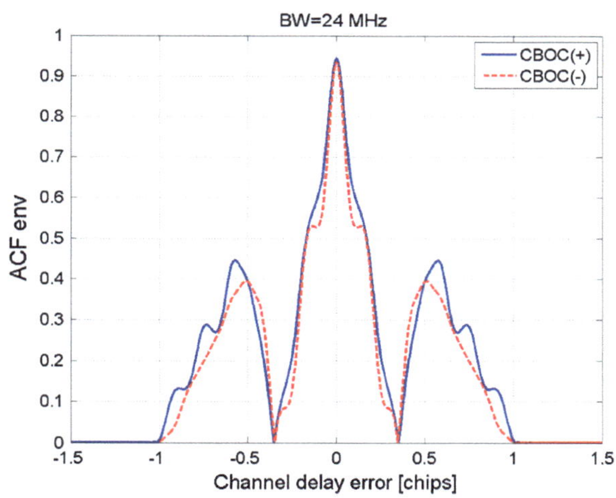

Fig. 3.15 Autocorrelation envelopes for a CBOC(+) and a CBOC(−)-modulated ideal code. 24.552 MHz double-sided bandwidth

$$\begin{aligned}R_{CBOC(+)}(\tau) &= w_1^2 \sum_{i=0}^{1}\sum_{i_1=0}^{1}\sum_{l=0}^{5}\sum_{l_1=0}^{5}(-1)^{i+i_1}\delta\left(\tau - i\frac{T_c}{2} + i_1\frac{T_c}{2} - l\frac{T_B}{12} + l_1\frac{T_B}{12}\right)\otimes\Lambda_{\frac{T_c}{12}}(\tau)\otimes R_s(\tau)\\ &+ w_2^2\sum_{i=0}^{11}\sum_{i_1=0}^{11}(-1)^{i+i_1}\delta\left(\tau - i\frac{T_c}{12} + i_1\frac{T_c}{12}\right)\otimes\Lambda_{\frac{T_c}{12}}(\tau)\otimes R_s(\tau)\\ &+ w_1w_2\sum_{i=0}^{1}\sum_{i_1=0}^{11}\sum_{l=0}^{5}(-1)^{i+i_1}\delta\left(\tau - i\frac{T_c}{2} + i_1\frac{T_c}{12} - l\frac{T_c}{12}\right)\otimes\Lambda_{\frac{T_c}{12}}(\tau)\otimes R_s(\tau)\\ &+ w_1w_2\sum_{i=0}^{1}\sum_{i_1=0}^{11}\sum_{l=0}^{5}(-1)^{i+i_1}\delta\left(\tau + i\frac{T_c}{2} - i_1\frac{T_c}{12} + l\frac{T_c}{12}\right)\otimes\Lambda_{\frac{T_c}{12}}(\tau)\otimes R_s(\tau)\end{aligned} \tag{3.19}$$

For CBOC(−), the weight w_2 should be replaced with $-w_2$ in the above equation. It is to be noted that such expressions have a very simple implementation in Matlab. Code examples can be found in (Strang and Borre 2012).

The PSD of CBOC(+) and CBOC(−) signals has been derived in (Lohan 2010) and is:

$$G_{CBOC(+)}(f) = \frac{1}{T_c}\left|\frac{\sin(\pi f T_c)}{\pi f}\left(w_1 e^{-j\pi f\frac{T_c}{2}}\tan\left(\frac{\pi f T_c}{2}\right) + w_2\tan\left(\frac{\pi f T_c}{12}\right)\right)\right|^2 G_b\left(e^{j2\pi f T_c}\right) \tag{3.19}$$

$$G_{CBOC(-)}(f) = \frac{1}{T_c}\left|\frac{\sin(\pi f T_c)}{\pi f}\left(w_1 e^{-j\pi f\frac{T_c}{2}}\tan\left(\frac{\pi f T_c}{2}\right) - w_2\tan\left(\frac{\pi f T_c}{12}\right)\right)\right|^2 G_b\left(e^{j2\pi f T_c}\right) \tag{3.20}$$

It was shown that Galileo CBOC signals may be processed not only with a CBOC receiver, but also with a SinBOC(1,1) with a maximum penalty of –0.9 dB in reception power (Hein et al. 2006).

AltBOC

In AltBOC-modulation case, four real signals (or two complex signals) are modulated onto the two phases of orthogonal sub-carriers (Artaud et al. 2008; Margaria et al. 2008; Sleewaegen et al. 2004). Assuming that the two complex signals are denoted via $s_1(t)$ and $s_2(t)$, the Non-Constant Envelope (NCE) AltBOC(m,n)-modulated signal can be written as (Lohan et al. 2006a, b):

$$\begin{aligned} s_{AltBOC(m,n),NCE}(t) &= p_{T_B}(t) \otimes \sum_{i=0}^{N_B-1}(-1)^i\delta(t - iT_B) \otimes s_1(t) \\ &+ p_{\frac{T_B}{2}}(t) \otimes \sum_{k=0}^{1}\sum_{i=0}^{N_B-1}(-1)^{i+k}\delta\left(t - iT_B - k\frac{T_B}{2}\right) \otimes s_2(t) \end{aligned} \tag{3.21}$$

Where $T_B = \frac{T_c}{N_B}$ and $N_B = \frac{2m}{n}$.

The expression of an ACF of a non-constant envelope AltBOC-modulated signal is given in (Lohan et al. 2006a, b) and examples are shown in Fig. 3.16.

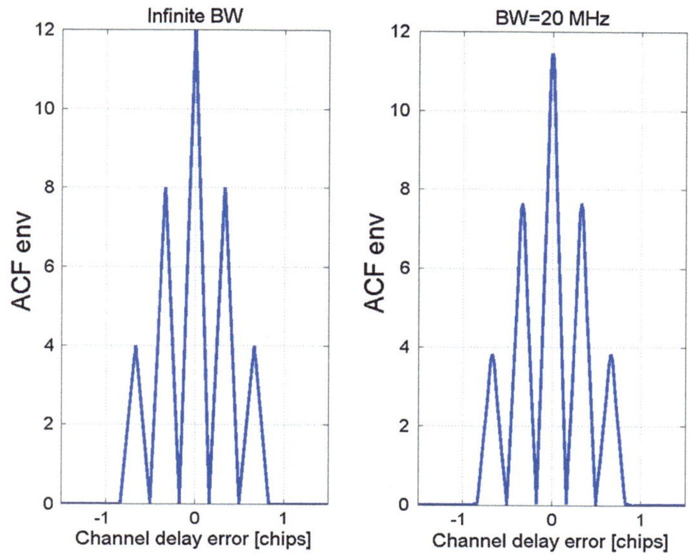

Fig. 3.16 Autocorrelation envelopes for a AltBOC(15,10)-modulated ideal code. *Left* infinite bandwidth, *Righ* 20 MHz double-sided bandwidth

$$\begin{aligned} R_{AltBOC(15,10),NCE}(\tau) &= \sum_{i=0}^{2}\sum_{k=0}^{2}\sum_{i_1=0}^{1}\sum_{k_1=0}^{1}(-1)^{i+i_1+k+k_1}\delta\left(\tau - i\frac{T_c}{3} + i_1\frac{T_c}{3} - k\frac{T_B}{6} + k_1\frac{T_B}{6}\right) \\ &\otimes \Lambda_{\frac{T_c}{6}}(\tau)\otimes R_s(\tau) \\ &+ \sum_{l=0}^{2}\sum_{m=0}^{2}\sum_{p=0}^{1}\sum_{p_1=0}^{1}(-1)^{l+m}\delta\left(\tau - l\frac{T_c}{3} - m\frac{T_c}{3} - p\frac{T_B}{6} + p_1\frac{T_B}{6}\right)\otimes\Lambda_{\frac{T_c}{6}}(\tau)\otimes R_s(\tau) \end{aligned} \tag{3.22}$$

The PSD of the NCE AltBOC(15,10) can be modeled as (Rebeyrol et al. 2005; Lohan et al. 2006a, b):

$$G_{AltBOC(15,10),NCE} = \frac{4}{T_c}\left(\frac{\sin\left(\pi f\frac{T_c}{6}\right)\cos(\pi f T_c)\tan\left(\pi f\frac{T_c}{3}\right)}{\pi f}\right)^2\left(\frac{G_{b_1}\left(e^{j2\pi f T_c}\right)}{\cos^2\left(\frac{\pi f T_c}{6}\right)} + \frac{G_{b_2}\left(e^{j2\pi f T_c}\right)}{\sin^2\left(\frac{\pi f T_c}{6}\right)}\right) \tag{3.23}$$

where $G_{b_1}(e^{j2\pi f T_c})$ and $G_{b_2}(e^{j2\pi f T_c})$ are the PSDs of the transmitted discrete-time symbol trains $\{b_{1n}\}_n$ and $\{b_{2n}\}_n$, corresponding to the signal waveforms $s_1(t)$ and $s_2(t)$, respectively.

In order to avoid the issue of a non-constant-envelope, (Ries et al. 2003) proposed a constant-envelope method, where the sub-carrier waveforms are chosen such that the sum and difference of complex values always lie on the unit circle in the complex plane. The PSD of the **constant envelope AltBOC(15,10)** has been derived in (Rebeyrol et al. 2005), under the assumption of $G_{b_1}(e^{j2\pi fT_c}) = 1, G_{b_2}(e^{j2\pi fT_c}) = 1$ and it is equal to:

$$G_{AltBOC(15,10),CE} = \frac{4}{T_c}\left(\frac{\cos(\pi fT_c)}{\pi f \cos\left(\pi f \frac{T_c}{3}\right)}\right)^2 \left(\cos^2\left(\frac{\pi fT_c}{3}\right) - \cos\left(\frac{\pi fT_c}{3}\right) - 2\cos\left(\frac{\pi fT_c}{3}\right)\cos\left(\frac{\pi fT_c}{6}\right) + 2\right) \tag{3.24}$$

More details on Galileo signals and modulations can be found also in Chap. 2 of the book (Strang and Borre 2012) and in the PhD thesis (Avila-Rodriguez 2008).

References

Artaud G, Lestarquit L, Issler JL (2008) AltBOC for dummies or everything you always wanted to know about AltBOC. In: Proceedings of the ION GNSS 2008, Savannah, GA, September 2008

Avila-Rodriguez JA (2008) On generalized signal waveforms for satellite navigation, PhD thesis, Munich, June 2008

Avila-Rodriguez JA, Wallner S, Hein G (2006) MBOC: the new optimized spreading modulation recommended for Galileo E1 OS and GPS L1C, ESA Navitec 2006, Noordwijk, The Netherlands, 11–13 Dec 2006

Avila-Rodriguez JA, Wallner S, Hein G (2006) How to optimize GNSS signals and codes for indoor positioning In: ION GNSS 19th international technical meeting of the satellite division, Sept 2006

Balaei AT, Akos DM (2011–2012) Cross correlation impacts and observations in GNSS Receivers, NAVIGATION 58(4):323–333

Betz J (1999) The offset carrier modulation for GPS modernization. In:Proceedings of ION technical meeting, Cambridge, Massachusetts June 1999, pp 639–648

Betz J (2000) Design and performance of code tracking for the GPS M code signal. In: CDROM Proceedings of ION meeting, Anaheim, CA, Sept 2000

Betz J (2001) Binary offset carrier modulations for radionavigation. Navig J Inst Navig 48 (4):227–246

Braasch M, Dierendonck AJ (1999) GPS receiver architectures and measurements. In: Proceedings of IEEE, pp 48–64

Collins JP (2012) Assessment and development of a tropospheric delay model for aircraft users of the global positioning system, technical report no 203, Department of geodesy and Geomatics Engineering, University of New Brunswick, Canada, 1999. http://gge.unb.ca/Pubs/TR203.pdf. Accessed Sep 2012

Dovis F, Presti LL, Fantino M, Mulassano P, Godet J (2008) Comparison between Galileo CBOC Candidates and BOC(1,1) in Terms of Detection Performance. Int J Navig Observ 793868:9. doi:10.1155/2008/793868

European GNSS (Galileo) (2010) Open service—signal in space interface control document, Feb 2010

Galileo Open Service (2008) Signal in space interface control document (OS SIS ICD). European Space Agency/European GNSS Supervisory Authority, Georgia

Gerein N, Olynik M, Clayton M, Auld J, Murfin T (2005) Early results are in GPS world. (November issue)

Éamonn P. Glennon, Dempster AG (2005) A novel GPS cross correlation mitigation technique. In: Proceedings of the 18th international technical meeting of the satellite division of the institute of navigation (ION GNSS 2005), Long Beach, CA, September 2005, pp 190–199

GPS Interface Control Document (ICD-GPS-200D) (2004) IRN-200C-004, U.S. Air Force. Accessed 7 Dec 2004
Hein G, Irsigler M, Rodriguez JA, Pany T (2004) Performance of GALILEO L1 signal candidates. In: CDROM proceedings of european navigation conference GNSS, May 2004
Hein G, Avila-Rodriguez J, Ries L, Lestarquit L, Issler J, Godet JJ, Pratt T (2005) A candidate for the GALILEO L1 OS optimized signal. In: ION GNSS 18th international technical meeting of the satellite division, Long Beach, CA, pp 833–845, September 2005
Hein G, Avila-Rodriguez J, Wallner S, Pratt A, Owen J, Issler J, Bezt J, Hegarty C, Lenahan S, Rushanan J, Kraay A, Stansell T (2006) MBOC: the new optimized spreading modulation recommended for GALILEO L1 OS and GPS L1C. In: Proceedings of IEEE/ION PLANS (2006), pp 883–892
Hoult N, Aguado E, Xia P (2008) MBOC and BOC(1,1) performance comparison. J Navig 61:613–627
GPS Space Segment/User Segment L5 Interfaces (2003) IS-GPS-705, U.S. air force. Accessed 24 Nov 2003
GPS Space Segment/User Segment L1C Interfaces (2006) Draft IS-GPS-800, U.S. Air Force. Accessed 19 Apr 2006
Kaplan ED, Hegarty KJ (2006) Understanding GPS: principles and applications, 2nd edn. Artech House, USA
Lohan ES, Renfors M (2007) Correlation properties of multiplexed binary offset carrier (MBOC) modulation. In: Proceedings of european wireless conference, 2007
Lohan ES, Lakhzouri A, Renfors M (2006a) Binary-offset-carrier modulation techniques with applications in satellite navigation systems. Wirel Commun Mobile Comput 7(6):767–779
Lohan ES, Lakhzouri A, Renfors M (2006b) Complex double-binary-offset-carrier modulation for a unitary characterization of GALILEO and GPS signals. IEE Proc Radar Sonar Navig 153 (5):403–408
Lohan, ES (2010) Analytical performance of CBOC-modulated Galileo E1 signal using sine BOC (1,1) receiver for mass-market applications. In: Proceedings of ION-PLANS 2010 3–5 May 2010. Palm Springs, CA
Macchi F, Petovello M, Lachapelle G (2008) Consequences of MBOC approximation by BOC modulation, In: International technical meeting of the satellite division ION GNSS. Savannah, GA
Margaria D, Dovis F, Mulassano P (2008) GALILEO AltBOC signal multiresolution acquisition strategy. IEEE Aeros Electron Syst Mag 23(11):4–10
Morton Y, Miller M, Tsui J, Lin D, Zhou Q (2007) GPS civil self-interference mitigation during weak signal acquisition. IEEE Trans Signal Process 55(12):5859–5863
Nunes FD, Sousa FMG (2012) GNSS near-far mitigation through subspace projection without phase information. IEEE Trans Aerosp Electron Syst 48(3):2746–2755
O'Driscoll C, Murphy CC (2007) A simple technique for the detection of multiple access interference in the parallel acquisition of weak GPS signals. In: Proceedings of the 20th international technical meeting of the satellite division of the institute of navigation (ION GNSS 2007), Fort Worth, TX, pp 282–291, September 2007
Proakis J (2008) Digital communications. McGraw Hill, New York
Rappaport T (1996) Wireless communications: principles and practice, 2nd edn. Prentice Hall, NJ
Rebeyrol E, Macabiau C, Lestarquit L, Ries L, Issler JL, Boucheret ML, Bousquet M (2005) BOC power spectrum densities. In: ION NationalTechnical Meeting (ION-NMT), 2005
Richert T, El-Sheimy N (2005) Ionospheric modeling the key to GNSS ambiguity resolution, GPS World, pp. 35–40. http://www.antcom.com/pdf/GPS7.06.05e.pdf. Accessed June 2005
Ries L, Legrand F, Lestarquit L, Vigneau W, Issler JL (2003) Tracking and multipath performance assessments of BOC signals using a bit-level signal processing simulator. In: ION GPS/GNSS, Portland, OR, USA, pp 1996–2010
Seco-Granados G, Lopez-Salcedo JA, Jiménez-Baños D, López-Risueño G (2012) Signal processing challenges in indoor GNSS, IEEE Signal Process Mag, 29(2):108–131

Simon MK, Omura JK, Scholtz RA, Levitt BK (1994) Spread spectrum communications handbook: revised edition. McGraw-Hill Inc, New York

Sleewaegen M, de Wilde W, Hollreiser M (2004) GALILEO AltBOC receiver. In: CDROM proceedings ENC GNSS 2004, Rotterdam, The Netherlands, May 2004

Strang G, Borre K (2012) Algorithms for Global Positioning. Wellesley-Cambridge Press, USA

Weiler RM, Paul B, Stephen H, Martin U (2008) The effect of cosine phased BOC modulation on the GNSS receiver search process. J Navig, 61:591. doi:10.1017/S0373463308004876

Winkel JO (2013) Spreading codes for a satellite navigation system, ESA Patent application WO/2006/063613. http://patentscope.wipo.int/search/en/WO2006063613. Accesssed May 2013

Part II
Receiver, the Core Galileo Positioning Technology

Chapter 4
Advanced RF Front-End Design Issues

Ernesto Perez and Marco Detratti

Abstract This chapter aims at providing general design guidelines which could be helpful to any designer facing the challenge of the implementation of a Radio Frequency Front End (RF FE) for multiple GNSS applications. Even if some considerations will be focused on specific cases for the sake of simplicity, we will try to provide a comprehensive analysis of the main issues that have to be taken into account to carry out a proper design. While most of the discussion in this Chapter is valid for any GNSS receivers, the parts specific to Galileo will be emphasized when applicable.

Introduction

Every GNSS receiver comprises two main blocks: radio front-end and baseband processing unit, as shown in Fig. 4.1. The front-end is responsible of actually receiving the signal from space and conditioning it in a way the baseband unit is able to process it mathematically to obtain the satellite observables and, eventually, the receiver's position and timing. These two blocks can be integrated in a single IC or separated in different implementations, including multi-chip front-ends and even complete computers as baseband processors in SDR (Software-Defined Radio) receivers (Principe et al. 2011).

Radio front-ends must carry out the following tasks:

- Amplification: GNSS signals are very weak and its amplitude must be increased before the receiver can actually process it.

E. Perez
Acorde, Cantabria, Spain

M. Detratti (✉)
European GNSS Agency, Prague, Czech Republic (formerly with Acorde)
e-mail: Marco.DETRATTI@gsa.europa.eu

J. Nurmi et al. (eds.), *GALILEO Positioning Technology*,
Signals and Communication Technology 182, DOI 10.1007/978-94-007-1830-2_4

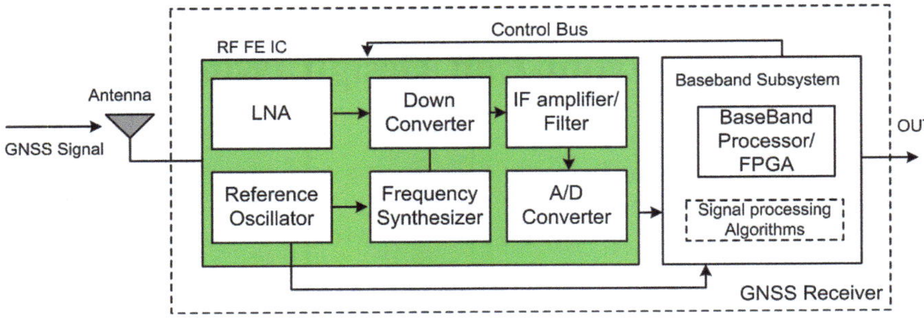

Fig. 4.1 Simplified GNSS receiver block diagram

- Filtering: ideally GNSS signals and only GNSS signals should be processed. Filters along the front-end chain make sure the interference from diverse sources is minimized.
- Frequency conversion: GNSS signals are transmitted at certain frequencies to allow their propagation through space, but these carriers must be shifted at the receiver.
- Digitization: GNSS signals must be sampled and quantized to be processed digitally. Analog-to-digital converters (ADCs) are the interface between the analog sections of the front-end and the baseband processor.

Designing an RF front-end targeting GNSS applications represents a challenge even for an experienced RF designer. Many factors have to be taken into account, starting from the fact that the circuit should be capable of receiving very weak radio navigation signals, well below the thermal noise floor, and thus very susceptible to interference.

The main drawback of currently available GNSS platforms is the poor sensitivity performance in typical multipath environments, as could be deep urban or indoor locations. Recently, to overcome this limit, the GPS modernization plan has added new, more sophisticated, civil codes. In addition to this, the European satellite navigation system Galileo will provide new signals many of which will be freely available as well. Once Galileo will be operational, the vast majority of user receivers sold will be capable of operating with multiple navigation systems. Users will benefit from receiving signals from multiple bands and services with improved accuracy, reliability, and availability. Multi-constellation multi-frequency receivers will become widespread.

Receiver operation is always affected by nearby communication devices (personal computers, mobile phones, radio transmitters,...) which should be carefully considered as they can seriously compromise performance given the very low power levels of navigation signals. The situation is even worse if the receiver has to be integrated in a multi-mode device (as is the case of smartphones). Hence, not only low-cost, low power consumption and smallest form factor, but also compatibility at the radio level with the appropriate standards should be addressed at the design stage.

GNSS RF Front-End Design

The front-end must not only satisfy the minimal functionality of signal conditioning through amplification, frequency conversion, filtering and sampling, but this must be done in such a way as to minimize the degradation of the carrier-to-noise density ratio (C/N_0) caused by the presence of potentially interfering signals. The following sections will provide an overview of the main figures of merit of radio front-ends and the design considerations associated to them.

Bandwidth

The available channel bandwidth should be adjusted to the strict minimum required by the signal, in order to improve out-of-band rejection while respecting the integrity of the full transmitted signal. In practice, however, this bandwidth has a strong impact on the design of a receiver.

A wider channel means more demanding design of the front-end and particularly the lower frequency sections: variable-gain amplifiers, active filters and analog-to-digital converters. Higher sampling frequencies means not only more power consumption from the ADCs but also in the low-level blocks of the baseband processor, correlators and tracking loops, with greater computational complexity.

On the other hand, wideband signals have some important advantages. They allow greater accuracy in the pseudorange estimation with a sharper autocorrelation, and potentially improve multipath performance (Principe et al. 2011). Figure 4.2 illustrates the normalized autocorrelation function of an ideal (i.e., infinite bandwidth) pseudorandom BPSK such as GPS L1 C/A (dashed line) and the autocorrelation of the same signal when the channel bandwidth is restricted to the main

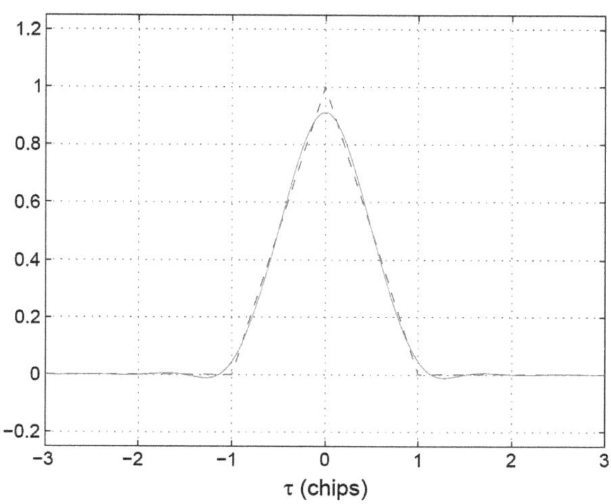

Fig. 4.2 Autocorrelation function of a band-limited BPSK PRN sequence

spectral lobe (continuous line). The power loss is minimal yet the correlation function is clearly smoothed because of this reduced bandwidth.

Therefore channel bandwidth is a trade-off mainly between performance and power consumption.

Noise Figure and Sensitivity

The ultimate goal of a GNSS receiver is to extract accurate position and time information from weak satellite signals. Sensitivity is a measure of the capability of the receiver to acquire and track these signals, and is affected by a combination of different parameters of the receiver including the baseband processing characteristics, thus resulting in a figure of merit not exclusively defined by the front-end.

From an RF point of view the effect of the FE on the sensitivity is a worsening of the signal-to-noise ratio (SNR) due to the front-end's own noise sources and other physical effects such as phase noise or nonlinearity.

Noise Figure (NF) is a measure of the degradation of the SNR caused by the front-end. Baseband sensitivity is typically expressed as a carrier-to-noise ratio or C/N_0, that is, the logarithmic difference between the power of the signal and the noise spectral density for which the signal can acquired and tracked with a certain probability, directly related to the integration interval of the correlators. If a baseband processor is said to be able to successfully acquire GNSS signals with a $C/N_{0,\,min}$, sensitivity can then be expressed in terms of minimum signal power, as typically done by receiver manufacturers:

$$P_{sens} = N_0(dB) + C/N_{0,min} + NF \tag{4.1}$$

For example, for a nominal spectral density at 290 K of −174 dBm/Hz, $C/N_{0,min}$ of 29 dB-Hz and a noise figure of 3 dB, we would have a sensitivity of −142 dBm. Therefore there is a dB-per-dB correspondence between NF and receiver sensitivity.

NF is the logarithmic expression of the *noise factor*, F.

$$NF(dB) = 10\log_{10}(F) \tag{4.2}$$

Friis's formula is used to obtain the combined noise factor of N cascaded stages, each of them with its own gain G_i and noise factor F_i (Friis).

$$F_{total} = F_1 + \frac{F_2 - 1}{G_1} + \frac{F_3 - 1}{G_1 G_2} + \cdots + \frac{F_N - 1}{\prod_{i=1}^{N-1} G_i} \tag{4.3}$$

There are two fundamental consequences of this formula: NF can never be lower than that of the first stage, and the higher the gain of a stage the lower the impact of the NF of any subsequent stage is. For this reason receivers designers make sure the first block is a *low noise amplifier* (LNA) with sufficient gain and low NF to keep the total NF of the front-end low. In many cases this first LNA is integrated with the antenna itself in what is known as an *active antenna*.

In addition to electrical noise, the quantization process inherent to the analog-to-digital conversion further degrades the SNR with this rounding of the analog amplitude to discrete values. This is known as *quantization noise*. Quantization noise depends fundamentally on the number of bits of the ADC and the quantization levels with respect to the signal amplitude; the closer the discrete amplitude of the digitized signal is to the analog input, the lower this noise is.

Quantization noise is traced mathematically as additive, but it is intrinsic to the discretization of the signal and Friis's formula does not apply; the SNR degradation caused by the ADCs must always be added to the NF of the combination of the analog blocks.

Under the idealization of white Gaussian noise this SNR degradation is approximately 1.96 dB for 1 bit, 0.55 dB for 2 bits and 0.17 dB for 3 bits. Above 3 bits the improvement in terms of NF is barely noticeable, so very few low-cost low-power receivers have greater precision than that. Such ADCs provide however greater protection against interference and are therefore preferred in professional applications where immunity to jamming or accidental interference is of utmost importance (Abdizadeh et al 2012). Multibit ADCs require an automatic gain to adjust the signal to its optimal level for quantization. Conversely, 1-bit ADCs behave as a simple comparator so they require no gain control, as the last amplification stages operate in saturation; these front-ends are simpler to design and very low power so they are preferred in many low cost applications, but perform poorly in the presence of interference.

Since quantization reshapes noise and GNSS signals are below this noise floor (meaning one can have the same spectral distribution of noise for different signals), more accurate characterization of SNR degradation by quantization usually relies on numerical simulations (Balaei et al. 2008).

Linearity

Electromagnetic compatibility is fundamental in a world of ubiquitous wireless communication devices, and GNSS receivers are no exception; precisely because of their inherent vulnerability to interferers and the vast spectrum of applications their integration allows, robustness against external signals is crucial. All systems transmitting at frequencies close to the bands of interest are potential sources of interference for a GNSS receiver, and even small leakages out of their allocated bandwidth or harmonic content of the transmitted signal represent a threat. The receiver must withstand these sources, so their power levels and frequencies must be considered together with the GNSS bands in the design process. Front-end linearity and out-of-band rejection determine this interference robustness to a large extent.

The linearity performance of the radio FE should ensure the correct behaviour of the GNSS receiving chain in the targeted environment. In this section we will refer to one of the most common of these multi-band scenarios, cellular handsets, but each application faces different challenges. GNSS is clearly the most vulnerable radio system within a terminal because of the inherently weak radio signals coming from the satellite; being a receiver means this coexistence with other devices is mostly a matter of whether the GNSS subsystem can withstand the presence of interferers or not.

In order to specify the required radio performance for a GNSS receiver to work properly in a multi-functional user terminal, it is of utmost importance the evaluation of the inter-system isolation. The amount of interference coming from other services occupying the frequency spectrum can interact and degrade the quality of the desired signal.

The inter-system isolation is determined primarily by the antenna characteristics and by the multiple channel select filter, which is usually present at the input of a multi-frequency receiver. Hence, to evaluate the required system performance it is necessary to know and characterize not only the signals present in the working environment, but also the performance of specific commercially available products that could be used to implement the targeted receiver. We will focus our attention on the situation depicted in Fig. 1.1, in which separate antennas for GNSS and WCS are used rather than a single-feed multiband antenna because of this situation can be more representative of a generic interference environment.

Taking as an example a tight integration scenario as in a mobile phone, the communication services which can strongly interact with a positioning system and its main characteristic are summarized in Table 4.1.

We will focus mainly in the effects on GPS/Galileo L1/E1 and L5/E5a, but the idea can be easily extrapolated to other GNSS bands.

Out-of-Band Linearity

The strongest and closest interferers are DCS1800 and E-GSM respectively which can radiate up to +30 dBm output power [3GPP08] with a frequency separation of only 135 and 262 MHz from L1/E1 and L5/E5a respectively. Due to the circuits' non linearity these strong out-of band signals can generate higher order products which can fall into the desired GNSS bands degrading reception performance. The goal is to design the RF FE so that its non linear performance (in terms of intermodulation) is sufficient to prevent such degradation. As the two signals have the same maximum output power, we can state that the worst case intermodulation will be generated by the DCS1800 jammer, which has lower separation with L1/E1.

What is important to evaluate, in order to properly specify the RF requirement, is the amount of power that can effectively enter in the GNSS receiver. This will be mainly determined by the characteristics of the selected GNSS antennas and RF selection filters (Fig. 4.3) which are needed to prevent too high level of out-of-band

Table 4.1 Maximum output power of WCS transmitters

Standard	Modulation	Frequency (MHz)	Max. radiated handset power (dBm @antenna output)
IS-95, CDMA2000	QPSK	824–849, 1,850–1,910	+23
E-GSM	GMSK	880–915	+30
DCS1800	GMSK	1,710–1,785	+30
PCS	QPSK	1,850–1,910	+30
UMTS	HPSK	1,920–1,980	+24

linearity requirement which will be not compatible with the low power requirements of a handset. As far as the antenna is concerned isolation values in the order of 20 dB can be obtained with proper antenna placement, but for sake of generality we can consider a value of 15 dB which can be easily obtained). However this value of isolation would not be enough to guarantee reduced FE power consumption and it is a good practice to ensure at least an overall out-of-band attenuation at the blocking frequency (and before the LNA) in the order of 40 dB which could be reached with an additional low loss filter at the input of the LNA (not shown in the figures), or by designing a frequency selective LNA matching circuitry by placing a proper LC filtering network at the input.

If this condition is met (and in our assumptions we suppose that this could be achieved with standard components and proper PCB design) the amount of unwanted jammer power which can enter the receiver in the worst case is:

$$P_{leak} = -10\,\text{dBm} \tag{4.4}$$

As in the GNSS standards specific intermodulation tests are not defined, what we should also do is to define proper intermodulation tests matched to the specific

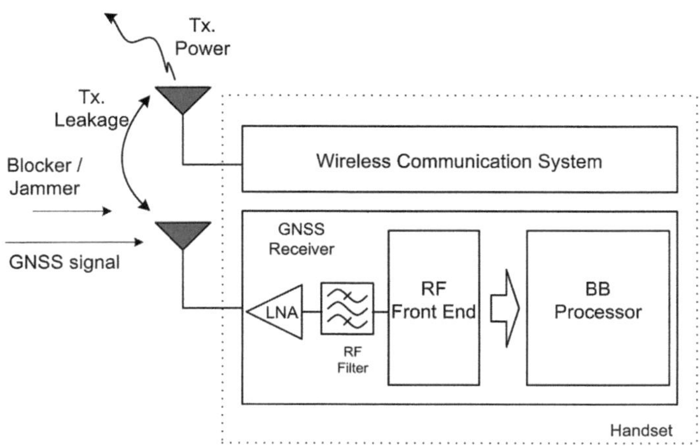

Fig. 4.3 GNSS coexistence with other signals within a cellular handset

application. We assume an additional out-of-band interference with $P_{Block} = -45$ dBm power (which represents a potential −30 dBm/100 kHz UMTS spurious emission in the 1–1.7 GHz band, entering the receiver through the 15 dB isolation path provided by the antenna), producing a third-order inter-modulation product (IM3) in the GNSS frequency band. In addition to this assumption, a maximum tolerated intermodulation power (P_{IM3}) should be defined in a manner similar to what done in other WCS standards. In this case we assume that the intermodulation distortion product could be specified to be 3 dB below the FE maximum input acceptable noise power in order to result in minimum NF (and sensitivity) degradation. Clearly this parameter depends on the selected RF BW. The scenarios we will consider as a worst case are around 6 and 15 MHz RF 3 dB BW for E1 and E5a channels respectively (to ensure some design margin, however these values should be confirmed in the RF FE design phase and will be always kept above the minimum values of 3 and 13 MHz [GRAMMAR D2.2 2012, GRAMMAR D2.3 2013, GRAMMAR D3.1 2012]. These values correspond to −106.2 and −102.2 dBm input referred noise power and thus (assuming $NF_{RF} = 2.5$ dB) −103.7 and −99.7 dBm maximum acceptable input noise power to maintain the targeted sensitivity. When inter-modulations products are present we can use a 3 dB weighting factor similarly with what is done in common communication standard specification documents, resulting in a worst case P_{IM3} below −106.7 dBm.

Now, for two tones spaced f_1 and f_2 from the fundamental such that $\pm 2f_1 \pm f_2$ falls in band but with different amplitudes, we have:

$$P_{IM3}[\text{dBm}] = 2P_{f_1}[\text{dBm}] + P_{f_2}[\text{dBm}] - 2IIP3[\text{dBm}] \tag{4.5}$$

Thus obtaining for the IIP3 the following expression:

$$IIP3[\text{dBm}] = \frac{2P_{f_1}[\text{dBm}] + P_{f_2}[\text{dBm}] - P_{IM3}[\text{dBm}]}{2} \tag{4.6}$$

In our case this results in IIP3 = +3.25 dBm which is still quite high for being achieved with typical low power CMOS implementations. However in the above calculations it has been supposed the inter modulating interference affected only by the antenna isolation, while in reality also the additional filtering required to achieve the 40 dB attenuation at the DCS frequency will affect the modulating signals. Adding, for example half of this attenuation to the modulated interferer the requirement on the third order intermodulation becomes IIP3 = −9.25 which would more reasonable for such implementations (GREAT 2008) and will be assumed as a baseline specification.

A remark should now be done in order not to assume these calculations as stringent requirements but mainly as design suggestions. In fact, in the IIP3 calculation it has been neglected the fact that the expressions used have been derived based on pure sinusoidal signals. It has been assumed that the jammers are the GSM or UMTS-like signals considered as CW signals with power equal to a specified value. In real conditions, these tones are modulated signals. The IP3 of a nonlinear building block with respect to a CDMA or GSM signal is different from the IIP3 for sinusoidal signals and should be tested directly on the implemented circuit to see how it really affects circuit performance. These simplifications make the above computations of the required IIP3 only an estimation and common practices in the characterization of real GNSS receivers is to apply real intermodulation signals and see if the resulting IMD3 product has less power than the maximum allowed.

The circuit that suffers from this specification the most is the input LNA, which should be linear enough to avoid the propagation of the distortion across the RF chain. The out-of-band intermodulation requirement for the stages following the LNA, and the RF filter preceding the down-converter (shown in Figs. 4.3 and 4.6, which is required in order to not saturate the baseband circuitry and ADCs and to achieve realistic out-of-band phase noise specifications), will be strongly relaxed, because the out-of-band blocker levels are further attenuated (even more at the down-converter output). If we consider for example a filter selectivity L_{SEL} = 30 dB at the PCS interferer and half of this attenuation at the additional out-of-band blocker (easily obtained with commercial low cost SAW filter (Chastellain et al. 2006)), we obtain for the mixer:

$$\mathrm{IIP3}_{\mathrm{MIXERout-of-band}} = \frac{P_{leak} + 2P_{Block} - P_{IM3}}{2} \approx -27\,\mathrm{dBm} \qquad (4.7)$$

And this does not represent a serious problem. Therefore, in the stages following the LNA the most limiting factor for the out-of-band non-linear performances will be the maximum signal level that should be tolerated at a specific point. This could be determined only once the complete gain value and distribution across the receiving chain is fixed. Once known this value, we can obtain an approximate IIP3 specification from the theoretical expression of IIP3 ≈ $P_{1\ \mathrm{dB}}$ + 9.6 dB.

In-Band Linearity

Taking into account the very low power level of GNSS signals in band performance of the FE are not as critical as the out-of-band. However some consideration can be done to further justify this assumption. In band jamming signals are typically caused by harmonics from displays, micro-controller, bus systems, or spurious WCS emission in the GNSS bands for which the correspondent standards do not pose specific recommendations.

Regarding the first class of in band jammers no specific requirement could be derived as it strongly depends on the physical implementation. However, specific counter measures to minimize the effects could include:

- Maintaining a good grounding concept in the design
- Shielding
- PCB Layout optimisation
- EMI Filtering
- Proper Placement of the GNSS antenna

For the second source of in-band jammers (WCS spurious) we should check if the level of signals which can enter the receiver (and could not be filtered as they would appear at the same GNSS frequency) could potentially degrade receiver performance. According to GSM and UMTS standards, the spurious emissions within a mobile station in the frequency range between 1 and 12.75 GHz should meet the following limitations: (Table 4.2)

These signals can be viewed as in-band CW interferers in order to apply the same expressions (approximations…) used in this chapter. In this case the power should be referred to a single Hz, hence spurious of GSM and UMTS would convert into −97 dBm/Hz and −90 dBm/Hz respectively. Two of these signals spaced for example 3 and 6 MHz from the wanted signal, could generate a third-order non linearity into the desired channel. This unwanted emission could not be filtered out, and the resulting receiver IIP3 requirement is given by (4.8):

$$\mathrm{IIP3}_{\mathrm{in-band}} = \frac{P_{GSM} + 2P_{UMTS} - P_{IM3}}{2} \approx -85\,\mathrm{dBm} \tag{4.8}$$

And this IIP3 requirement does not pose any stringent requirement on the GNSS radio FE to be developed.

This IIP3 requirement does not seem to be too stringent at least from the RF circuitry point of view. In the following calculations, the linearity of the RF front-end is therefore set to a value slightly higher than the overall receiver requirement (for example we can suppose $\mathrm{IIP3}_{\mathrm{RF}}$ = −24 dBm, since several implementations that fulfil this requirement have been already published) to allow the calculation of the analog specification. In this case, actually, analog circuitry needs to be linear enough to satisfy the in-band requirement, since the in band blockers are not attenuated at mixer output. Here, the specifications for the analog baseband circuit are calculated using a model for the RF front-end. This is sufficient at the receiver block-planning phase; the circuit designers are responsible for fulfilling the specifications for the analog baseband block. In reality, the architecture of the baseband part is subject to changes during the circuit design and probably cannot be fixed at an early phase. The receiver-system designer should have a basic knowledge of circuit design and its possibilities, and of related state-of-the-art implementations, to avoid unrealistic specifications. In addition, feedback from circuit designers can be used to reconsider and optimize the specifications of the different building blocks of the receiver. However, the specifications for the analog part can be estimated with hand calculations.

In practice, there are no matched interfaces between the down-conversion mixers, analog baseband block, and ADCs or inside the analog baseband circuit. Therefore, the use of dBm values, i.e. the power referred to 1 mW should be avoided at on-chip interfaces, since the impedances at these interfaces are not

Table 4.2 WCS spurious emissions in GNSS bands

E-GSM, DCS, PCS	−47 dBm/100 kHz(idle mode)
IS-95, CDMA, UMTS	−30 dBm/1 MHz

typically well defined or known. The signal levels and linearity parameters in dB are converted into the corresponding levels in dBV at voltage mode interfaces: P_{TGSM} of −57 dBm corresponds to −70 dBV, P_{TCDMA} of −40 dBm to −53 dBV and P_{IMD} of −104 dBm to −117 dBV in a single ended 50 Ω system. If we assume that only the analog circuitry produces intermodulation, the required analog IIP3 becomes:

$$\mathrm{IIP3_{analog}}[\mathrm{dBV}] = (3\mathrm{P_{int}} - \mathrm{P_{IM3}})/2 + 20\log_{10}(\mathrm{A_{v,RF}}) \tag{4.9}$$

where $A_{v,RF}$ is the linear voltage gain of the RF input section, which has to be added (in dBV) to both in band interferer and intermodulation product. Substituting the above calculated value we obtain:

$$\mathrm{IIP3_{analog}} = +2\,\mathrm{dBV} \tag{4.10}$$

Assuming an interferer power equal to the arithmetic mean between P_{TGSM} and P_{TCDMA} (i.e. −61 dBV) and a quite realistic voltage gain of 35 dB. In practice, the values of all signal-related parameters are simulated or measured at the output (considering, for example, the peak-to-peak output voltage for which the device compresses, output voltage for 1 dB compression). Thus, they must be referred to the input of the block by dividing them with the passband gain of the block. The input and output interfaces of the analog baseband circuit may affect the results. Therefore, these interfaces must be properly modelled in the circuit simulations and measurements.

From equations above, we get the level of intermodulation distortion generated in the RF front-end once known the $IIP3_{RF}$:

$$P_{IM3,RF} = 3P_{\mathrm{int}} - 2IIP3_{RF} \tag{4.11}$$

The distortion product of the RF front-end and baseband circuit can be summed either coherently or non-coherently. In the former case, the distortions are summed in voltages, which give a worst case estimate for the IIP3 parameters of both blocks. In the latter case, the powers of the distortion components are summed. The coherent summation is used here to be on the safe side. The distortion generated in the RF front-end in V_{RMS} referred to the input of the RF front-end is:

$$v_{IM3,RF}\sqrt{10^{\frac{(3P_{\mathrm{int}}-2IIP3_{RF})}{2}} \cdot 1\,\mathrm{mW} \cdot 50\,\Omega} \tag{4.12}$$

A similar expression holds for $v_{IM3,\,RX}$ of the overall receiver distortion, and the maximum value of intermodulation distortion due to the analog circuitry becomes:

$$v_{IM3,analog} = v_{IM3,RX} - v_{IM3,RF} \tag{4.13}$$

which corresponds to the following $P_{IM3,analog}$ in dBV, referred to the input of the RF front-end:

$$P_{IM3,analog} = 20\log_{10}(v_{IM3,RX} - v_{IM3,RF}) \tag{4.14}$$

Thus, we have:

$$IIP3_{analog}[dBV] = (3P_{int} - P_{IM3,analog})/2 + 20\log_{10}(A_{v,RF}) \tag{4.15}$$

Assuming $IIP3_{RX} = -25$ dBm, $IIP3_{RF} = -24$ dBm, we get $P_{IM3,analog} = -120.7$ dBV and $IIP3_{analog} = +3.85$ dBV. The only mean to reduce the in-band requirement is to increase isolation between antennas, and this requires further investigation and measurements on the specific products and the configuration which will be finally used.

Second Order Distortion

In direct conversion receivers the DC-offset, 1/f noise and second-order intermodulation can be important sources of noise. Generally, in a low-IF architecture we do not have the DC component problem and, assuming we adopt adequate precautions (i.e. the value of the IF is chosen sufficiently high, few MHz being enough), flicker noise and also second order intermodulation products do not affect the behaviour of the receiver. This second order products are generated when a modulated signal enters the receiver (in care of a pure tone the second order distortion is a tone at DC) and are located close to DC. However in the case of a very low IF receiver or when the IF frequency could not be selected too high in order to avoid a high power consumption and the IF bandwidth extends close to DC, the second-order intermodulation, introduces undesirable spectral components at baseband and can degrade receiver sensitivity.

All the signals entering the receiver with a frequency separation Δf which falls within the IF filter BW are potentially generating second order intermodulation distortion products in the receiver BW (Fig. 4.4).

Clearly the biggest problem appears with in band interfering signals which pass through the entire receiver (and amplified) without being attenuated by the filters and frequency selectivity of the receiver building blocks. This problem can be avoided if the IF frequency is chosen to be higher than the RF filter bandwidth, but this is not our case, in which the IF should be less than 10 MHz (to save power) and the RF filter bandwidth could be greater than 15 MHz in the worst case (E5a reception).

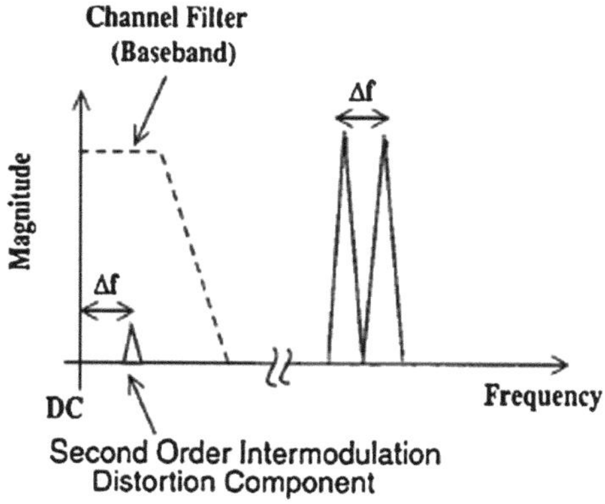

Fig. 4.4 Second order intermodulation mechanism

To fix this specification, once again we must fix a maximum acceptable input referred noise power, in this case due to second-order intermodulation products. Proceeding in the same way as in the case of third-order intermodulation, we have:

$$P_{IM2} \leq -106.7\,\text{dBm} \tag{4.16}$$

It can be shown that [LD09]:

$$IIP2 = 2 \cdot P_{int} - P_{IM2} - 6\,\text{dB} \tag{4.17}$$

Substituting $P_{int} = -45$ dBm (the maximum modulated interferer value which can enter the RX filter bandwidth, the required second-order intermodulation performance for the Front-end is:

$$IIP2 \geq +10.7\,\text{dBm} \tag{4.18}$$

Assuming a LNA gain between 15 and 20 dB we obtain a required $IIP2_{MIX} \approx +30$ dBm, which represent a value which can be easily obtained with standard mixer topologies. However it should be stressed how this parameter could not be measured on the complete receiver and should require a separate implementation of the down converter. We can assume that due to the very low IP2 requirements, this specification does not pose any kind of limitations in the performance of the receiver.

Desensitization

The term desensitization refers to the effect of a single strong interfering signal present at the receiver input. The effect of this unwanted signal can reduce the small-signal gain at the fundamental frequency. This, in turns, reduces the C/N_0 at the input or, equivalently, increases the noise figure. The phenomenon of gain compression caused by large interfering signal acting on the amplifier nonlinearity is well-known and the basic relations are summarized here for completeness. If the desired signal ($s_{in}(t) = A \cdot \sin(\omega t)$) is present at the input of the receiver together with a strong interferer ($A_{block} \cdot \sin(\omega_{block} t)$, with $A_{block} \gg A$), at the output, we have:

$$s_{out}(t) \approx \left(\alpha_1 + \frac{3}{2}\alpha_3 A^2_{Block}\right) \cdot A\sin(\omega t) + \cdots \tag{4.19}$$

where α_1 is the small-signal voltage gain and α_3 is the coefficient of the third-order non linearity, normally $\alpha_3 < 0$. Thus,

$$IIP3_{(linear)} = \sqrt{\frac{4}{3}\frac{\alpha_1}{\alpha_3}} \tag{4.20}$$

We can impose the condition that the interferer causes a reduction of −D dB on the nominal gain (D dB desensitization), and obtain:

$$20 \cdot \log_{10}\left(1 + \frac{3}{2}\frac{\alpha_1}{\alpha_3} \cdot A^2_{Block}\right) = -D \tag{4.21}$$

Solving with respect to the blocker power (A_{block} represents the linear amplitude), we have:

$$P_{Block}[\text{dBm}] = 10\log_{10}\left(1 - 10^{\frac{-D}{20}}\right) + IIP3[\text{dBm}] - 3\,\text{dB} \tag{4.22}$$

From (4.22) we can estimate the power of a strong blocking signal which causes D dB desensitization on the fundamental gain. In the case of 3 dB desensitization it can be shown that 4.22 can be rewritten as:

$$P_{block} \approx P_{1\,dB} + 1.2\,\text{dB} \tag{4.23}$$

The condition to be imposed is related to the selectivity of the receiver on the interferers of the various WCSs present in the cellular environment so that (4.23) is satisfied for a fixed value of IIP3 and D:

$$\text{Selectivity} = P_{Tleak} - P_{Block} \tag{4.24}$$

In the worst case we have a blocker at $f_{offset} = 135$ MHz with a $P_{TLeak} = +2$ dBm. In Fig. 4.5 there are plotted the required values of selectivity preceding the LNA as a function of desensitization for different values of IIP3. Clearly, the higher the IIP3 of the LNA, the lower the required selectivity (at the cost of higher power consumption).

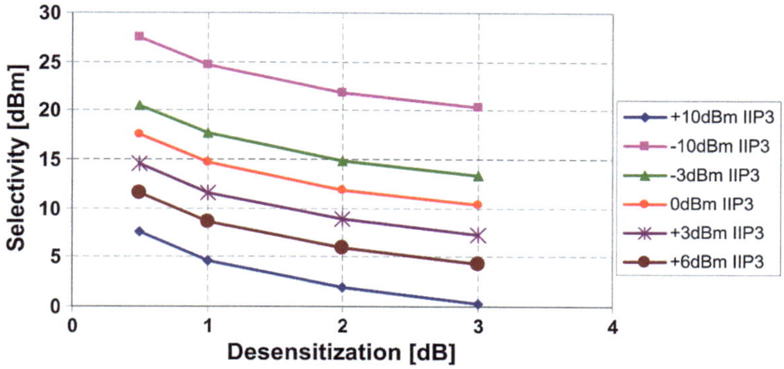

Fig. 4.5 Required front end selectivity as a function of the desensitization level for different values of IIP3 and supposing a blocker with +2 dBm power

Gain

The required FE gain depends on the dynamic range of the ADC and noise figure. The combined amplification of the whole receiver must be enough to make sure the standard deviation of the signal matches the optimal level to minimize the quantization noise.

For a given quantization function there is an optimal signal level σ_N that minimizes the quantization noise. In GNSS receivers this signal is dominated by noise, and the nominal noise power density at the input of an ideal antenna is N_0. This level is increased not only by the gain of the system G but its noise figure NF as well. Given this noise power is applied to a load R we can immediately obtain the gain required to amplify this noise integrated over the receiver bandwidth so it matches this signal level σ_N, by factoring the equivalent noise bandwidth of the channel BW:

$$N_0 + 10\log_{10}(BW) + NF + G + 10\log_{10}(R) = 20\log_{10}(\sigma_N) \tag{4.25}$$

$$G = 20\log_{10}(\sigma_N) - N_0 - NF - 10\log_{10}(R) - 10\log_{10}(BW) \tag{4.26}$$

When the ADCs are simple 1-bit converters, under the assumption of ideal comparators, theoretically no gain would be needed, since an ideal comparator by definition implicitly has infinite gain. In practice however the comparator is a saturating amplifier with limited gain and non-zero offset, so the combined gain of the RF FE and the ADC must be high enough to provide clean transitions—accurate quantization.

This optimal σ_N must be maintained during normal operation, so the gain control VGA must provide enough margin to keep the total gain around this target G in every configuration case for the application (passive or active antenna, external LNA, etc.) despite tolerances, additional losses, or simply different bandwidth choices in the case of reconfigurable receivers.

Frequency Stability

Under ideal circumstances the RF signal is down-converted to an IF plus a small deviation due to the Doppler shift. This frequency shift around IF is estimated and compensated by the BB. The local oscillator or oscillators within the receiver used for the frequency conversion also introduce an additional random offset, the accuracy of which depends mostly on a reference given by a crystal oscillator.

A GNSS receiver typically has to deal with a Doppler shift approximately between −10 and 10 kHz (Principe et al. 2011). If for example an oscillator with an accuracy of 1 ppm (parts-per-million) was used to down-convert GPS L1 to an IF of 3 MHz (f_{LO} = 1572.42 MHz), this search interval would be increased by about 1.57 kHz by each side, so a receiver with these values would have to search for SVs approximately between −11.57 and 11.57 kHz.

This frequency deviation from its nominal value is also modified by variations in the supply voltage, temperature, acceleration… which is generally specified by the manufacturer of the reference. The acquisition engine must be consistent with the uncertainty of this carrier.

Besides these static and slow-varying errors, the reference is also subject to faster frequency drifts. The slowest of these components are usually referred to as short-term stability. Even if a receiver is prepared to compensate this error, if the frequency varies too rapidly it may not be possible for the PLL to track it correctly. The combined effect of the variation of the Doppler shift and the local oscillator frequency must be low enough not to shift this IF beyond the tracking capabilities set by the PLL dynamic response.

Phase Noise

Even though technically speaking all random frequency variations are phase noise, the term phase noise is usually preferred for the fast-varying components. Even if a PLL is able to produce the desired output frequency, fast random variations of the phase during the correlation can reduce the SNR, thus reducing the effective sensitivity of the receiver.

As with quantization, the effect of phase noise can be treated as additive noise as well, although it is in general difficult to model accurately without using numerical methods. Some closed-form expressions exist for simplified statistical models (Pérez et al. 2010), but the main conclusion is phase noise limits the maximum practical integration period as the phase error increases over time, which means the SNR may reach a point where it no longer increases with the correlation time. This is of great importance in high-sensitivity designs relying on long coherent integration. The impact of phase noise on GNSS baseband correlation has been also studied in (Thombre et al. 2014).

Frequency Planning

The choice of the reference and local oscillator frequencies is not a trivial one, as one source of interferers is the receiver itself. Leaving aside the mix with potential external interferers, the harmonics of the reference as well as their mixing products must be taking into account.

If the reference is f_{XTAL}, even if measures to reduce coupling are taken it is expected to find significant harmonics at $N \cdot f_{XTAL}$ where N is any integer value. If this reference is internally divided by M to be used in the frequency synthesis, additional harmonics at $N \cdot f_{XTAL}/M$ are to be expected as well, even if generally weaker. For this reason a good frequency choice will be to try to make sure none of these harmonics, particularly those at $N \cdot f_{XTAL}$, falls within the bands of interest.

Additionally, for each frequency conversion additional harmonics may enter the signal path. For example, if an 8 MHz reference is used in an L1/E1 receiver (1575.42 MHz) and there is a first conversion to generate an intermediate frequency of 400 MHz, the 197th harmonic of the reference will enter the RF chain ($8 \times 197 = 1{,}576$ MHz) while the 50th would interfere at IF. Harmonics are less and less problematic as the signal progresses through the front-end being amplified, but for the first stages it can be critical. Choosing these frequencies carefully can reduce the filter and isolation requirements significantly.

GNSS RF Front-End Architectures

In ideal operating conditions a receiver considers exclusively the signals for which it is designed, and its characteristics in the processing of the signal are the unique factors which determine receiver requirements. Theoretically there are literally infinite equivalent implementations of radio-front ends that perform exactly the same transformations over a given group of signals. Besides design complexity, constraints in what is technically and economically achievable for each building block will drive the architectural choices. Some of the most common front-end architectures will be described in the following sections along with their trade-offs.

Single-Frequency Architectures

GNSS-based devices (especially targeting mass-market applications) usually require compact, low-power, low-cost solutions. In order to achieve this, designers must aim at a receiver as close to a single-chip circuit as possible: external components must be kept to a minimum, specially the very area-demanding passive filters. Single-frequency receivers achieve satisfactory performance as demonstrated by the millions of such devices that are produced every year. We will first study front-end implementations for these receivers, and then propose architectures for multi-frequency implementations in the following section.

Direct Conversion

The simplest radio front-end architecture is that of a direct conversion receiver (Fig. 4.6).

Once the input signal is amplified by a low-noise stage (one or more LNA blocks, one of which can be part of an active antenna) it is down-converted to DC with an in-phase/quadrature mixer. No explicit image rejection is needed, and the resulting signal is simply filtered with a low-pass function to avoid aliasing of out-of-band noise and interferers, and sampled.

The main drawbacks of these direct-conversion receivers mainly are in the low frequency sections of the front-end: voltage offsets due to process mismatches and high noise around 0 Hz. Voltage offsets require additional circuitry that effectively acts as a very sharp high-pass filter (Zheng et al. 2009). Flicker noise can be reduced through circuit optimization but it is mainly technology-dependent, so the fabrication process sets a lower boundary (Lundberg). It has been proposed the use of regular AC-coupling to avoid these problems even in direct-conversion receivers at the cost of signal integrity, but the simplicity of this solution does seem to outweigh the performance losses (Shaomin and Namgoong 2005).

Although the term *direct-conversion* refers to a single frequency conversion from RF to DC, the more general *zero-IF* (*intermediate frequency*) is often used, especially when there are several mixing stages but the resulting output is still centered at 0 Hz.

Low-IF

So-called low-IF architectures share the simplicity of direct conversion front-ends without their main disadvantages: the front-end topology can be kept essentially the

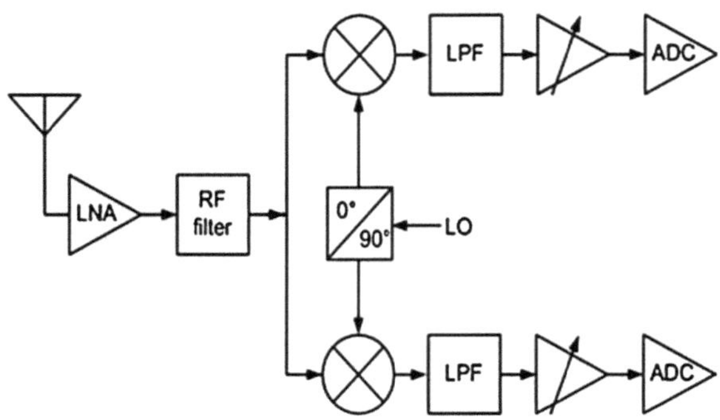

Fig. 4.6 Direct conversion receiver

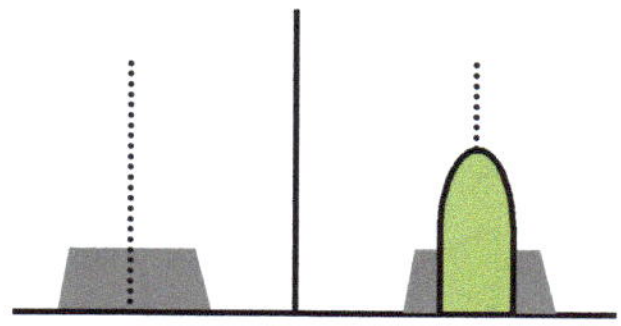

Fig. 4.7 Real filtering of a complex signal

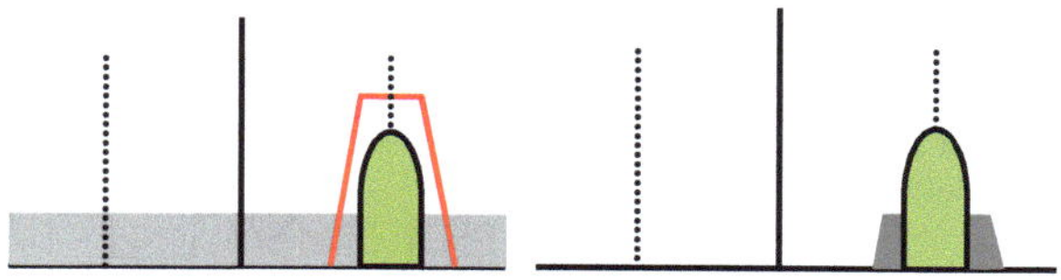

Fig. 4.8 Complex filtering of a complex signal

same but the signal is down-converted to an intermediate frequency low enough to allow inexpensive sampling while sufficiently high to cancel DC offsets with simple high-pass filters. Most commercial receivers follow this approach.

There are two main strategies for filtering and sampling. The simplest approach, illustrated in Fig. 4.7, is to use low-pass or band-pass filters around this complex I/Q signal (in green) and its image. The resulting—still complex—signal is hen sampled and processed in the digital domain.

Alternatively, a *complex filter* can be used, affecting the complex signal as a whole by selectively removing its image, as in Fig. 4.8.

The implementation of this kind of analog filter is more sophisticated (Behbahani et al. 2001) but allows sampling at lower frequencies, as the image, now removed, does not have be preserved in the process. Thus this sampling can be *real* (only one front-end output per channel instead of two, real and *imaginary*).

Subsampling and Direct Sampling

In some cases the last frequency conversion is not performed by a mixer but by the ADCs themselves, by sampling at frequencies below IF, which effectively reduces the output IF by means of aliasing (Barth 2011). This requires a careful choice of frequencies and filter design to avoid unintended aliasing of noise or even between signal replicas. It is a very convenient method when the signal bandwidth is relatively narrow, allowing a relative flexibility between intermediate and sampling frequencies, so it is not rare to find GPS front-ends using this sampling strategy. The main disadvantage is this can increase the performance requirements of the ADCs compared to choosing a lower IF, as their analog bandwidth and jitter characteristics have to allow accurate aliasing.

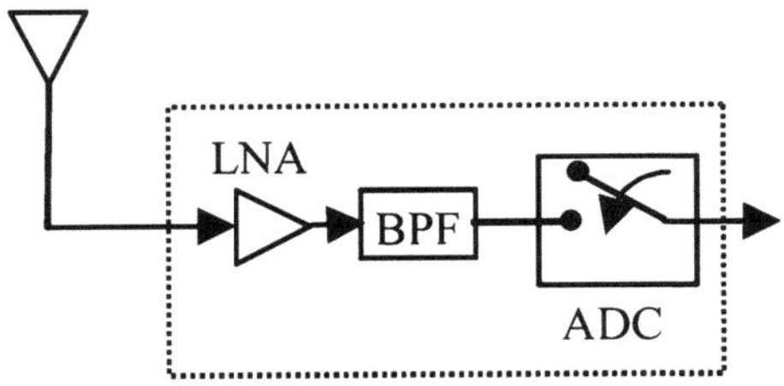

Fig. 4.9 Direct sampling front-end

Fig. 4.10 Dual-channel front-end with completely independent chains

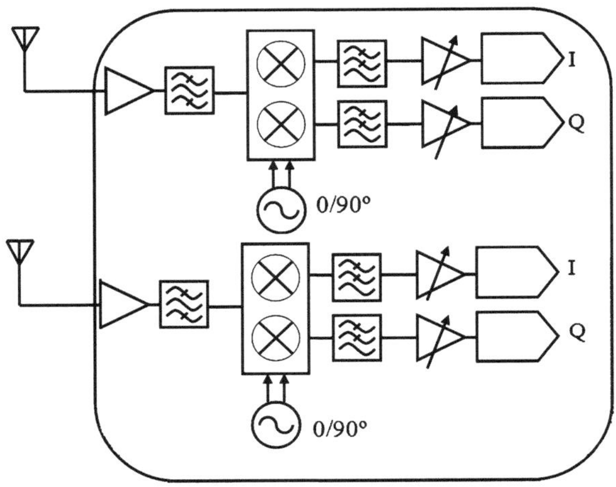

In the recent years, with the notable developments in ADC design and digital technologies, more applications have become suitable candidates for direct sampling architectures where the signal is sampled directly at RF. Subsampling receivers require a very specific ADC design, which is the most important part of such systems, up to the point that the performance of the receiver is ultimately set by these blocks. These architectures have been proposed for GNSS systems (Barth et al. 2012) but so far both professional and low-cost mass market receivers opt for more conventional and technically convenient approaches regarding sampling (Figs. 4.9, 4.10, 4.11 and 4.12).

Multi-frequency Architectures

Although current state of the art already provides excellent accuracy for many applications using exclusively single-frequency receivers, either in stand-alone operation or with *augmentation* services to reduce positioning errors (mainly by means of ionospheric error models or differential corrections), multi-frequency

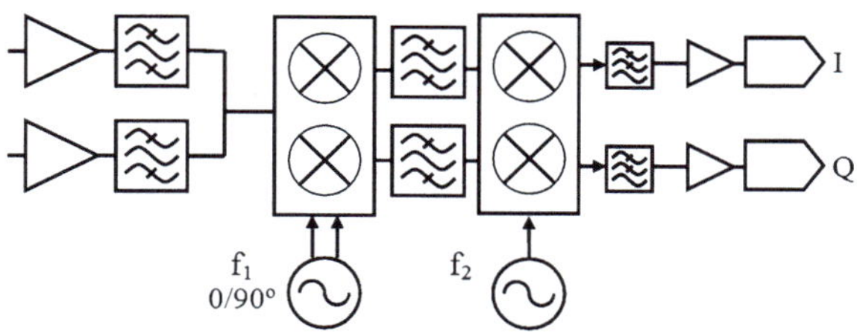

Fig. 4.11 Double-conversion front-end with shared stages

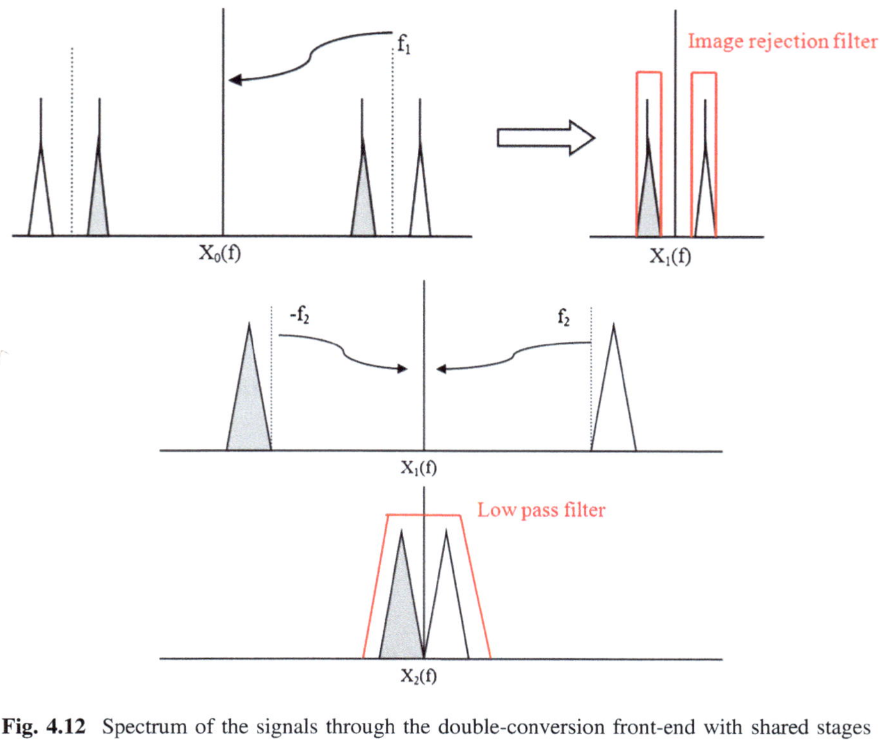

Fig. 4.12 Spectrum of the signals through the double-conversion front-end with shared stages

devices offer an even greater degree of accuracy, often in the millimeter range. In addition, these receivers are generally more robust as they can rely on single band operation when subject to interference.

The following sections will show some architectural options to implement multi-frequency radio front-ends. These front-ends can be used with either dedicated or multi-band antennas followed; multi-band or wideband antennas shared between channels are in general more challenging to design and less robust to interference, as the RF bandpass filters are usually placed *after* the LNA to optimize sensitivity, but offer the potential advantage of compactness. Note that for high-accuracy positioning the antenna should be the same for all channels, or the offsets between the phase center of each antenna should be accounted for.

Front-End with Dedicated Channels

The simplest approach for multi-channel front-ends is the integration of several single-frequency front-ends in the same system. This allows reuse or centralization of certain subsystems such as a clock generation or control while keeping separate chains (Qi et al. 2011).

Since each chain almost completely independent, this architecture offers the best performance and robustness; for example, if a strong jammer is present in one of the bands, the receiver may still operate normally with the remaining channels, as long as their respective RF sections offer sufficient out-of-band rejection.

In the case of single-chip receivers, the integration of several chains in the same ASIC is not as critical for GNSS as it is in other applications, as inter-channel coupling is less important in DSSS receivers. In some cases certain blocks can be shared, such as local oscillators in double-conversion front-ends where the intermediate frequency is the same for both channels.

Double-Conversion Front-End with Shared Stages

It is possible to use low-pass IF filters to accommodate two or more complex signals by using two or more frequency conversions. The idea is to mix the signals with very specific frequencies in order to place the different bands close to each other but at different halves of the spectrum.

In the front-end of the figure only two frequencies are considered. The first conversion shifts both signals to the same IF, which then go through a regular band-pass filter that will remove the image of the *next* conversion.

A second conversion places both signals at a low IF, and since their images are filtered, there is no overlap. The resulting complex I/Q signal can now be sampled, and each signal will be in its respective side of the spectrum.

This kind of architecture can be extended to several bands with some differences, the main idea being essentially the same: through frequency conversions the signals are brought to adjacent bands and they are sampled together. It is also possible to sample them together at the same side of the spectrum therefore using only *real* IF sections or perform the second conversion in the digital domain as seen in (Botteron et al. 2005).

The main disadvantage is the increased complexity of the design, as multiple conversions and/or synthesizers can introduce many spurs and potential inter-modulation problems. In an interference-hostile environment the saturation or distortion of a channel can lock down the whole front-end, which is a severe drawback for secure applications.

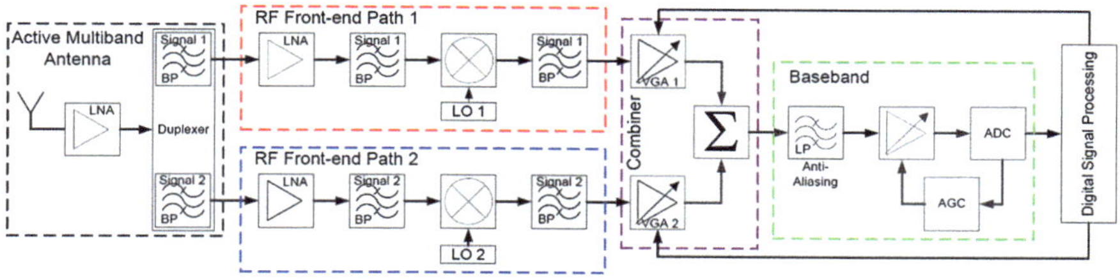

Fig. 4.13 Dual-frequency overlay front-end

Overlay Front-End

The previous architecture can be made more robust by using two separate RF chains as shown in Fig. 4.13:

Two or more bands are shifted into adjacent or even overlapping intervals of the spectrum to share the lower frequency sections (Rugamer et al. 2011). Since the RF sections are essentially independent, interference in one of the bands can be mitigated by switching the corresponding channel off.

The main disadvantages are design complexity, potential noise floor degradation (sensitivity loss) when the conversion frequencies are chosen to make the signals overlap (for example to reduce bandwidth requirements in the low frequency block), and lower resistance to interference in all the shared sections.

Front-End with Additional Frequency Translation

Another strategy to share resources is to have a common front-end at a fixed RF (or moderately high IF) and use different conversion stages for each band. An example of such strategy is shown in Fig. 4.14.

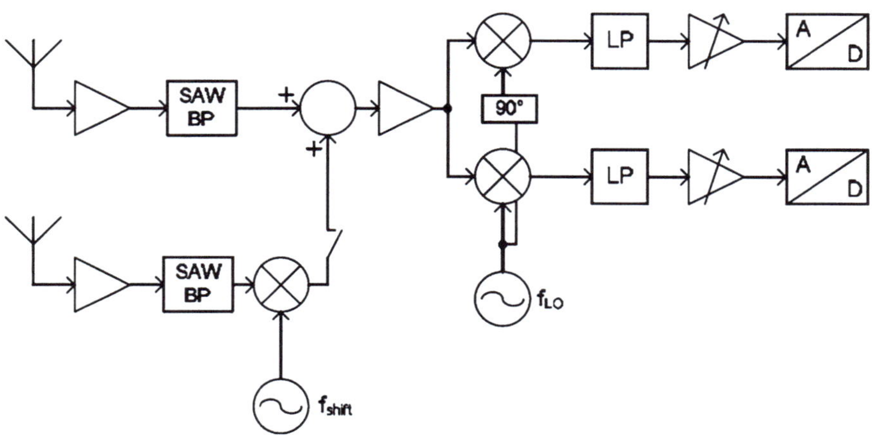

Fig. 4.14 Dual-frequency front-end with frequency translation

The front-end of the figure has a simple single-conversion single-frequency front-end that optionally can take the output of another down-conversion stage that takes the signal from another band. The way these two bands are combined can be diverse: they can overlap (increasing noise floor but still allowing separate demodulation because of signal orthogonality), they can be placed in adjacent to each other, or they can even be switched. The advantage is again a major sharing of resources between the channels, and because of this one of the major disadvantages is a higher vulnerability to interference. In some cases the performance and/or design complexity can be higher than other alternatives.

References

Abdizadeh M, Curran J, Lachapelle G (2012) Quantization effects in GNSS receivers in the presence of interference. In: ION ITM 2012

Balaei AT, Dempster A, Akos D (2008) Quantization degradation of GNSS signal quality in the presence of CW RFI. In: IEEE 10th international symposium on spread spectrum techniques and applications, 2008

Barth C, Linscott IR, Inan US (2012) A 9 mW direct RF sampling GPS receiver front-end in 0.13 μm BiCMOS. In: IEEE International Symposium on Circuits and Systems, 2012

Barth C (2011) Subsampling GPS receiver front-end. Ph.D. thesis, 2011

Behbahani F, Kishigami Y, Leete J, Abidi A (2001) CMOS mixers and polyphase filters for large image rejection. IEEE J Solid-State Circuits 36(6):873–887

Botteron C, Chastellain F, Farine PA (2005) A low-power RF front-end architecture for an l1/l2cs gps receiver. In: ION_GNSS 18th international technical meeting of the satellite division, Long Beach, CA, pp 628–634, 13-16 Sept 2005

Chastellain F, Botteron C, Waelchli G, Zamuner G, Manetti D, Farine (2006) A Galileo E1b,c RF front-end optimized for narrowband interferers mitigation. In: GNSS 19th international technical meeting of the satellite division, pp 1069–1075, 2006

Friis HT (1944) Noise figures of radio receivers. In: Proceedings of the IRE, July 1944

GRAMMAR D2.2 (2012) GRAMMAR project deliverable. Radio front-end design report (including receiver architecture and design justification)—Interim. http://www.kn-s.dlr.de/grammar/documents/deliverables.html

GRAMMAR D2.3 (2013) GRAMMAR project deliverable. Radio Front-end design report. http://www.kn-s.dlr.de/grammar/documents/deliverables.html

GRAMMAR D3.1(2012) GRAMMAR project deliverable. Report on advanced algorithms design, simulation, prototyping and evaluation—Interim. http://www.kn-s.dlr.de/grammar/documents/deliverables.html

GREAT (2008) Project deliverable. Final report for WP4000, EHE-08-3056-D, April 2008

Kaplan ED, Hegarty CJ (2006) Understanding GPS: principles and applications, 2nd edn. Artech House, London

Lundberg KH(2002)Noise sources in bulk CMOS. http://web.mit.edu/klund/www/papers/UNP_noise.pdf

Pérez E, Thombre S, Valkama M, Lohan S, Syrjälä V, Detratti M, Hurskainen H, Nurmi J (2010) Local oscillator phase noise effects on GNSS code tracking. InsideGNSS Magazine 52–62 (November 2010)

Principe F, Bacci G, Giannetti F, Luise M (2011) Software-defined radio technologies for GNSS receivers: a tutorial approach to a simple design and implementation. Int J Navig Observ 2011 (979815):27

Qi N, Xu Y, Chi B, Xu Y, Yu X, Zhang X, Wang Z (2011) A dual-channel gps/compass/galileo/glonass reconfigurable GNSS receiver in 65 nm cmos. In: Custom integrated circuits conference (CICC), 2011 IEEE, pp 1–4

Rugamer A, Mongredian C, Urquijo S, Rohmer G (2011) Optimal path-control for dual-frequency overlay GNSS receivers. In: International conference on localization and GNSS (ICL-GNSS), 2011, Tampere

Shaomin H, Namgoong W (2005) An AC-coupled direct-conversion receiver for global positioning system. In: IEEE international symposium on circuits and systems, 2005

Thombre S, Tchamov NN, Lohan ES, Valkama M, Nurmi J (2014) Effects of radio front-end PLL phase noise on GNSS baseband correlation. ION J Navig 1:13–21

Zheng Y, Yan J, Xu Y (2009) A CMOS VGA with DC offset cancellation for directconversion receivers. IEEE Trans Circuits Syst—I: Regul Papers 56(1):103–113

Chapter 5
Advanced Acquisition and Tracking Algorithms

Elena Simona Lohan

Abstract This chapter focuses on the signal processing in the acquisition and tracking stages in a Galileo receiver. The acquisition stage refers to the stage where the presence of one or several Galileo satellites is identified in the sky and some coarse estimates of the propagation delay (namely Line of Sight delay) and Doppler frequency shift corresponding to the identified satellites are obtained. The tracking stage refers to the block where fine estimates of LOS delay and carrier frequency and phase are obtained. We start this section with an overview of the challenges in acquiring and tracking a Galileo signal, by emphasizing the difference with GPS. We then present a compact overview of the acquisition methods used more widely, in any DS-SS system, and which can be directly applicable to Galileo signals, in the so called 'ambiguous' mode. Afterwards, we focus on the Galileo-specific processing in acquisition, namely the unambiguous signal processing which takes into account the ambiguities of most of the proposed Galileo signals and tries to better cope with them. The next parts are dedicated to tracking: first focusing on an overview of generic tracking methods valid for any DS-SS system, and then discussing unambiguous tracking methods and multipath impact on tracking. This chapter is not offering an exhaustive review of all possible algorithms and related challenges in acquisition and tracking, but it tries to cover some of the main aspects, in such a way that the reader becomes familiarized with the basic challenges and the available solutions. Many parts of this chapter illustrate the Author's own understanding and own views on the problems addressed, and the notations may be sometimes different from what may be found in other GNSS-related books. The Author however tries to clarify these situations and gives references to other literature sources that can complement the Readers' understanding.

E.S. Lohan (✉)
Department of Electronics and Communications Engineering,
Tampere University of Technology, Tampere, Finland
e-mail: elena-simona.lohan@tut.fi

J. Nurmi et al. (eds.), *GALILEO Positioning Technology*,
Signals and Communication Technology 182, DOI 10.1007/978-94-007-1830-2_5

BOC-Related Ambiguities and Their Impact on Acquisition and Tracking

As in any DS-SS system, both acquisition and tracking processes are based on the correlation of the incoming signal with the reference pseudorandom code, typically using the same modulation as at the transmitter side or a simplified variant of it (e.g., we can correlate a CBOC-modulated signal with a SinBOC(1, 1)-modulated reference code, as shown for example in (Lohan 2010) if we want to decrease the receiver complexity).

Now, to understand better the differences between Galileo and GPS, we can compare for example the ACF of a CBOC(−)-modulated signal (used for Galileo E1 pilot signal) with the ACF of a BPSK(1)-modulated signal (corresponding to the GPS C/A code). This comparison is shown in Fig. 5.1 for infinite bandwidth cases (more shapes were illustrated in Chap. 3). We notice additional notches within ±1 chip interval from the maximum peak. These notches are also called 'ambiguities' and they are illustrated in more detail in Fig. 5.2 (this time for a CBOC(+) modulated code as used for Galileo E1 data signal and a 24.552 MHz bandwidth).

Such ambiguities pose two kinds of problems, according to whether we are in the acquisition or in a tracking stage.

In the acquisition stage, if a high time-bin step is used (for example 0.5 chips step, as typically in GPS C/A code), we might lose the main correlation peak, as illustrated in Fig. 5.2. Indeed, in the acquisition stage, correlations are performed with different code phases of the reference code, and the jump from one tentative code phase to the next one is typically called the time-bin step. The notches in a CBOC(−)-modulated waveform shown in Fig. 5.2 are about 0.7 chips apart. For certain channel delays, it may happen that the correlations with various code phases of the reference signal occur close to the notches (as illustrated in Fig. 5.2) and thus the main correlation peak may be missed in the acquisition process, especially when the channel noises are taken into account.

In order to deal with the ambiguities in the acquisition stage, the most straightforward solution is to decrease the time-bin step of scanning different code phases, but this means an increased acquisition time (since more code phases are to be tested). Alternatively, we can use the so called unambiguous acquisition methods that are presented later on in this chapter.

In the tracking stage, the problem that appears is the problem of false lock peaks, as illustrated in Fig. 5.3. Indeed, the two additional sidelobes in the correlation function of a SinBOC(1, 1) or a CBOC-modulated signal have strong peaks, and the tracking process may converge wrongly to one of the false peaks. The problem is even more severe if we have higher order BOC modulations (e.g., CosBOC(15, 2.5) with 22 additional side peaks: 11 at each side around the main peak). The number of additional (or false) peaks within ±1 chip from the main (correct) peak equals to $2N_B - 2$, with N_B being the BOC modulation order defined in Chap. 3, and those who pose the greatest challenge are the two nearest false peaks to the main peak.

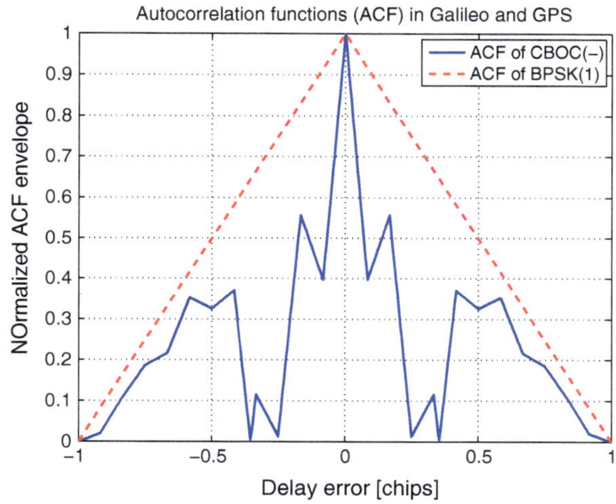

Fig. 5.1 Illustration of ACF shapes for Galileo E1 pilot code and GPS C/A code. Infinite BW case

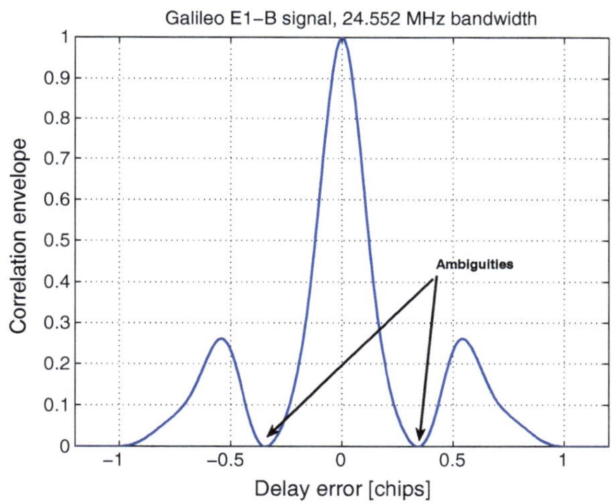

Fig. 5.2 Illustration of the ambiguities in a CBOC(+)-modulated signal with 24.552 MHz *double-sided front-end* bandwidth

The delay resolution interval is defined by the width of the main lobe of the autocorrelation function, as illustrated in Fig. 5.4.

The delay resolution (or the width of the main lobe of the ACF envelope) is shown in Table 5.1 for the Galileo modulations. The term delay resolution is used here because it represents also the minimum spacing between two resolvable channel paths. Also the maximum recommended time-bin step in the basic acquisition process (when no supplementary ambiguity removal algorithms are used) is shown in this table, using the rule of thumb of a maximum one fourth of the delay resolution (by analogy with GPS C/A code, where the typical time-bin step is 0.5

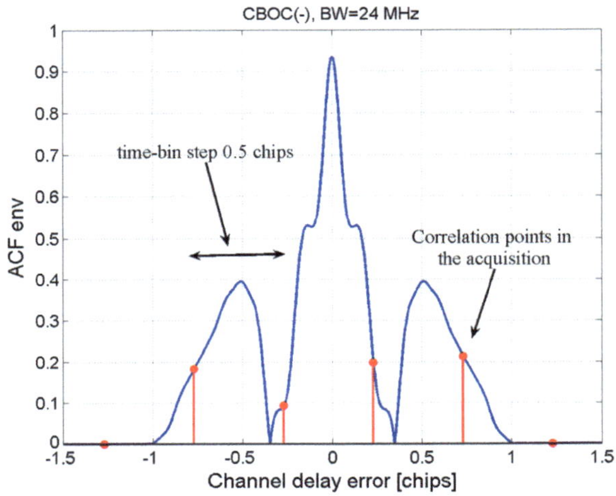

Fig. 5.3 Ambiguity problem in the acquisition stage for large time-bin steps. Here, CBOC(−) modulation and a time-bin step of 0.5 chips are used

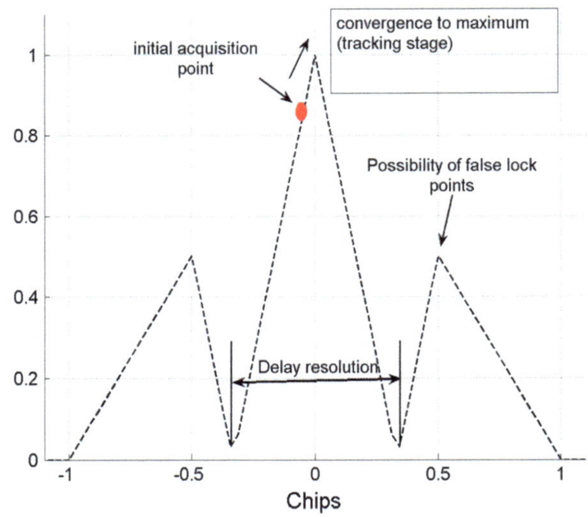

Fig. 5.4 Example of false lock points in the tracking stage, SinBOC(1, 1)-modulated signal

chips). These values hold for infinite bandwidth. For smaller receiver bandwidths, these values slightly increase with the decrease in the receiver bandwidth.

The maximum correlation loss L_{corr} in the acquisition stage is directly proportional to the time offset τ between signal and reference for a given signal, the time bin step $(\Delta\tau)_{bin}$ and the receiver double-sided bandwidth B_W:

$$L_{corr}[dB] = 10\log_{10}\left(\frac{B_W|H(0)|^2}{\int\limits_{-\infty}^{\infty}|H(f)|^2 df}\right) + 10\log_{10}\left(\frac{R(\tau + (\Delta\tau)_{bin})}{R(0)}\right) \tag{5.1}$$

Table 5.1 Delay resolution, distance to the strongest false lock peak in tracking and maximum recommended time-bin step

Modulation	Delay resolution [chips]	Distance between main peak and the nearest zero in the ACF envelope [chips]	Maximum recommended time—bin step in ambiguous acquisition under infinite bandwidth [chips]
BPSK	2	–	0.5
CBOC(+)	0.68	0.34	0.175
CBOC(−)	0.68	0.34	0.175
CosBOC (15, 2.5)	0.08	0.04	0.02
CosBOC (10, 5)	0.22	0.11	0.055
AltBOC (15, 10)	0.32	0.16	0.08

Above, H(f) is the IF filter frequency response and $R(\tau)$ is the code autocorrelation function, including the modulation (e.g., BPSK; SinBOC, etc.).

Survey of Traditional Acquisition Methods

By the term 'traditional' here we mean any 'ambiguous' acquisition method currently existing in the research literature for DS-SS signals. In ambiguous methods, the ambiguities due to BOC family are not dealt with. Since GALILEO signals are basically DS-SS signals, any traditional or ambiguous DS-DCMA acquisition method can be applied for GALILEO, with the only constraint of having a sufficiently low time-bin step (recommended equal or smaller to 0.175 chips, as shown in Table) in order to achieve a sufficiently high detection probability.

In the acquisition stage, the purpose is to obtain some coarse estimates of the channel delay and frequency shift. This is achieved through correlation of the incoming signal $r_{\text{mod}}(t), \text{mod} = CBOC, BPSK, AltBOC$, etc with a reference code with a certain tentative delay $\tilde{\tau}$ and tentative Doppler frequency $\tilde{f_D}$:

$$R\left(\tilde{\tau}, \tilde{f_D}\right) = r_{\text{mod}}(t) \otimes \left(s_{ref}(t - \tilde{\tau})e^{j2\pi\tilde{f_D}t}\right),$$
$$\tilde{\tau} = 0, (\Delta\tau)_{bin}, 2(\Delta\tau)_{bin}, 3(\Delta\tau)_{bin}, \ldots$$
$$\tilde{f_D} = -3(\Delta f_D)_{bin}, -2(\Delta f_D)_{bin}, -(\Delta f_D)_{bin}0, (\Delta f_D)_{bin}, 2(\Delta f_D)_{bin}, 3(\Delta f_D)_{bin}, \ldots \tag{5.2}$$

The received signal $r_{\text{mod}}(t)$ is the transmitted modulated signal $s_{\text{mod}}(t)$ (employing one of the modulation types used in GALILEO), plus the channel effects (multipaths, additive white noise, Doppler and IF shifts and other sources of interferences that are here lumped into the noise term). Indeed, a simplified base-band model of the received signal is the following:

$$r_{\text{mod}}(t) = \left(s_{\text{mod}}(t) \otimes h_{chan}(t)e^{j2\pi f_D t} + n(t)\right) \otimes h_{filt}(t) \tag{5.3}$$

where $h_{chan}(t)$ is the channel impulse response (e.g., including fading multipath effect, see for example the models in (Rappaport 1996)), f_D is the sum between the Doppler shift and the IF frequency, $n(t)$ is the additive noise (e.g., Gaussian noise sources plus other interferences such as multi-access and narrowband interferences) and $h_{filt}(t)$ is the impulse response of the bandwidth limiting filter at the receiver (if absent, then $h_{filt}(t) = \delta(t)$).

The tentative delays $\tilde{\tau}$ of the reference code scan the whole delay axis (or delay uncertainty), namely all the code phases, with a certain time-bin step $(\Delta\tau)_{bin}$. The tentative Doppler frequencies scan the whole frequency axis (or frequency uncertainty) within a minimum and maximum expected frequency shift (typically ±10 kHz for Galileo (Campana et al. 2008), though the exact maximum value can vary), with a certain frequency-bin step $(\Delta f_D)_{bin}$ (typically $(\Delta f_D)_{bin} = 1/\text{N}_\text{c}$, where N_c is the coherent integration length in ms). A time-frequency uncertainty space is illustrated in Figure. Each correlation value corresponding to a tentative delay-Doppler pair ($\tilde{\tau}$, $\tilde{f_D}$) is mapped to a time-frequency bin. Several bins put together will form a time-frequency window.

For example, the number of timing hypotheses (or the time uncertainty space) per frequency bin for the acquisition of Galileo E1 OS signal when a time-bin step of 0.175 is 23382 code phases (or $\lfloor 4092/(\Delta\tau)_{bin} \rfloor$).

The division into bins and windows is essential for understanding the division into serial, hybrid and parallel search algorithms described in the next section. The correct time-frequency bin is the bin corresponding to the correct time delay (or LOS delay) and the correct frequency shift (e.g., incorporating Doppler spread and shift and the IF shift).

An example of the time-frequency correlation mesh (or correlation values) for a correlation window is shown in Fig. 5.5.

The steps of the acquisition process are illustrated in Fig. 5.7, and they will be briefly reviewed in the next sections:

1. *Correlation stage* in this stage correlations with the reference code with certain tentative code delays and Doppler frequencies are formed. Possible coherent and non-coherent integrations are applied. The motivation for using both coherent and non-coherent integration is described in detail in Corazza and Pedone (2007), Van Diggelen (2009). Briefly, coherent integration deals better with noise reduction than the non-coherent one, but it is limited by the data bit transitions, residual Doppler shifts and clock mismatches. Thus, additional non-coherent integration can further decrease the noise level, with the cost of some non-coherent integration loss in comparison with the coherent integration case.

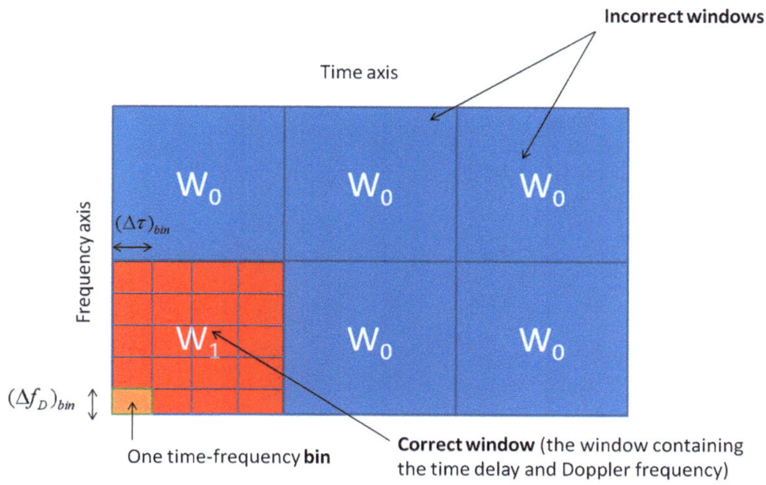

Fig. 5.5 Illustration of the time-frequency uncertainty space in the acquisition process

The coherent integration length is denoted here via N_c (N_c is the number of ms of coherent integration) and the non-coherent integration length is denoted via N_{nc}. The block before the non-coherent integration can be either an envelope detector or a squared envelope detector (the later one is usually preferred in hardware implementations). The total integration length (coherent and non-coherent) is $N_c N_{nc} T$ where T is the code epoch interval in Galileo (e.g., $T =$ 4 ms for E1 OS signals).

2. *Search stage* in this stage, one or several time-frequency bins are used together in order to form a so-called decision statistic, which will inform us whether the signal is present or absent in a certain time-frequency bin (i.e., at a certain tentative delay-Doppler pair $(\tilde{\tau}, \tilde{f_D})$). The decision statistic can be for example the maximum signal energy in a time-frequency window.
3. *Detection stage* in this stage, the test statistic formed previously is compared with a threshold. If it is higher than this threshold, then acquisition is declared and the tracking stage is enabled. If the test statistic is smaller than this threshold, then the search continues for other delay-Doppler pairs. The outputs here are the coarse estimates of the code delay and frequency shift $\tilde{\tau}, \tilde{f_D}$. The accuracy of these estimates depends on the time and frequency—bin steps. For example, in the absence of multipaths or other error sources, the time accuracy after the acquisition stage is bounded by $(\Delta\tau)_{bin}/2$ and the frequency accuracy is bounded by $(\Delta f_D)_{bin}/2$ (Fig. 5.6).

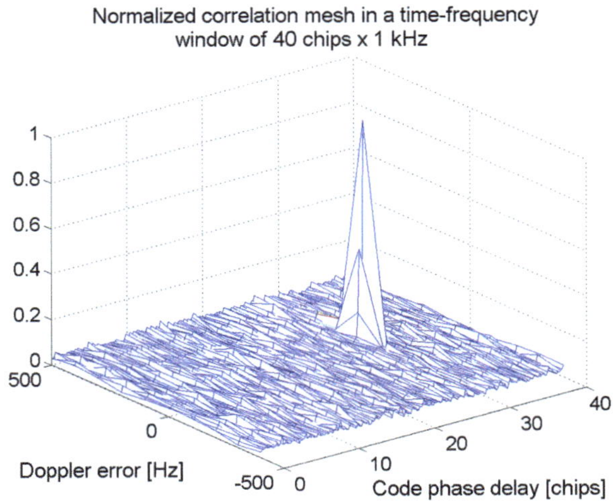

Fig. 5.6 Illustration of a time-frequency correlation mesh in a time-frequency window of 40 chips and 1 kHz, $(\Delta\tau)_{bin} = 0.175$ chips, $(\Delta f_D)_{bin} = 100$ HZ. Correct LOS delay in this example is at $\tilde{\tau} = 28.93$ chips and correct frequency is at $\tilde{f}_D = 15$ HZ

Correlation Stage

The correlation block from Fig. 5.7 can be implemented either in time domain (also called matched filter) or in frequency domain, by taking into account the fact that the convolution in time domain is equivalent with the multiplication in frequency domain. The correlation block performs a complex correlation (both the received signal and the reference code are complex-valued signals).

Examples of time-domain correlation architectures can be found in Lakhzouri et al. (2004), Sorokin et al. (2006), Pany et al. (2010), Van Diggelen (2009).

The concept of frequency-based correlation has been introduced in Cheng et al. (1990) for the acquisition of DS-SS signals. Since then, it has been widely applied to GNSS signals (Akopian 2001; Borre et al. 2007; Chibout et al. 2007; Lin et al. 1999; Van Nee and Coenen 1991; Sagiraju et al. 2006; ODriscoll and Murphy 2005).

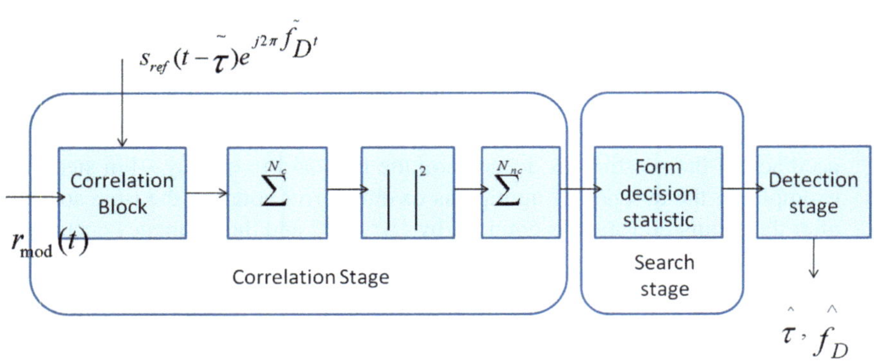

Fig. 5.7 Block diagram of the acquisition process

Regarding these frequency domain or Fast Fourier Transform (FFT)-based correlation structures, there are several acquisition methods proposed in the literature, such as: the classical FFT-based acquisition, also called sometimes N + N ms acquisition method (Chibout et al. 2007), the half-bit acquisition method proposed in Psiaki (2001) and employed when we want to avoid data bit transitions, and the double block zero padding method, proposed also in (Chibout et al. 2007) in order to reduce the acquisition time.

A comparison in terms of complexity between time-domain and frequency-domain approaches can be found in Lakhzouri et al. (2004). It was shown in (Lakhzouri et al. 2004) that frequency-domain based correlations are better choices in the acquisition stage in terms of complexity. Also, frequency-domain correlations are the architectures of choice for software-defined GNSS receivers (Borre et al. 2007).

A more recent comparison of time-domain and frequency-domain approaches in terms of acquisition sensitivity can be found in (Pany et al. 2010). The correlator outputs can be further coherently averaged in order to reduce the additive white noise. As previously mentioned, the coherent integration length is limited by the presence of the navigation data bit transitions, by the stability of the local oscillator at the receiver, by the changes in channel Doppler characteristics, and by the residual Doppler error in the acquisition process. That is why a non-coherent averaging is also sometimes employed, especially in high-sensitivity receivers, in order to increase the total integration gain and to boost the signal-to-noise ratio (Van Diggelen 2009).

Search Stage

In the search stage a decision statistic is formed based on one or several correlation bins. According to the number of bins used for one test statistic, the search stage can be divided into three classes:

- *Serial search* only one bin is used (e.g., see the orange small bin from Fig. 5.5); the test statistic (or decision statistic) is the correlation output in one bin after the integration periods. Polydoros and Simon (1984), Jovanovic (1988), Borre et al. (2007).
- *Hybrid search* several time-frequency bins are used together in order to form a test statistic (e.g., see the red large window from Fig. 5.5). The decision statistic can be, for example, equal with the maximum among the correlation bins within one window. The window size depends on the number of available parallel correlators that can be used. In the hybrid search, there are at least two windows in the whole time-frequency uncertainty space (see Fig. 5.5). Burian et al. (2006), Pajala et al. (2005), Lohan et al. (2004a, b), Zhuang (1996), Povey (1998).
- *Fully parallel search* this is a particular case of the hybrid search; in here the decision statistic is taken on the whole time-frequency uncertainty space and

there is only one window in the whole time-frequency uncertainty space. Zhuang (1996), Zheng (2010), ODriscoll (2007). For illustration, the fully parallel search would correspond to the whole time-frequency space of Fig. 5.5.

In order to form the final decision statistic which will be the input to the Detection Stage, the following generic procedure is used: assume that there are N bins in each time-frequency window (e.g., N = 1 for serial search and N > 1 for hybrid and fully parallel search) and assume that the outputs of the correlation stage (after the non-coherent integration) are denoted with Z_i, i = 1,..., N (where each index *i* corresponds in fact to a certain pair of tentative delay and Doppler estimates), then, the decision statistic *Z* will be formed as a function $f(\cdot)$ of all the Z_i variables:

$$Z = f(Z_1, Z_2, \ldots, Z_N) \tag{5.4}$$

In Fig. 5.8 an example of a time-frequency window is shown, with bins denoted by i, i = 1,...,N.

The function $f(\cdot)$ can be, for example, the maximum among these variables, or a ratio between the strongest maximum and the next maximum at a certain interval apart, and so on. Several different ways of forming the decision statistics in hybrid and fully parallel search have been investigated for example in Corazza (1996), Pajala et al. (2005).

The decision statistic can be based on a single signal at time (e.g., either data or pilot) or can use a combination of data and pilot signals. The combining methods can be either non-coherent or coherent. At the current moment, only few research papers have addressed the issue of how to combine the data and pilot signals in forming the decision statistic. The general approach is the non-coherent one (Borio and LoPresti 2008): the correlation outputs from the two signals (e.g., data and pilot) can be squared, non-coherently integrated and then summed together in order to form the decision variable.

Alternatively, coherent combining methods have been proposed in Borio and LoPresti (2008), Mattos (2005), Yang et al. (2004) for both E1 and E5 GALILEO signals. The main idea of coherent combining methods is to add, respectively subtract the pilot and data signals and to form two decision variables Z_1 and Z_2, and then to form a decision variable, based, for example, on the maximum between the two Z_i outputs. According to the results reported in Borio and LoPresti (2008), the

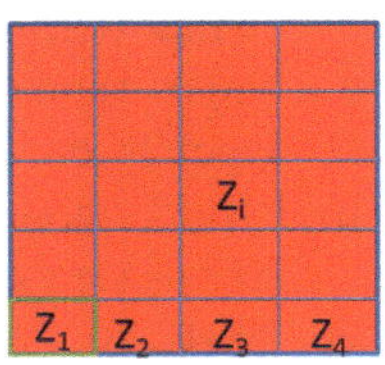

Fig. 5.8 Example of a time-frequency window and the corresponding correlation values $Z_i, i = 1, \ldots, N$ used in forming a decision statistic

coherent combining scheme slightly outperforms the non-coherent combining scheme, but only if the signal level is strong enough. At moderate and low Carrier to Noise Ratio densities (e.g., C/N_0 below 35 dB-Hz), the non-coherent and coherent combining gave the same performance results.

Detection Stage

In the detection stage, the goal is to compare the detection statistic Z obtained in the search stage with a threshold γ in order to decide whether the signal is present or absent. The signal is present when we correlate the incoming signal with a reference signal having the correct delay and Doppler frequency. This translates to the test statistic being higher than a threshold: $Z \geq \gamma$. When either the reference delay or the reference Doppler frequency is wrong, then the signal will be 'absent'. This translates to the test statistic being smaller than a threshold: $Z < \gamma$.

The general problem of detecting a signal in the background noise has been extensively presented in (Kay 1998). The detection can be done in a single stage or in multiple stages, with various coherent and non-coherent integration times. The single-stage detection is known as *single-dwell detection* (Simon et al. 1994; Povey 1998) and the multiple-stage detection is known as *multi-dwell detection* (Simon et al. 1994; Kang and Lee 2003; Lohan et al. 2004a). The concept of multi-dwell approach is illustrated in Fig. 5.9 (it encompasses the case of single-dwell detection for K = 1).

Each dwell in Fig. 5.9 is characterized by the so-called dwell time τ_d, which is simply the integration time in the correlator part $\tau_d = N_c N_{nc} T$. Then a decision statistic Z is formed at a certain tentative code phase estimate and Doppler

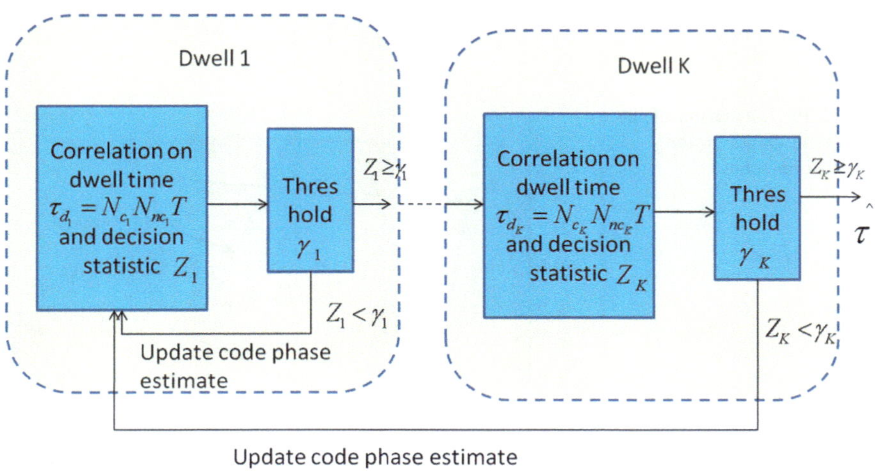

Fig. 5.9 Multi-dwell structure in code acquisition

frequency estimate $\tilde{\tau}, \tilde{f_D}$ and this decision statistic is compared with a threshold. If the decision statistic is higher than the threshold, then we go to the next dwell, with a higher dwell time, in order to cross-check the decision taken at the previous dwell. At each dwell (or decision stage), the same steps are under-taken. The final delay and Doppler estimates are obtained after the last dwell. Ideally, the more number of dwells, the better tradeoff between a high detection probability and a low false alarm probability we can get, though this might not be always the case (Lohan et al. 2004b). Moreover, a higher number of dwells means an increased complexity of the acquisition stage. The multi-dwell structures have not been extensively studied in GNSS systems. The main references found with respect to multi-dwell architectures are: Sheen and Chiou (1999), Aguado et al. (2004), Konovaltsev et al. (2006a).

Each stage (or dwell) is characterized by a certain detection probability $P_{d_k}, k = 1, \ldots, K$ (i.e., the probability that the decision statistic is higher than the threshold when we are in the correct bin) and a certain false alarm probability $P_{fa_k}, k = 1, \ldots, K$ (i.e., the probability that the decision statistic is higher than the threshold when we are in an incorrect bin). The overall detection and false alarm probabilities are the product of the dwell probabilities: $P_d = \prod_{k=1}^{K} P_{d_k}$ and $P_{fa} = \prod_{k=1}^{K} P_{fa_k}$, respectively.

The relationship between the detection probability, false alarm probability and the detection threshold is schematically illustrated in Fig. 5.10. The two curves plotted in there stand for the probability density functions under a correct bin and an incorrect bin hypothesis, respectively. The higher the threshold, the smaller the false alarm probability is, but also the smaller the detection probability.

A performance criterion for the acquisition stage is the Mean Acquisition Time (MAT) (Polydoros and Weber 1984a, b) which refers to the average time needed to be spent in the acquisition stage, until the delay and frequency estimates are obtained. Naturally, different combinations between search and detection mechanisms will

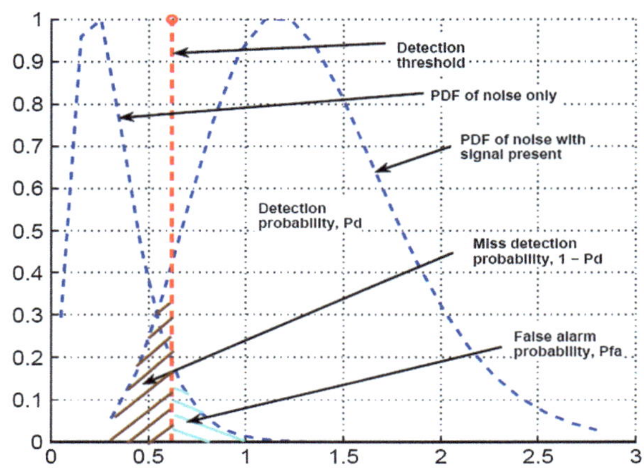

Fig. 5.10 Illustration of the threshold choice in the acquisition stage

provide different MAT. Other performance criteria than MAT exist as well. Some examples can be found in (Omahony 2010).

Typically, the tradeoff is between the complexity of the acquisition unit (e.g., number of parallel correlators) and the acquisition times. Generic methods to compute MAT for DS-SS signals can be found in Povey (1998) , Kang and Lee (2003). For GALILEO, MAT values have been analyzed in Lohan et al. (2004a), Konovaltsev et al. (2006a, b), Villanti et al. (2007). Some techniques to improve MAT both at the search and at the detection stage have been discussed in (Suwansantisuk and Win 2007).

Multipath Mitigation Methods in the Acquisition Process

The problem of the acquisition in multipath channels has usually been treated as the problem of acquisition with multiple correct bins (Glisic and Katz 2001; Kong and Nam 2010). Acquisition methods focused on multipath channels for Galileo signals have not been extensively studied so far; usually the multipath mitigation task is moved to the tracking stage. Among the few methods currently found in the literature that are trying to mitigate the multipath error already at the acquisition stage, we mention: the Differential Correlation methods (Schmid and Neubauer 2004; Pajala et al. 2006) and the Monte Carlo Markov Chain (MCMC) method (Brahim et al. 2008), based on an approximation of Maximum A Posteriori (MAP) estimate.

Also, it was shown in (El-Natour et al. 2005) that the multipath effect on the acquisition stage is not so significant (as in the tracking stage), in the sense that the signal can be still acquired within ±1 chip delay error in most of the studied multipath scenarios.

Unambiguous Acquisition

In order to deal with the ambiguities of the traditional BOC correlation, several unambiguous methods have been proposed in the literature. Unambiguous processing is specific to GALILEO, modernized GPS and any signal that uses a split-spectrum modulation, such as BOC family. The underlying idea of any unambiguous approach is to convert, through some filtering or other transforms, the ambiguous correlation-shape into a 'Binary Phase-Shift Keying (BPSK)-like' shape, thus removing the notches and side peaks within ±1 chip from the maximum peak correlation peak. They can be divided into the following main algorithms:

1. *Betz & Fishman* (*B&F*) algorithm or sideband correlation (Betz 1999; Fishman and Betz 2000): in this method, the spectral sidebands of the signal (respectively of the code) are processed individually. The block diagram of the Dual-Side Band (DSB) B&F method is shown in Fig. 5.11. The Single-Side Band (SSB)

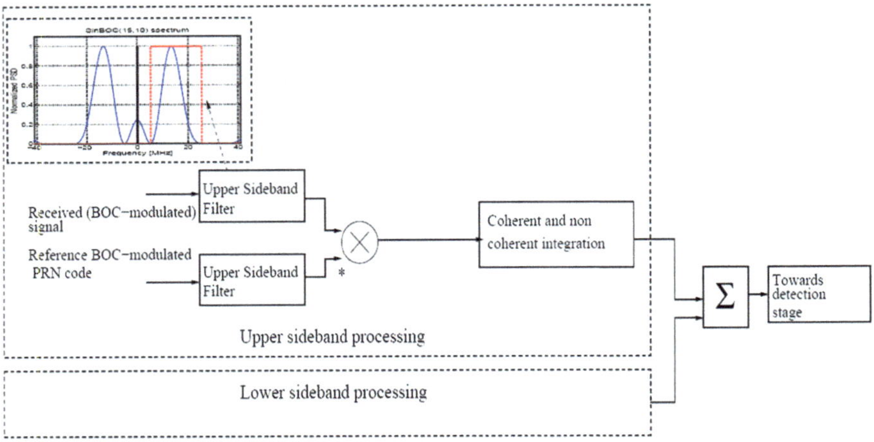

Fig. 5.11 Block diagram of Betz and Fishman (B&F) unambiguous acquisition method (here, the frequency spectrum is shown for sine BOC(15, 10))

B&F method keeps only one of the bands (either upper or lower) when forming the decision statistic. Both the received signal and the reference code (assumed to be real) are filtered and their upper (or lower) bands are correlated, then added non-coherently. This is the most encountered unambiguous acquisition method in current GNSS literature.

2. *Martin & Heiries* (*M&H*) approach or BPSK-like techniques (Martin et al. 2003; Heiries et al. 2004): the block diagram is shown in Fig. 5.12. Here both main side lobes of the received signal are filtered and correlated with a shifted version of the PRN reference code, up-sampled to the same rate as the incoming

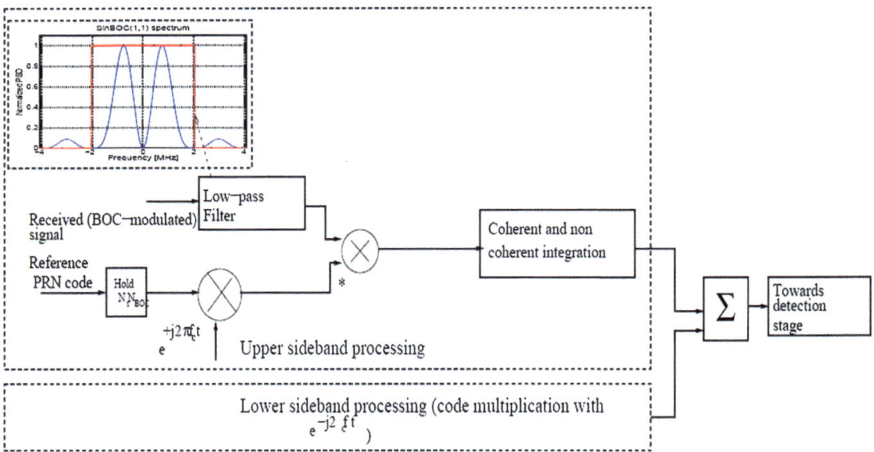

Fig. 5.12 Block diagram of Martin & Heiries (M & H) unambiguous acquisition method (here, the frequency spectrum is shown for sine BOC(1, 1))

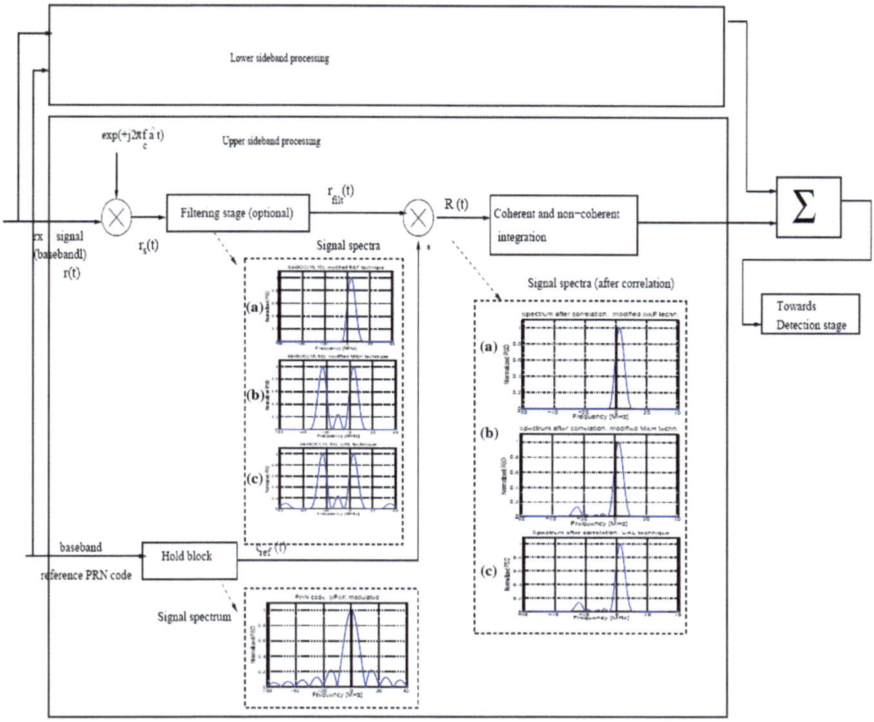

Fig. 5.13 Block diagram of unambiguous acquisition methods of lower complexity than B&F method (here, the frequency spectrum is shown for sine BOC(15, 10)). **a** Modified B&F, **b** Modified M&H, **c** UAL

signal. The drawback for this method is that it is not working properly for odd BOC modulation orders (this does not apply to GALILEO signals that have even BOC modulation orders). We recall that N_B is the BOC modulation order. For example, for sine BOC(1, 1), $N_B = 2$, and for MBOC variants $N_B = 12$ (Lohan et al. 2006a, b).

3. *Modified B&F method* (Lohan et al. 2008): Its principle is shown in Fig. 5.13, the upper signal spectrum (a). The received signal is shifted with $\pm\hat{a}f_c$in frequency domain, then the main lobe is selected (either upper or lower), and the resulting signal is correlated with the reference PRN code (brought at the signal rate). The shifting factor $\hat{a}$ depends on the BOC modulation order. For sine BOC (1, 1) and CBOC signals, $\hat{a} = 1$.
4. *Modified M&H method* (Lohan et al. 2008): Its principle is shown in Fig. 5.13, the middle signal spectrum (b). The difference with the modified B&F approach is that now we select both main lobes of the incoming signal (not only one of them). The performance of the modified M&H method is exactly the same as that of M&H method, but it has a lower complexity.

5. *Unsuppressed Adjacent Lobes* (*UAL*) *method* (Lohan et al. 2008): its principle is shown in Fig. 5.13, the middle signal spectrum (c). In here, the filtering part is removed completely, in order to decrease the implementation complexity. It has slightly worse performance than the modified B&F and modified M&H approaches, but it offers the lowest complexity among all the unambiguous acquisition approaches.

All the above methods have both a DSB and a SSB implementation, depending whether we keep both upper and lower bands (and combine them non-coherently) or only one of them.

Recent studies on unambiguous approaches on MBOC signals showed that unambiguous methods can offer up to 4 dB C/N_0 enhancements in the acquisition process (Samad and Lohan 2009), providing that the receiver bandwidth is sufficiently high. The best performance is achieved with B&F method, which has also the highest complexity, while the lowest complexity unambiguous method, namely UAL. For low receiver bandwidths however, recent studies showed that there is no benefit in using unambiguous acquisition, and the severe bandwidth limitation (e.g., to 3 or 4 MHz double-sided bandwidth) act in such a way to alleviate the ambiguities (Lohan 2011).

An example of the performance of unambiguous double-sideband approaches compared with the performance of unambiguous acquisition is shown in Fig. 5.14 for infinite bandwidth, serial search, and CosBOC(15, 2.5) modulation. More examples regarding the performance of unambiguous approaches can be found in Lohan et al. (2008), Samad and Lohan (2009), Lohan (2011).

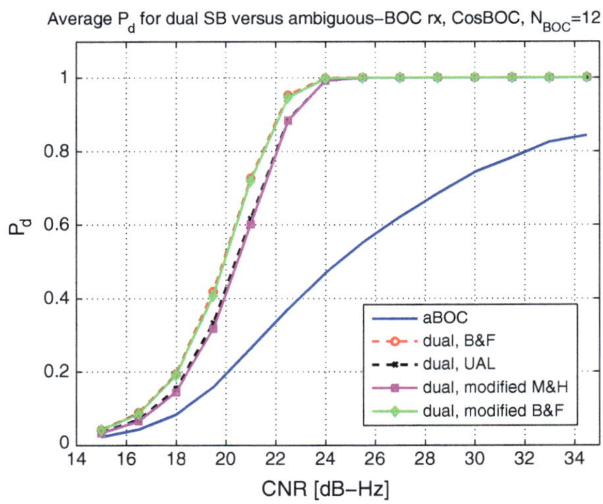

Fig. 5.14 Average detection probability for CosBOC(15, 2.5) modulation with ambiguous and unambiguous processing. Serial search, $P_{fa} = 10^{-3}$, $N_c = 30$ ms, $N_{nc} = 6$

Survey of Traditional Tracking Methods

The next sub-sections deal with code and carrier tracking algorithms. Once the acquisition stage is completed, the receiver enters in the tracking mode. Tracking has usually two parts: code tracking and carrier tracking. The tracking mechanisms in GNSS have been described in detail in Kaplan and Hegarty (2006), Parkinson and Spilker (1996), Misra and Enge (2011), Borre et al. (2007). In here, we target at a literature review of the main code and carrier tracking methods that can be used, in general, for any DS-SS system, and, in particular, for GALILEO signals.

Code Tracking

Again, by 'traditional' here we refer to any 'ambiguous' tracking method currently existing in the research literature for DS-SS signals, including both well-established methods and the state-of-art methods. Many of them can also be used as a basis for the unambiguous tracking methods discussed later on in this chapter. A classification of code delay tracking methods according to the Author is shown in Fig. 5.15 and each class is discussed in the following sections. Many of these methods have been derived to deal better with multipath propagation, and thus they have some multipath mitigation properties as well. Later on in this chapter there is also a

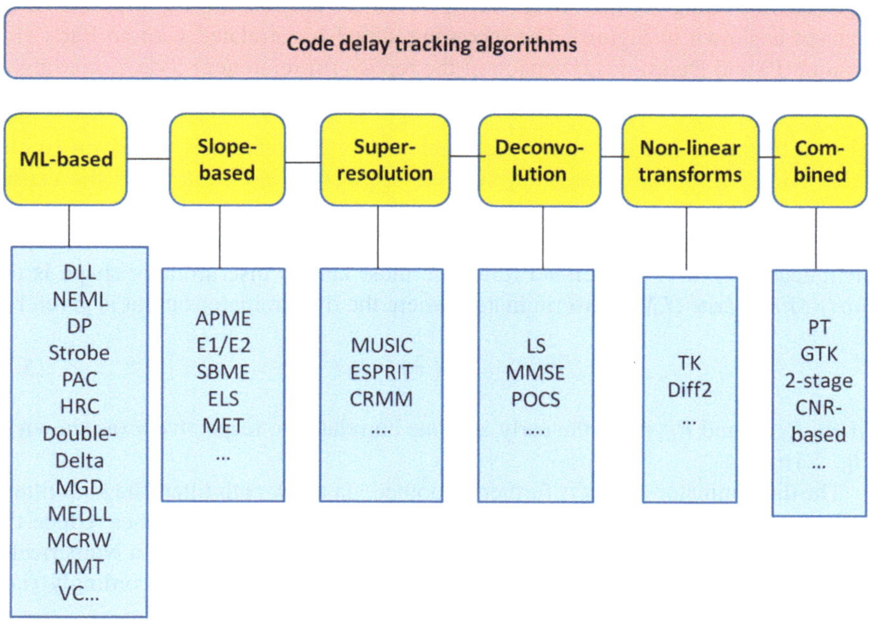

Fig. 5.15 Classification of code delay tracking methods

section dealing particularly with multipath effects and multipath mitigation in the context of GNSS code tracking.

Maximum Likelihood-Based Approaches

The idea behind any *Maximum Likelihood* (*ML*) estimator is to determine the parameters that maximize a likelihood function, which is the joint probability density function (PDF) of the sample data. This estimation method does not require a priori information and assumes that the unknown parameters are constant over an observation period, typically hundreds of milliseconds or multiple seconds for high-sensitivity receivers (Sahmoudi and Landry 2008). A straightforward implementation of an ML estimator is typically very complex, as it requires the receiver to measure the received signal cross-correlation function for each reflected path with multiple correlators and to process these measurements with complex algorithms. That is why, in practice, some sub-optimal implementations or ML-based approaches are preferred, with a lower number of correlators. The simplest of all ML-based approaches is the classical *Delay Locked Loop* (*DLL*) (Kaplan and Hegarty 2006; Parkinson and Spilker 1996). In here, at most three complex correlation branches (or, equivalently 6 real correlators) are employed: an early replica, a prompt-replica, and a late replica. An error curve is created based on the early, late and sometimes also prompt replicas and the code phase of the replica code is adjusted according to this error curve. A block diagram of the classical DLL concept is shown in Figure . The incoming signal is correlated with an Early (E), Prompt (P) and Delayed (D) version of the replica (or reference) code, according to a certain tentative delay estimate $\tilde{\tau}$ (initial value comes from the acquisition stage). Also, a tentative Doppler shift $\tilde{f_D}$ is employed in the reference waveform (initial value comes from the acquisition and sub-sequent values come from the carrier phase and carrier frequency loops). The correlation outputs, possibly coherently and non-coherently averaged, are fed into a discriminator function, which combines them according to a pre-defined rule. The most known discriminator shape is the *Early Minus Late* (*EML*) discriminator, where the discriminator output is given by:

$$D(\tilde{\tau}) = R_E(\tilde{\tau}) - R_L(\tilde{\tau}) \tag{5.5}$$

where $R_E(\cdot)$ and $R_L(\cdot)$ are the early and late correlations, respectively (as shown in Fig. 5.16).

The discriminator output is further smoothed via a low path filter (the loop filter), with a small filter bandwidth B_L(typically few Hz), in order to further reduce the noise, and the output of the loop filter is used as the control input of a Numerically Controlled Oscillator (NCO), which adjusts the reference code accordingly (i.e., updates the code delay $\tilde{\tau}$ in such a way to minimize the discriminator error, in absolute value).

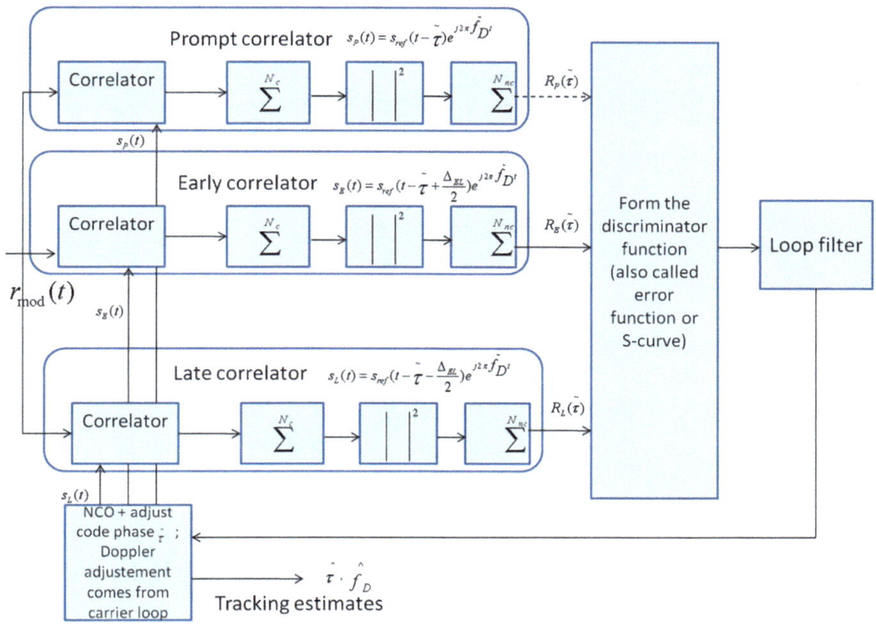

Fig. 5.16 Block diagram of the classical DLL tracking loop (non-coherent version)

Other variants of classical DLLs are the normalized EML and the *Dot-Product* (*DP*) discriminator (Borre et al. 2007).

According to the spacing Δ_{EL} between early and late correlations, the DLL can be wide ($\Delta_{EL} = 1$chip) or narrow ($\Delta_{EL} < 1$chip). The narrow correlator concept has been introduced in 1992 (Dierendonck et al. 1992) as a DLL variant better coping with noise, and, since then, it has proved to give moderate to good results also in the presence of multipaths and GALILEO ambiguities. An example of the narrow correlators is the *Narrow Early-Minus-Late* (*NEML*) code tracker (Dierendonck et al. 1992).

If we introduce two additional early and late correlators (a so-called 'very early', spaced at $-\Delta_{EL}$ from the prompt one and a 'very late', spaced at Δ_{EL} from the prompt one), we get the family of double-delta correlators: the *High Resolution Correlator* (*HRC*) (McGraw and Braasch 1999), the *strobe correlators* (Garin and Rousseau 1997), *Pulse Aperture Correlator* (*PAC*) (Fenton et al. 2004), etc.

The discriminator function for double-delta correlators mentioned above is of the form:

$$D(\tilde{\tau}) = \left(R_E(\tilde{\tau}) - R_L(\tilde{\tau})\right) - a\left(R_{VE}(\tilde{\tau}) - R_{VL}(\tilde{\tau})\right) \tag{5.6}$$

where $R_{VE}(\cdot)$ and $R_{VL}(\cdot)$ are the very early and very late correlations, respectively, and a is a weight factor (typically $a = 0.5$). The differences between various double-delta correlators are usually minor.

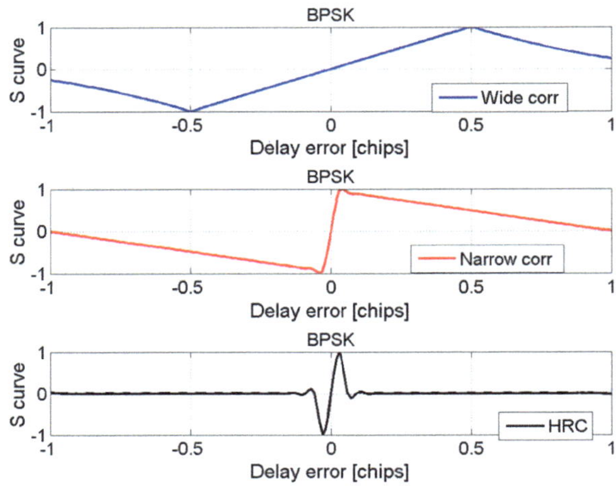

Fig. 5.17 Example of S-curves for BPSK-modulated codes and various discriminator types. Single path channels

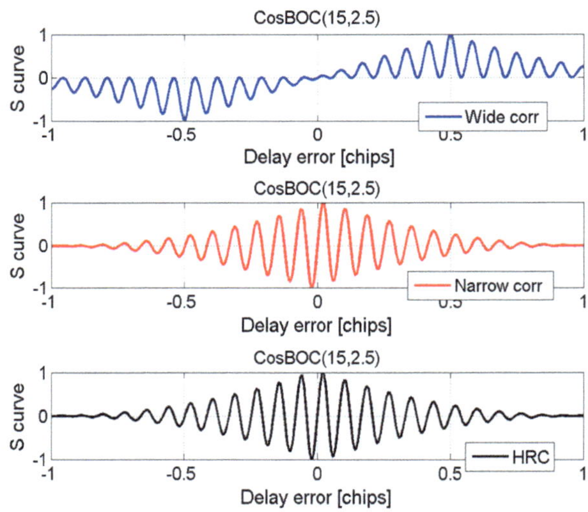

Fig. 5.18 Example of S-curves for CosBOC(15, 2.5)-modulated codes and various discriminator types. Single path channels

Examples of S curves or discriminator curves for wideband correlator ($\Delta_{EL} = 1$ chip), narrowband correlator ($\Delta_{EL} = 0.02$ chips) and HRC ($\Delta_{EL} = 0.02$ chips) are shown in Figs. 5.17 and 5.18 for BPSK and CosBOC(15, 2.5)-modulations, respectively. A double sided bandwidth of 24.552 MHz was used in both cases with Butterworth filtering. For BPSK case, the pull-in range (i.e., the linear region around zero crossing) is smaller when we move from a larger time-bin step (wideband correlator) to a smaller time-bin step (narrow correlator) or to HRC case. This means a better resistance to multipaths, but also a higher likelihood of faster loss of lock in noisy situations. For CosBOC(15, 2.5), the wideband correlator is not sufficient, since the zero crossing at correct delay is barely visible. The narrow correlator and HRC correlator in this particular case have similar performance.

An extension of the double-delta correlator concept has been presented in (Hurskainen et al. 2008) under the form of *Multiple Gate Delay* (*MGD*) structures. In there, the number and the spacing of the correlators can vary and can be optimized according to the estimated channel profile, in order to achieve a better robustness against multipaths. The optimization criterion in MGD is built according to the multipath errors. A least-squares based optimization of the discriminator (or S curve) has been proposed in (Pany et al. 2005). The same idea of using multiple correlators as in MGD is used, but the weighting coefficients used in forming the discriminator are optimized in the minimum square error sense (by minimizing the squared-absolute-valued error between the multipath distorted discriminator and an ideal, reference discriminator).

The *Modified Correlator Reference Waveforms* (*MCRW*) concept introduced in (Zhdanov et al. 1999) makes use of a modified replica code at the receiver, in such a way to provide a cross-correlation function with inherent resistance to errors caused by multipath. An example belonging to this family of MCRW is the *gated correlator* (McGraw and Braasch 1999), which consists of periodically blanking the received signal between code chip transitions. However, due to the modified reference correlation function, some losses in the signal to noise ratio occurs and the receiver is typically more sensitive to noise.

The *Multipath-Estimating Delay-Lock Loop* (*MEDLL*) (Van Nee 1992) used for example in NovAtel receivers also uses several correlators (e.g., 6–10) in order to determine accurately the shape of the multipath-corrupted correlation function. Instead of the discriminator function from Figs. 5.17 and 5.18, a multipath estimation unit is incorporated in a loop. In this multipath estimation unit we determine via extensive maximum likelihood search the best combination of LOS and multipath components (i.e., amplitudes, delays, phases and number of multipaths) which would have produced the measured correlation function. The *Reduced Search Space Maximum Likelihood* (*RSSML*) estimator (Bhuiyan et al. 2009) is a reduced-complexity implementation of MEDLL, based on some lookup tables.

The *Multipath Mitigation Technology* (*MMT*) introduced in (Saarnisaari 1996) and studied later also in (Weill 2002) is also based on a direct computation of log-likelihood functions, similar with MEDLL. However, unlike in MEDLL, a coordinate transforms is performed and the complexity of the search process is substantially reduced.

Another ML-based technique to deal with multipath propagation was proposed by Novatel (Fenton and Jones 2005) under the name of *Vision Correlator* (*VC*). By distinction with all the above correlators, VC is a chip-level delay estimator, where the multipaths are searched directly in the transitions of the received signal (before its correlation with the incoming code). In order to deal with low signal levels of the spread signal, coherent integration is performed on several successive chips and an equivalent chip transition waveform is built and compared with some theoretical reference. The theoretical references are built for several multipath scenarios, and then, the "best fit" is selected. In (Fenton and Jones 2005), VC was shown to have better multipath resistance than the narrow correlators and double-delta correlators. However, the results were shown only at rather high signal levels and there is still an open issue whether the Vision Correlator is able to deal with low and moderate carrier-to-noise ratios. VC has been built based on MMT, as noticed also in (Blanco-Delgado 2011).

Slope-Based Approaches

Slope-based approaches are based on the idea of using some information from certain slopes of the ACF (usually close to the maximum ACF peak) and trying to infer multipath information based on such slope information. An example is shown in Figure for a CBOC(+) ACF under infinite BW. The slope in this example can be calculated as:

$$\tan(\theta) = \frac{b-a}{d-c} \tag{5.7}$$

where a, b, c and d are the points illustrated in Fig. 5.19 and they depend on the chosen correlation values.

The *A Posteriori Multipath Estimation* (*APME*) of Septentrio (Sleewaegen and Boon 2001) relies on the a posteriori estimation of the multipath tracking error. It incorporates a narrow correlator plus an independent estimation module. The estimation module uses some slope information, according to the correlation function with and without multipaths.

The *Early-Late Slope* (*ELS*) correlator (Irsigler and Eissfeller 2003; Bhuiyan et al. 2008), also known as *Multipath Elimination Technique* (*MET*), is based on two correlator pairs at both sides of the correlation functions with parameterized

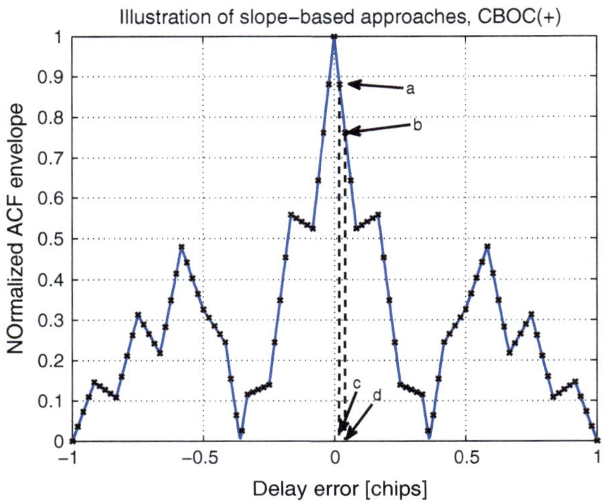

Fig. 5.19 Illustration of slope-based approaches principle

spacing (similarly with MGD and HRC cases). Based on these correlation pairs, the slopes of the correlation function can be computed and then, their intersection point can be found. A pseudorange correction is then performed. The MET is based on the symmetry of the autocorrelation function of the codes which is lost in the presence of multipath. Simulation results performed in (Irsigler and Eissfeller 2003) showed that the ELS is outperformed by HRC and by the strobe correlators in multipath environments. Improvements of ELS technique have been more recently proposed in (Heiries et al. 2006).

Another slope-based correlator structure is the *Early 1/Early 2 (E1/E2) tracker*, initially proposed in (Mattos 1996) and later analyzed also in (Irsigler and Eissfeller 2003) In E1/E2 structure, two correlators are located on the early slope of the correlation function (with an arbitrary spacing). The corresponding amplitudes are compared with the amplitudes of an ideal reference correlation function (no multipath) and some delay correction factor is computed based on the measured amplitudes and reference amplitudes.

A more recent *Slope-Based Multipath Estimator (SBME)* has been proposed in (Bhuiyan et al. 2010a) also attempts to compensate the multipath error contribution of a narrow correlators tracking loop by utilizing the slope information of an ideal normalized correlation function. The main difference with APME stays in the way of computing the multipath error contribution. Its reported results are better than those of narrow correlators and HRC in multipath environments.

Super-Resolution Approaches

The super-resolution or subspace based algorithms involve the decomposition of the space spanned by the observation vector (i.e., the vector formed by the received signal samples) into several subspaces, usually a noise subspace and a signal subspace. Furthermore, these algorithms use the orthogonality property between noise subspace and signal subspace in order to estimate the channel parameters. The subspace approaches involve eigenvector decomposition of high-order matrices. The main advantage of the subspace-based methods is their increased resolution in the parameter estimates. The main disadvantage is their increased complexity with respect to other existing delay code trackers for DS-SS signals. The most known subspace-based methods which have been employed in delay estimation of DS-SS signals are the *Multiple Signal Classification (MUSIC)* (Jakobsson et al. 1998; Chu and Mitra 1999), the *Estimation of Signal Parameters via Rotational Invariance techniques (ESPRIT)* (Jakobsson et al. 1998), and the *Expectation Maximization (EM)* algorithm (Antreich et al. 2005). Also the *Space-Alternating Generalized Expectation-maximization (SAGE)* algorithm described for example in Fessler and Hero (1994), Antreich et al. (2005) is a reduced complexity implementation of EM.

The super-resolution approaches have been but little investigated so far in GNSS literature, mainly due to their complexity and reported sensitivity to noise. Recent efforts in this direction are the studies of the *Complexity Reduced Multipath*

Mitigation (*CRMM*) subspace techniques for GPS and GALILEO (Groh and Sand 2008; Groh 2011).

Deconvolution Approaches

Resolving the multipath components can also be seen as a deconvolution problem (Lohan et al. 2005) in which we try to estimate the non-zero elements of an unknown gain vector, modeling the channel complex coefficients at each possible time delay. The first non-zero component higher than a threshold will be the estimate of the LOS component. The classical solution of a deconvolution problem is the *Least Square* (*LS*) solution. However, the LS solution is highly sensitive to noise and it is not well suited to GNSS signals (Lohan et al. 2006c). Improved deconvolution approaches are the *Minimum Mean Square Error* (*MMSE*) estimation (Bhuiyan et al. 2009; Lohan et al. 2006c) and the *Projection Onto Convex Sets* (*POCS*) estimation (Kostic et al. 1992; Skournetou et al. 2009; Lohan et al. 2006c; Dragunas 2010; Dragunas and Borre 2011). Deconvolution approaches typically suffer of rather low robustness to noise. However, their full potential for GALILEO signals in multipaths is yet to be investigated.

Non-linear Transform-Based Approaches

A different class of delay trackers or estimators is based on a non-linear transform of the correlation function. A non-linear scheme based on the slope differential or *second-order derivative* (*Diff2*) of the correlation function has been proposed in (Bhuiyan 2006). It was shown there that this scheme has better multipath performance than narrow EML and strobe correlators.

Another non-linear transform that can be used is the nonlinear quadratic *Teager Kaiser* (*TK*) operator, first introduced for measuring the real physical energy of a system (Kaiser 1990) and later applied to multipath delay estimation in DS-SS systems (Lohan et al. 2005) and more specifically, in GALILEO systems (Lohan et al. 2006c; De Castro et al. 2007).

The non-linear *wavelet transform* has also been investigated in the context of GPS delay estimation in (Souza et al. 2008).

Combined Approaches

Combinations of two or several between the above-mention algorithms have also been studied in the literature with the scope of achieving better tradeoffs between performance and complexity. For example, *the two-stage Carrier-To-Noise-based*

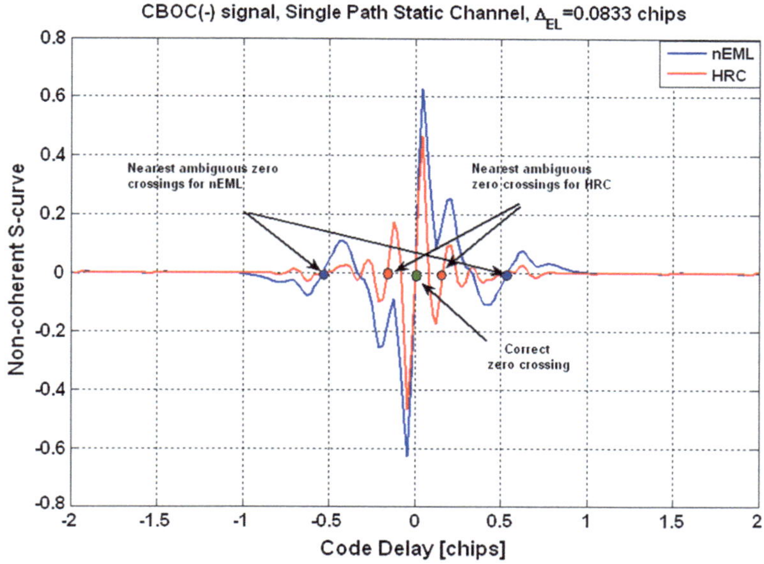

Fig. 5.20 Non-coherent S-curve for CBOC(−) modulated single path static channel with infinite receiver bandwidth. Illustration of the nearest ambiguous zero-crossing points for High Resolution Correlator (HRC) and narrow earlyminus-late correlator (nEML)

estimator proposed in (Bhuiyan et al. 2010b) combines a narrow correlators (at low to moderate C/N_0) with a High Resolution Correlator (at moderate to high C/N_0), thus alleviating or decreasing the problem of loss of lock specific to HRC and increasing the performance in multipaths. The motivation of this approach is given by the fact that nearest ambiguous zero crossings for HRC is much closer to that of a narrow correlators, implying that the probability of locking to any of the side peaks is much higher for HRC than that of narrow correlators. An illustration of this problem is shown in Fig. 5.20. In here, nEML stands for narrow early minus late correlator. The discriminator curves (or S curves) have a correct zero crossing (from below) at zero delay error and additional zero crossings from below at around ±0.17 chips for HRC and ±0.5 chips for nEML.

The *Peak Tracking* (*PT*) techniques described in (Bhuiyan et al. 2008) employs several non-linear transforms of the correlation function and forms a pool of competitive peaks, among which the final LOS delay estimate is selected according to a weighting algorithm, thus achieving an increased performance in multipath environments. The *Generalized Teager Kaiser* (*GTK*) estimator combines the deconvolution algorithms with the TK non-linear estimator (Hamila et al. 2005) and it manages to outperform both TK and deconvolution stand-alone approaches. The *Differential Teager-Kaiser* algorithm starts from TK and first-order derivatives ideas and offers a new closed-loop algorithm for GNSS code tracking (Lohan 2012). A *Fast Iterative Maximum Likelihood Algorithm* (*FIMLA*) combines the narrow correlator with a maximum likelihood search, thus reducing the complexity of the final tracking algorithm (with respect to ML and MEDLL approaches) and still preserving a better multipath accuracy than the narrow correlator (Sahmoudi and Landry 2008). The work in (Blanco-Delgado 2011) also focuses on ML estimators, this time in conjunction with interpolation methods in order to increase the estimation accuracy.

Multipath Mitigation and Code Tracking

Basically, all the code tracking algorithms mentioned in the previous sections can cope with multipath to a certain extent. Typically, the tradeoff is between complexity and performance, but sometimes even a simple structure such as the narrow correlator can outperform the more complex ones (e.g., in very noisy scenarios) (Bhuiyan et al. 2008; Bhuiyan and Lohan 2010). More advanced multipath mitigation structures also require antenna arrays at the GNSS receivers (Seco-Granados 2000; Ray 2000). Additionally, front-end multipath mitigation structures can be used, such as choke ring antennas (Tranquilla 1994) or antenna polarization and antenna pattern shaping (Williams et al. 2000), but they are out of the scope of this Chapter.

The simplest way to analyze the performance of an algorithm in the presence of multipaths is to look at the Multipath Errors (ME) under some simplified channel assumptions (such as very low or no noise, static channels, few paths) and to build the so called *Multipath Error Envelopes* (MEE) (Ray 2000; Bhuiyan et al. 2008) or *Running Average Error* (RAE) curves (Avila-Rodriguez et al. 2006a). The concept of multipath error is illustrated in Fig. 5.21, starting from an S-curve for a CBOC(−)-modulated signal, narrow correlators with $\Delta_{EL} = 0.08$ chips spacing and 24.552 MHz double-sided bandwidth (Butterworth filtering). In the absence of any multipath (i.e., single path channels), the S curve would cross the zero axis at exactly zero delay error. In the presence of multipaths (here 3, one LOS and 2 multipaths), the zero crossing and thus the delay estimation error is displaced, and we get a certain multipath error which is dependent on the channel profile. By comparing various MEs under similar channel profiles, we get the MEE. In a typical analysis of MEE, only 2 channel paths are assumed, having either the same phase (in-phase paths) or 180° shift (out-of-phase paths). The second path is typically assumed to be 3 dB or 6 dB weaker than the first one. The MEEs for these two extreme cases are built, and we may get some plots similar with those shown in Fig. 5.22.

Clearly, MEEs shapes depend on the assumed channel profile and on the receiver bandwidth, as well as on the delay tracker algorithm (or discriminator shape).

A unified comparison between all the code tracking algorithms in terms of their performance in multipath scenarios is still lacking in the current literature. A partial survey between several of the above-mentioned code tracking algorithms for GNSS signals can be found for example in (Lohan et al. 2006c) and (Bhuiyan and Lohan 2010).

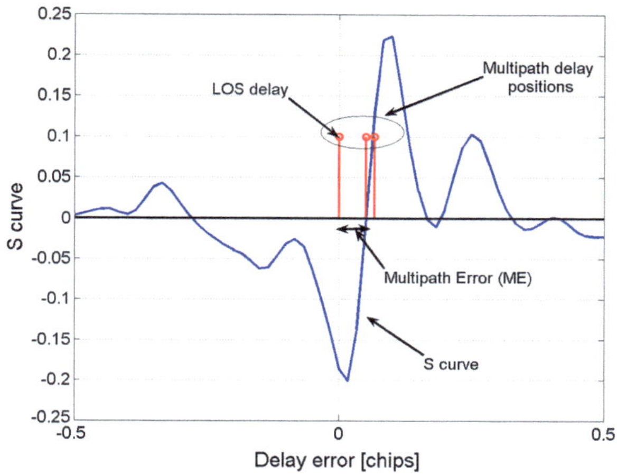

Fig. 5.21 Illustration of the Multipath Error (ME) concept, based on S-curve of a delay tracker

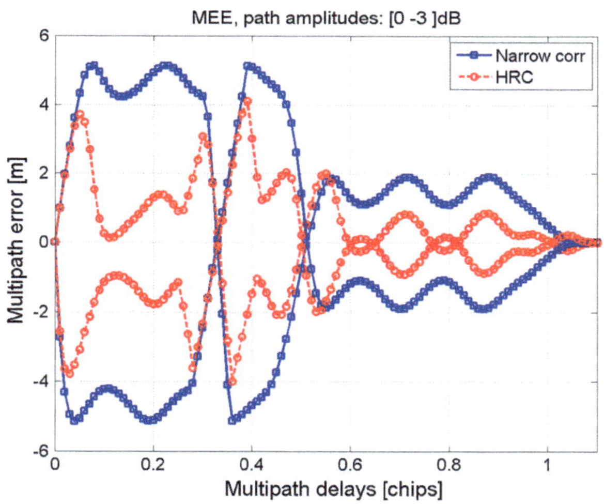

Fig. 5.22 Example of MEE curves for narrow correlator and HRC, CBOC(−) signal, 24.552 MHz double-sided bandwidth, $\Delta_{EL} = 0.08$ chips early-late spacing

Unambiguous Code Tracking

The unambiguous methods used in acquisition tend to 'destroy' the main correlation peak, while removing the ambiguities within one chip interval across the main peak. However, for code tracking purpose, we would like to preserve the narrow main correlation lobe of the BOC modulation, while removing or diminishing the adjacent side peaks that may create false peak convergence.

An innovative unambiguous tracking technique, that keeps the sharp correlation of the main peak, is proposed in (Julien et al. 2005). This approach uses two correlation channels, completely removing the side-peaks from the correlation function in case of a SinBOC(n, n)-modulated waveform. However, the Julien et al.

method is not applicable to other types of BOC modulation. A similar method with better multipath resistance has been proposed in (Heiries et al. 2005), under the name of *Sub-Carrier Phase Cancellation* (*SCPC*) technique. This time, the method was also tested for CBOC waveforms. The differential correlation approaches presented in (Schmid and Neubauer 2005) can also be used to a certain extent in order to diminish the side lobes or false peaks in the correlation envelope. The method presented in (Julien et al. 2005) has been also developed in (Julien et al. 2007) under the name of *ASPeCT*. It was noted in (Julien et al. 2007) that the ASPeCT method could also be used for the signal acquisition process, allowing for a comparable sensitivity to the ambiguous acquisition scheme, but with the need of twice as many correlators in order to achieve the same mean-time-to-first-fix compared to a ambiguous acquisition scheme.

More recently, a method called *Sidelobe Cancellation Method* (*SCM*) has been introduced in (Burian et al. 2006) and later developed in (Burian et al. 2007) in order to deal with any sine and cosine BOC-modulated waveform and to diminish the threat of false lock points in tracking. An illustration of SCM principle is shown in Fig. 5.23. A so-called subtraction waveform is generated based on the BOC-properties, in such a way to cover the side-peaks the closest to the main peak (the red dashed curve in the upper plot). The closed formula of the pulse to be subtracted for SinBOC and CosBOC modulations is given by Burian et al. (2006), Burian et al. (2007):

$$R_{sub}(\tau) = \sum_{i=0}^{N_B-1} \sum_{j=0}^{N_B-1} \sum_{k=0}^{N_{B_2}-1} \sum_{l=0}^{N_{B_2}-1} (-1)^{i \times j+k+l} \Lambda_{T_B}\left(\tau + i\frac{T_c}{N_B} - j\frac{T_c}{N_B} + k\frac{T_c}{N_B N_{B_2}} - l\frac{T_c}{N_B N_{B_2}}\right) \tag{5.8}$$

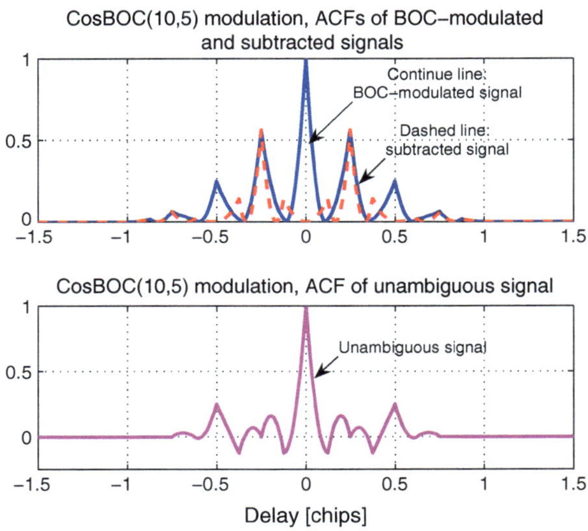

Fig. 5.23 SCM principle: examples of the ambiguous correlation function and subtracted pulse (*upper plot*) and obtained unambiguous correlation function (*lower plot*), in a single-path channel

where $T_B = \frac{T_c}{N_B N_{B_2}}$ is the minimum BOC sub-interval, $N_B = \frac{2m}{n}$ is the BOC modulation order, defined also in Chap. 3, and N_{B_2} is a flag equal to 1 for SinBOC case and equal to 2 for CosBOC case. For more details about this model, the reader is directed to (Lohan et al. 2006a).

This subtraction waveform needs first to be aligned to the maximum correlation peak, and then subtracted from it in order to obtain a correlation function free of the highest false peaks. This is the reason for which SCM method cannot be used alone in multipath channels, but needs to be combined with some other multipath mitigation methods. Examples can be found in (Burian et al. 2007), where it is shown that SCM in combination with narrow correlator provides a good multipath performance, while diminishing the threat of the false lock tracking points.

A comprehensive comparison between all existing unambiguous tracking techniques nowadays is still missing in the Galileo literature.

Carrier Phase and Frequency Tracking

The code tracking loop is used together with a carrier phase and frequency tracking loop, illustrated schematically in Fig. 5.24. The task of a carrier phase and frequency tracking loop is to estimate and track the Doppler frequency shift. For this purpose, a combination between a Phase Locked Loop (PLL) and a Frequency Locked Loop (FLL) is employed. The PLL is concerned with carrier phase tracking, while the FLL focuses on carrier frequency tracking. If the Doppler frequency variations are not very large, sometimes only the PLL loop is employed (Borre et al. 2007). The output of the complex correlator illustrated in Fig. 5.24 is input to the PLL and FLL discriminators, and the discriminator outputs are then combined in a loop filter. The basic carrier tracking loops (both PLL and FLL) have the same principles as DLL, but with a different discriminator. An example is shown in Fig. 5.25 (Borre et al. 2007).

A typical PLL discriminator function is the arctangent discriminator (Borre et al. 2007), while a typical FLL discriminator for GNSS is the four-quadrant arctangent discriminator. The loop filter can be individual for PLL and FLL or joint for the two (Kaplan and Hegarty 2006). Another example of joint DLL-PLL-FLL is illustrated in Figure. This example is based on the implementation of the open-source GALILEO E1-E5a Simulink simulator (Zhang and Lohan 2010).

The multipath effect on carrier tracking is much less (translated in meters) than the multipath effect on code tracking (usually it is in the order of cm for the carrier tracking and of the order of meters for code tracking). Some of the algorithms used for code tracking multipath mitigation also work for the carrier phase tracking in the presence of multipaths. Examples of such algorithms are the MEDLL (Van Nee 1992) and the deconvolution algorithms (Skournetou et al. 2009), which estimate also the multipath phases. Examples of carrier phase multipath effects and a discussion about mitigation carrier phase multipaths via antenna arrays can be found in

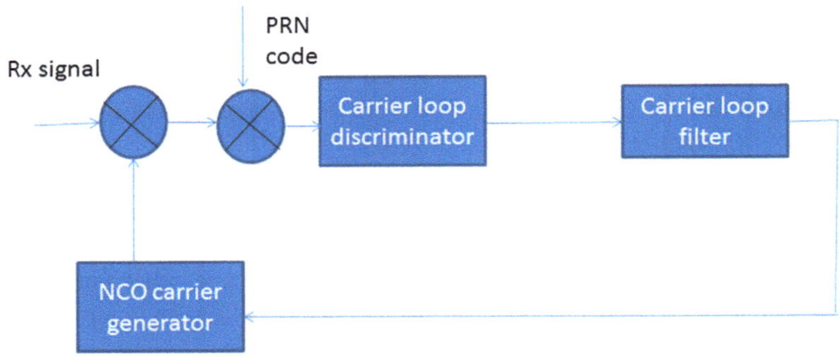

Fig. 5.24 PLL/FLL discriminators principle

(Ray 2000). The carrier tracking loops in the presence of multipaths have been investigated much less than the code tracking loops in the context of GNSS signals (Fantino et al. 2008) (Fig. 5.25).

Conclusions and Look Forward

Acquisition and tracking algorithms in GALILEO have to cope with the usual challenges in acquiring and tracking a DS-SS signal over a wireless signal (namely multipaths, dealing with the acquisition of long codes, relatively weak signal

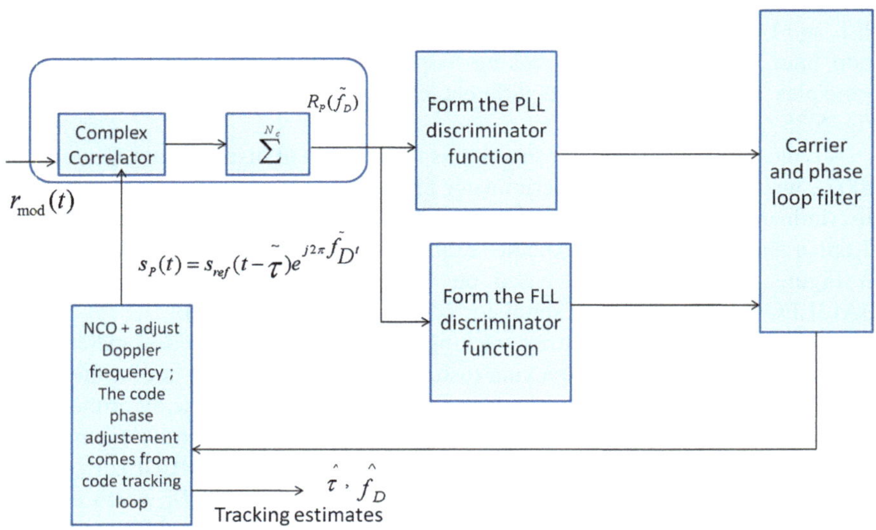

Fig. 5.25 Carrier phase and frequency tracking loop

powers, etc.) but also with an additional challenge due to the utilization of BOC-class modulations. This challenge is related to the presence of ambiguities in the correlation envelopes and manifests itself differently in acquisition and tracking stages. In acquisition stage, we would like to preserve a wide correlation main lobe while removing the notches, while in the tracking stage the narrow main lobe is desirable to be preserved, at the same time as the additional side lobes are removed. For these purposes, different unambiguous algorithms have been developed for the acquisition and tracking of GALILEO signals. The main ones currently existing in the GNSS literature have been summarized in this Chapter and some illustrative examples have also been showed. However, a comprehensive comparison between various unambiguous approaches for all BOC modulation classes currently proposed in GALILEO is still missing nowadays. Moreover, the performance of such unambiguous algorithms in the presence of multipaths and fading channels is still a topic of intense research interest.

References

Akopian D (2001) A fast satellite acquisition method. In: Proceedings ION-GPS 2001, Salt Lake City, pp 2871–2881

Antreich F, Esbri-Rodriguez O, Nossek JA, Utschick W (2005) Estimation of synchronization parameters using SAGE in a GNSS-receiver, estimation of synchronization parameters using SAGE in a GNSS-Receiver. In: Proceedings ION GNSS, Long Beach, CA, USA, Sept 2005

Avila-Rodriguez JA, Wallner S, Hein G (2006a) MBOC: the new optimized spreading modulation recommended for GALILEO E1 OS and GPS L1C, ESA Navitec 2006, Noordwijk, The Netherlands, Dec 11–13

Avila-Rodriguez JA, Wallner S, Hein G (2006b) How to optimize GNSS signals and codes for indoor positioning. In: ION GNSS 19th international technical meeting of the satellite division, Sept 2006

Betz J (1996) The offset carrier modulation for GPS modernization, In: Proceedings of ION technical meeting. Cambridge, Massachusetts, pp 639–648, June 1999

Bhuiyan MZH, Lohan ES (2010) Advanced multipath mitigation techniques for satellite-based positioning applications. Int J Navig Obs. doi:10.1155/2010/412393

Bhuiyan MZH, Lohan ES, Renfors M (2008) Code tracking algorithms for mitigating multipath effects in fading channels for satellite-based positioning. EURASIP J Adv Signal Process. doi:10.1155/2008/863629

Bhuiyan MZH, Lohan ES, Renfors M (2009) A reduced search space maximum likelihood delay estimator for mitigating multipath effects in satellite-based positioning. In: Proceedings of 13th international association of institute of navigation, 2009 (IAIN'09), Stockholm, Sweden

Bhuiyan MZH, Lohan ES, Renfors M (2010a) A slope-based multipath estimation technique for mitigating short-delay multipath in GNSS receivers. In: CDROM Proceedings of IEEE ISCAS, Paris, France

Bhuiyan MZH, Zhang J, Lohan ES (2010b) Enhanced delay tracking performance of a C/N0-based two-stage tracker for GNSS receivers, at the European Navigation Conference on Global Navigation Satellite Systems, ENC-GNSS 2010, Braunschweig, Germany

Bhuiyan MZH (2006) Analyzing code tracking algorithms for GALILEO open service signal. M. S. Thesis, Tampere University of Technology, Tampere, Finland

Blanco-Delgado N (2011) Signal Processing Techniques in Modern Multi-Constellation GNSS Receivers, PhD thesis, Technical University of Lisbon, 2011

Borio D, LoPresti L () Data and pilot combining for composite GNSS signal acquisition. Int J Navig Obs 2008(Article ID 738183):12. doi:10.1155/2008/738183

Borre K, Akos DM, Bertelsen N, Rinder P, Jensen SH (2007) A software-defined GPS and GALILEO receiver- A single frequency approach, Birkhäuser, 2007

Brahim F, Chonavel F, Rabaste O (2008) An MCMC algorithm for BOC and AltBOC signaling acquisition in multipath environments. In: IEEE/ION Proceedings of PLANS, 2008

Burian A, Lohan ES, Renfors M (2006) BPSK-like methods for hybrid-search acquisition of GALILEO signals. In: CDROM Proceedings of ICC 2006, Turkey, JunE 2006

Burian A, Lohan ES, and Renfors M (2007) Efficient delay tracking methods with sidelobes cancellation for BOC-modulated signals. EURASIP J Wireless Commun Networking, vol 2007 no 2, January 2007

Campana R, Gottifredi F, Valle V, Lombardo BF (2008) Different acquisition algorithms for the GALILEO L1 signal with BOC(1,1) modulation, satellite communications and navigation systems, signals and communication technology, 2008(2):341–352. doi:10.1007/978-0-387-47524-0_26

Cheng U, Hurd WJ, Statman JI (1990) Spread-spectrum code acquisition in the presence of doppler shift and data modulation. IEEE Trans Commun 38:241–250

Chibout B, Macabiau C, Escher AC, Ries L, Issler JL, Corrazza S, Bousquet M (2007) Comparison of acquisition techniques for GNSS signal processing in geostationary orbit, ION-NMT, San Diego, CA, Jan 2007

Chu LC, Mitra U (1999) Analysis of MUSIC-based delay estimators for direct-sequence code-division multiple-access systems. IEEE Trans Commun 47:133–138

Corazza GE, Pedone R (2007) Generalized and average likelihood ratio testing for post detection integration. IEEE Trans Commun 55(11):2159–2171

Corazza GE (1996) On the MAX/TC criterion for code acquisition and its application to DS-SSMA systems. IEEE Trans Commun 44(9):1173–1182

De Castro D, Diez J, Fernandez A (2007) High resolution multipath mitigation technique based on the teager-kaiser operator for GNSS signals. In: Proceedings of the 20th international technical meeting of the satellite division of the Institute of Navigation (ION GNSS 2007), Sept 2007, Fort Worth, TX

Dierendonck A, Fenton P, Ford T (1992) Theory and performance of narrow correlator spacing in a GPS receiver. J Inst Navig Fall 39(3):265–283

Dragunas K, Borre K () Multipath mitigation based on deconvolution. J Global Positioning Syst 10 (1):79–88. doi:10.5081/jgps.10.1.79

Dragunas K (2010) Indoor multipath mitigation. In: International conference on indoor positioning and indoor navigation, IPIN 2010. IEEE Press, pp 578–584

El-Natour H, Escher AC, Macabiau C, Boucheret ML (2005) Impact of multipath and cross-correlation on GPS acquisition in indoor environments. In: CDROM Proceedings of ION-NMT, San Diego, CA

Enrique Aguado L, Brodin GJ, Cooper JA (2004) Combined GPS/Galileo highly-configurable high-accuracy receiver. In: Proceedings of ION GNSS 2004, Sept 2004

Fantino M, Marucco G, Mulassano P, Pini M (2008) Performance analysis of MBOC, AltBOC and BOC modulations in terms of multipath effects on the carrier tracking loop within GNSS receivers, in 2008 IEEE/ION Position, Location and Navigation Symposium. Monterey, CA, pp 369–376, May 2008

Fenton P, Jones J (2005) The theory and performance of Novatel Inc's Vision Correlator. In: Proceedings of ION-GNSS, Long Beach, CA

Fenton P, Smith B, Jones J (2011) Theory and performance of the pulse aperture correlator. Technical report, http://www.novatel.com/customersupport/alltechpapers.html (active Mar 2011), NovAtel Inc., Sept 2004

Fessler JA (1994) Hero AO, Space-alternating generalized expectation-maximization algorithm. IEEE Trans Signal Proc 42(10):2664–2677

Fishman P, Betz J (2000) Predicting performance of direct acquisition for the M-code signal. In: Proceedings of ION NMT, pp 574–582

Garin L, Rousseau J (1997) Enhanced strobe correlator multipath rejection for code & carrier. In: Proceedings of ION GPS-97. Kansas City, MO, pp 559–568, Sept 1997

Glisic S, Katz MD (2001) Modeling of code acquisition process for rake receivers in CDMA wireless networks with multipath and transmitter diversity. IEEE J Select Areas Commun 19 (1):21–32

Groh I, Sand S (2008) Complexity reduced multipath mitigation in GNSS with the Granada bit-true software receiver. In: Proceedings of the IEEE/ION position location and navigation symposium (PLANS), Monterey, CA, USA, May 2008

Groh I (2011) Efficient time-variant synchronization in spread-spectrum navigation receivers. PhD Thesis, Universitas Miguel Hernández, Elche, Spain, Sept 2011

Hamila R, Lakhzouri A, Lohan ES, Renfors (2005) A highly efficient generalized Teager-Kaiser-based technique for LOS estimation in WCDMA mobile positioning. EURASIP J Appl Signal Process 5:698–708. doi:10.1155/ASP.2005.698, http://dx.doi.org/10.1155/ASP.2005.698

Heiries V, Oviras D, Ries L, Calmettes V (2004) Analysis of non ambiguous BOC signal Acquisition performance. In: CDROM Proceedings of ION GNSS, Long Beach, CA, Sept 2004

Heiries V, Avila-Rodriguez JA, Irsigler M, Hein G, Rebeyrol E, Roviras D (2005) Acquisition performance analysis of composite signals for the L1 OS optimized signal. In: Proceedings of ION GNSS, Long Beach, CA, pp 877–889, Sept 2005

Heiries V, Roviras D, Calmettes V, Ries L (2006) An enhanced correlation processing multipath mitigation technique for BOC signals. In: CDROM Proceedings of IEEE-PLANS, Apr 2006

Hurskainen H, Lohan ES, Hu X, Raasakka J, Nurmi J (2008) Multiple gate delay tracking structures for GNSS signals and their evaluation with simulink, systemc and VHDL. Int J Navig Observ Article ID 785695:17

Irsigler M, Eissfeller B (2003) Comparison of multipath mitigation techniques with consideration of future signal structures. In: CDROM Proceeding of ION-GNSS 2003, Portland, Oregon, Sept 2003

Jakobsson A, Swindlehurst AL, Stoica P (1998) Subspace-based estimation of time delays and Doppler shifts. IEEE Trans Sign Process 46:2472–2483

Jovanovic VM (1998) Analysis of strategies for serial-search spread-spectrum code acquisition-direct approach. IEEE Trans Commun 36(11):1208–1220

Julien O, Macabiau C, Cannon M, Lachapelle G (2005) BOC signal acquisition and tracking method and apparatus, US Patent Application Publication 2005/0270997 A1, Dec 2005

Julien O, Macabiau C, Cannon M, Lachapelle G (2007) ASPeCT: unambiguous Sine-BOC(n, n) acquisition/tracking technique for navigation applications. IEEE Trans Aerosp Electron Syst 43 (1):150–162

Kaiser JF (1990) On a simple algorithm to calculate the 'energy' of a signal. In: Proceedings of IEEE international conference on acoustics, speech, and signal processing (ICASSP '90), vol 1. Albuquerque, NM, USA, pp 381–384, April 1990

Kang BJ, Lee IK (2003) A performance comparison of code acquisition techniques in DS-CDMA systems. Wireless Pers Commun 25:163–176

Kaplan ED, Hegarty KJ (2006) Understanding GPS: principles and applications, 2nd edn. Artech House, Norwood

Kay SM (1998) Fundamentals of statistical signal processing. In: Detection Theory, vol 2, 1st edn. Prentice Hall, Feb 1998

Kong SH, Nam W (2010) A-GNSS sensitivity for parallel acquisition in asynchronous cellular networks. IEEE Trans Wireless Commun 9(12):3370–3378

Konovaltsev A et al (2006a) Acquisition tradeoffs for Galileo SW receiver, European navigation conference, ENC2006. Manchester, UK

Konovaltsev A, Denks H, Hornbostel A, Soellner M, Kaindl M (2006b) General approach to analysis of GNSS signal acquisition performance with application to GALILEO signals. In: 3rd ESA workshop on satellite navigation user equipment technologies, Proceedings CD-ROM. The European Space Agency (ESA), NAVITEC '2006, 2006-12-11—2006-12-13, Estec, Noordwijk, The Netherlands

Kostic ZZ, Sezan MI, Titlebaum EL (1992) Estimation of the parameters of a multipath channel using set-theoretic deconvolution. IEEE Trans Commun 40(6):1006–1011

Lakhzouri A, Lohan ES, Renfors M (2004) Reduced-complexity time-domain correlation for acquisition and tracking of BOC-modulated signals. In: CDROM Proceedings of ESA NAVITEC, Dec 2004

Lin D, Tsui J, Howell D (1999) Direct P(Y)-code acquisition algorithm for software GPS receivers. In: The Proceedings of ION international technical meeting, pp 363–368, Sept 1999

Lohan ES, Lakhzouri A, Renfors M (2004a) Selection of the multiple-dwell hybrid-search strategy for the acquisition of GALILEO signals in fading channels. In: CDROM Proceedings of IEEE PIMRC, Sept 2004

Lohan ES, Burian A, Renfors M (2004b) Acquisition of GALILEO signals in hybrid double-dwell approaches. In: CDROM Proceedings of ESA NAVITEC, Dec 2004

Lohan ES, Hamila R, Lakhzouri A, Renfors M (2005) Highly efficient techniques for mitigating the effects of multipath propagation in DS-CDMA delay estimation. IEEE Trans Wireless Commun 4(1):149–162

Lohan ES, Lakhzouri A, Renfors M (2006a) Binary-offset-carrier modulation techniques with applications in satellite navigation systems. Wireless Commun Mobile Comput 7(6):767–779

Lohan ES, Lakhzouri A, Renfors M (2006b) Complex double-binary-offset-carrier modulation for a unitary characterization of GALILEO and GPS signals'. IEEE Proc Radar, Sonar, Navig 153 (5):403–408

Lohan ES, Lakhzouri, Renfors M (2006c) Feedforward delay estimators in adverse multipath propagation for GALILEO and modernized GPS signals. EURASIP J Appl Signal Process 50971:19

Lohan ES, Burian A, Renfors M (2008) Low-complexity acquisition methods for split-spectrum CDMA signals. Wiley Int J Satell Commun 26:503–522

Lohan ES (2009) Differential teager-kaiser-based code tracking loop discriminator for multipath mitigation in a NSS Receiver, US patent 8,160,124 B1, assigned to qualcomm atheros. Inc., San Jose, Calif. (US), Filed on Oct 22, 2009, As Appl No 12/604, 15

Lohan ES (2010) Analytical performance of CBOC-modulated GALILEO E1 signal using sine BOC(1,1) receiver for mass-market applications. In: Proceedings of ION-PLANS 2010, Palm Springs, CA, 3–5 May 2010

Lohan ES (2011) Limited bandwidths and correlation ambiguities: do they co-exist in GALILEO Receivers? Positioning Journal, pp 14–21, Pub. Date: 2011-02-24, DOI:10.4236/pos.2011.21002

Martin N, Leblond V, Guillotel G, Heiries V (2003) BOC(x,y) signal acquisition techniques and performances. In: Proceedings of ION-GPS2003, Portland, OR, US, pp 188–198, Sept 2003

Mattos PG (1997) Multipath elimination for the low-cost consumer GPS. In: Proceedings of the 1997 10th international technical meeting of the satellite division of the institute of navigation, ION GPS-96, Part 1(of 2), Proceedings of ION-GPS, Institute of Navigation, vol 1, pp 665–671

Mattos PG (2005) Acquisition of the GALILEO OS L1b/c signal for the mass-market receiver. In: Proceedings of the 18th international technical meeting of the satellite division of the Institute of Navigation (ION GNSS '05), pp 1143–1152, Long Beach, Calif, USA, Sept 2005

McGraw G, Braasch M (1999) GNSS multipath mitigation using gated and high resolution correlator concepts. In: Proceedings 1999 NTM of the institute of navigation, San Diego, CA, pp 333–342, Jan 1999

Misra P, Enge P (2011) Global positioning system: signals, measurements, and performance, Revised Second Edition, Ganga-Jamuna Press, ISBN 0-9709544-1-7

O'Driscoll C, Murphy CC (2005) Performance analysis of an FFT based fast acquisition GPS Receiver. In: Proceedings of the 2005 national technical meeting of the institute of navigation, San Diego, CA, pp 1014–1025, Jan 2005

O'Driscoll C (2007) Performance analysis of the parallel acquisition of weak GPS signals. PhD Thesis, National University of Ireland, Jan 2007

O'Mahony NM (2010) Variable dwell time verification strategieS for CDMA acquisition with application to GPS signals. PhD Thesis, National institute of Ireland, Dec 2010, online at (Active on Sep 2012). http://plan.geomatics.ucalgary.ca/papers/thesis_niamhom_final_dec10.pdf

Pajala E, Lohan ES, Renfors M (2005) CFAR detectors for hybrid-search acquisition of GALILEO signals. In: CDROM Proceedings of ENC-GNSS, Munich, July 2005

Pajala E, Lohan ES, Huovinen T, Renfors M (2006) Enhanced differential correlation method for the acquisition of GALILEO signals. In: 10th IEEE Singapore international conference on communication systems, ICCS, Oct 2006

Pany T, Irsigler M, Eissfeller B (2005) Optimum coherent discriminator based code multipath mitigation by S-curve shaping for BOC(n, n) and BPSK signal. In: CDROM Proceedings of ENC-GNSS, Munich, Germany, July 2005

Pany T, Gohler E, Irsigler M, Winkel J (2010) On the state-of-the-art of real-time GNSS signal acquisition—a comparison of time and frequency domain methods. Int Conf Indoor Positioning Indoor Navig (IPIN). doi:10.1109/IPIN.2010.5647537

Parkinson BW, Spilker JJ (1996) Global positioning systems: theory and applications. American Institute of Aeronautics and Astronautics, Inc

Polydoros A, Simon M (1984) Generalized serial search code acquisition: the equivalent circular state diagram approach. IEEE Trans Commun 32(12):1260–1268

Polydoros A, Weber CL (1984) A unified approach to serial serach spread-spectrum code acquisition-Part I: general theory. IEEE Trans Commun COM-32(5):542–549

Polydoros A, Weber CL (1984) A unified approach to serial search spread spectrum code acquisition—Part II: a matched-filter receiver. IEEE Trans Commun COM-32(5):550–560, May 1984

Povey GJ (1998) Spread spectrum PN code acquisition using hybrid correlator architectures'. Wireless Pers Commun 8:151–164

Psiaki M (2001) Block acquisition of weak GPS signals in a software receiver., ION GPS

Rappaport T (1996) Wireless communications: principles and practice, 2nd edn. Prentice Hall, Upper Saddle River

Ray JK (2000) Mitigation of GPS code and carrier phase multipath effects using a multi-antenna system. PhD Thesis, University of Calgary

Saarnisaari H (1996) ML time delay estimation in a multipath channel. In: IEEE 4th international symposium on spread spectrum techniques and applications proceedings, vol 3, pp. 1007–1011, Sept 1996

Sagiraju PK, Agaian S, Akopian D (2006) Reduced complexity acquisition of GPS signals for software embedded applications. IEEE Proc Radar, Sonar, Navig 153:69–78

Sahmoudi M, Landry R (2008) Probabilities and multipath: mitigation techniques using maximum-likelihood principles. Inside GNSS journal, Nov/Dec 2008. http://www.insidegnss.com/node/921. Accessed Mar 2011

Samad MF, Lohan ES (2009) MBOC performance in unambiguous acquisition. In: Proceedings of ENC-GNSS2009, May 2009, Naples, Italy

Schmid A, Neubauer A (2004) Performance evaluation of differential correlation for single shot measurement positioning. In: CDROM Proceedings of ION-GNSS

Schmid A, Neubauer A (2005) Differential correlation for GALILEO/GPS receivers. In: Proceedings of IEEE international conference on acoustics, speech, and signal processing (ICASSP), vol 3, pp 953–956

Seco-Granados G (2000) Antenna arrays for multipath and interference mitigation in GNSS receivers. PhD Thesis, Universitat Politecnica de Catalunya (UPC), Spain

Sheen WH, Chiou S (1999) Performance of multiple-dwell pseudo-noise code acquisition with I-Q detector on frequency-nonselective multipath fading channels. Wireless Netw 5:11–21

Simon MK, Omura JK, Scholtz RA, Levitt BK (1994) Spread spectrum communications handbook: revised edition., McGraw-Hill Inc, New York

Skournetou D, Sayed AH, Lohan ES (2009) A deconvolution algorithm for estimating jointly the line-of-sight code delay and carrier phase of GNSS signals. In: Proceedings of ENC-GNSS, Naples, Italy

Sleewaegen JM, Boon F (2001) Mitigating short-delay multipath: a promising new technique. In: CDROM Proceedings of ION-GPS, Sept 2001

Sorokin H, Lohan ES, Takala J (2006) Memory-efficient time-domain correlation for BOC-modulated signals. In: CDROM Proceedings of ESA NAVITEC, Nordwijk, Netherlands

Souza EM, Monico JFG, Polezel WGC, Pagamisse A (2008) Spectral analysis and low-frequency multipath mitigation for kinematic applications. In: 2008 IEEE/ION position, location and navigation symposium. Monterey, CA, pp 413–417, Sept 2008

Suwansantisuk W, Win MZ (2007) Multipath aided rapid acquisition: optimal search strategies. IEEE Trans Inf Theor 53(1):174–193

Tranquilla JM (1994) Analysis of a choke ring ground plane for multipath control in Global Positioning System (GPS) applications. IEEE Trans Antennas Propag 42(7):905–911

Van Diggelen F (2009) A-GPS, assisted GPS, GNSS and SBAS, Artech House

Van Nee DJR, Coenen AJRM (1991) New fast GPS code-acquisition technique using FFT. Electron Lett 27(2):158–160

Van Nee DJR (1992) The multipath estimating delay lock loop. In: IEEE second international symposium on spread spectrum techniques and applications. Yokohama, Japan, pp 39–42

Villanti M, Palestini C, Pedone R, Corazza G (2007) Robust code acquisition in the presence of BOC modulation for future GALILEO receivers. In: Proceedings IEEE International Conference on Communications (ICC), pp 5801–5806

Weiler RM, Blunt P, Hodgart S, Unwin M (2008) The effect of cosine phased boc modulation on the GNSS receiver search process. J Navig 61:591–611 (Cambridge University Press, 2008)

Weill R (2002) Multipath mitigation using modernized GPS signals: How good can it get? In: Proceedings ION GPS 2002. Portland, OR, pp 493–505, Sept 2002

Williams JH, Davis R, Rosario EN (2000) Multipath mitigation performance of planar GPS adaptive antenna arrays for precision landing ground stations. MITRE publication, http://wwwsrv1.mitre.org/work/tech_papers/tech_papers_00/williams_multipath/williams_multipath.pdf (active Sept 2012)

Yang C, Hegarty C, Tran M (2004) Acquisition of the GPS L5 signal using coherent combining of I5 and Q5, In: Proceedings of the 17th international technical meeting of the satellite division of the institute of navigation (ION GNSS '04). Long Beach, California, USA, pp 2184–2195, Sept 2004

Zhang J, Lohan ES (2011) Galileo E1 and E5a link-level performances in single and multipath channels. In: Proceedings of PSATS 2011, Feb 2011, Malaga, Spain, (open-source Simulink Galileo simulator. http://www.cs.tut.fi/tlt/pos/Software.htm)

Zhdanov A, Veitsel V, Shodzishsky M, Ashjaee J (1999) Multipath error reduction in signal processing. In: Proceedings ION GPS 1999, Nashville, TN, Sept 1999, pp 1217–1223

Zheng Y (2010) A software-based frequency domain parallel acquisition algorithm for GPS signal. In: IEEE 2010 international conference on anti-counterfeiting security and identification in communication (ASID), p 298, July 2010

Zhuang W (1996) Noncoherent hybrid parallel PN code acquisition for CDMA mobile communications. IEEE Trans Veh Technol 45(4):643–656

Chapter 6
Baseband Hardware Implementations for Galileo Receiver

Heikki Hurskainen and Jari Nurmi

Abstract In Galileo receiver the main functions of baseband are to find the signals of interest, to decode the navigation data when found and to measure signals transmission time from the tracking channels. Both of these baseband processes, namely acquisition and tracking, described in Chap. 5, require a large amount of computational power in comparison of the rest of the processes involved in reception. Due to this computation requirement, the baseband functionality is still mostly implemented in hardware, even if the trend is towards software realizations. SW Galileo receivers are described later in this book. Quite many of the hardware baseband solutions created for GPS signals are usable also for Galileo reception, but there exists few distinct features in Galileo signals, such as the memory based codes, pilot signals and usage of composite BOC subcarrier, that require special attention in hardware implementation but are more favorable towards software receivers. This chapter provides the reader with introduction to the baseband of satellite navigation receiver, its functionality and specific solutions for the challenges set by the aforementioned properties of Galileo signals.

Baseband Processes and Hardware

Chapter 3 has already discussed briefly the fundamentals of data decoding in the signal reception. This section goes a bit deeper to the structures where the decoding is actually performed inside of the receiver.

Figure 6.1 illustrates a generic receiver block diagram, where along the signal chain from the left are: antenna, radio front end, baseband processing, and finally the navigation computation. For simplification only blocks for a single tracking channel are illustrated inside of the baseband part of the receiver.

H. Hurskainen (✉) · J. Nurmi
Department of Electronics and Communications Engineering,
Tampere University of Technology, Tampere, Finland
e-mail: heikki.hurskainen@gmail.com

J. Nurmi et al. (eds.), *GALILEO Positioning Technology*,
Signals and Communication Technology 182, DOI 10.1007/978-94-007-1830-2_6

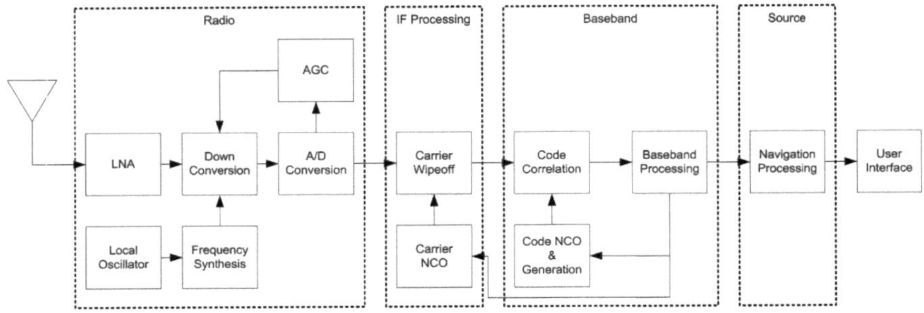

Fig. 6.1 A block diagram of the GNSS receiver

Signals are received through antenna and forwarded to analog processing in GNSS radio front end, where the signals are been down-converted to zero or intermediate frequency (IF) and sampled to a digital stream. The radio output can be represented as in Eq. 6.1. Where the incoming stream $s(t)$ consists of signals amplitude A_X, spreading code $C_X(t)$, navigation data $D_X(t)$ and the carrier wave. The carrier component usually contains the remaining intermediate frequency f_{IF}, unknown Doppler frequency $f_{Doppler}$ and unknown carrier phase offset $\varphi_{carrier}$:

$$s(t) = A_X C_X(t) D_X(t) \cos(2\pi(f_{IF} + f_{Doppler}) + \varphi_{carrier}) \tag{6.1}$$

The next fundamental component in GNSS reception chain is the digital signal processing, usually referred to as the baseband processing. The (usually hardware) baseband processing is located between the analog to digital converter (ADC) and the navigation processor. In between these two domains the functionalities of IF processing and baseband are located. It is quite often when these two are merged as a single baseband block. In reality the division of these blocks is not always as straightforward but for understanding the nature of baseband processes this simplified division of receiver functionality could be followed for the rest of this section.

In the baseband processing, a part of the main objectives is to measure the range and range-rate and to demodulate the navigation data out of the received signal stream (Braasch and Dierendonck 1999). The range and range-rates are measured by multiple tracking processes, which try keeping the receiver locked to visible satellite signals. For each satellite a separate tracking process, or a channel when speaking of hardware implementation, is needed.

The tracking is a key process to achieve the range measurement since it takes care of the exact reception of the signal. For accurate reception tracking process creates the replicas of carrier, subcarrier and spreading code, which it uses to demodulate the navigation data out of the received stream (as illustrated in Fig. 6.1). Since the signals travel at the speed of light, a timing error of 1 ms equals to a ranging error of 300 km. Thus, it is quite obvious that the signal replicas need to be really accurate estimates of the received signal components. The rougher estimates are achieved as output of the acquisition process, where a massive search over all possible code delays and Doppler frequencies is performed, as explained in Chap. 5 . In the acquisition phase, the carrier phase can be neglected since typically the received signals amplitude is searched by combining the results of in-phase and quadrature phase reception branches. The tracking process continues the tuning of these three unknown estimates (i.e. code delay, Doppler, phase) in order to fulfill the needs of accurate positioning.

Traditionally, the baseband processes have been implemented as hardware. The reasoning for this comes from the computational demands of the processes. The core functionality of GNSS receiver baseband consumes a significant amount of simple calculus. Another demand for this computation is its hard real-time requirement. For example, for the tracking process it is not allowed to miss a single integration epoch or even a sample. The typical input streams are digitized outputs of ADC, widths ranging from 1 to 8 bits and rates from 2 to 50 MHz's.

In next two subsections we take a more detailed look to acquisition and tracking processes, now mostly from the hardware implementation point of view. The generic principles of acquisition and tracking were described in Chap. 5.

Acquisition Is the Process of Search

Before the receiver can lock to a visible satellite it needs some prior knowledge of the received signal. This information is achieved from a process called acquisition, where the large search space is swept through to solve the coarse estimates of the characteristics of visible signals. These characteristics include the identity of the satellite (each satellite has unique transmission, CDMA modulation), unknown signals (code) delay and Doppler frequency. Indeed, the acquisition can be seen as a 3-dimensional search over these variables. It is important to note that Galileo and global positioning system GPS are both CDMA based systems.

The main objective of the acquisition process is to find and separate the satellite signals from the noise. This is also the key challenge in acquisition since received signals have really low power, which is substantially lower than the power of environment noise. The minimum reception powers for Galileo signals in the face of the Earth are given in Table 6.1. For comparison, with a minimal 2 MHz receiver bandwidth the power of environmental noise achieves power levels around −138 dBW (Braasch and Dierendonck 1999). And, when the bandwidth is increased also the reception noise increases.

In the early days of GPS receivers, when the acquisition was performed serially, the tracking channels (illustrated in Fig. 6.3) were used to perform also the acquisition stage. In serial acquisition the frequency search space is swept through by alternating NCO variables. The tradeoff with the easy implementation is that going through all possibilities is a lengthy process since samples needs to be integrated over some period (at least 1 ms in GPS, minimum 4 ms in Galileo) to see whether the satellite is visible or not. This integration needs to be repeated to all

Table 6.1 Galileo and GPS signal levels in shared frequency bands [Galileo ICD], (GPS ICD 2003, 2004, 2006; US-EU 2007)

Galileo		
Galileo signal	Signal component	Total received minimum power (dBW)
E1	E1 SoL (composite)	−157.0
E5	E5a	−155.0
	E5b	−155.0
GPS		
GPS signal	Signal component	Total received minimum power (dBW)
L1	C/A	−158.5
	L1C (composite)	−157.0
	P(Y)	−161.5
L5	L5	−157.9

combinations of code delays and Doppler frequency bins. E.g. even in a simple case for current GPS signal the search would take approximately 5 s per signal (1 ms predetection integration time (PIT), 1,023 chips/3 correlators, 15 Doppler bins (±5 kHz, bin size = 2/3*PIT), but when looking for Galileo signals the search time would be over 392 s per signal (4 ms PIT, 2*4,092 chips due BOC/5 correlators per channel, approx. 60 Doppler bins). It is quite obvious that consuming more than 6 min per satellite is not meeting the requirements set by the users in modern world.

Acquisition process can be speeded up drastically by executing its' two dimensional search in a single dimension (by parallelizing the search in other). Acquisition can use the parallelism either in the time domain or in the frequency domain. The time domain approach is to extend the tracking structure (Fig. 6.3) to contain more correlators, maybe over full spreading code epoch period. This structure could be called a matched filter. In frequency domain approach the Fast Fourier Transform (FFT) is applied both in incoming stream and locally generated replica code. Since correlation (or convolution) in time domain is equal to a multiplication in frequency domain, the acquisition result is achieved simply by taking inverse FFT over the resulting multiplication, as illustrated in Fig. 6.2. The carrier is removed from the incoming stream and the result is transformed to the frequency domain. Then, a complex conjugate of the transformed replica code epoch is multiplied with the stream and after inverse transform the absolute value of resulting vector represents the autocorrelation function of the process in duration of one code epoch. For a more detailed description of FFT acquisition additional literature, for example (Borre et al. 2007) or (van Nee and Coenen 1991) is recommended.

As mentioned earlier, the first GPS receivers had serial acquisition implementations. Such an approach is really simple but it does not meet the current user requirements for acquisition speed and Time To First Fix (TTFF). Based on these reasons and on the fact there current receivers have more computational power

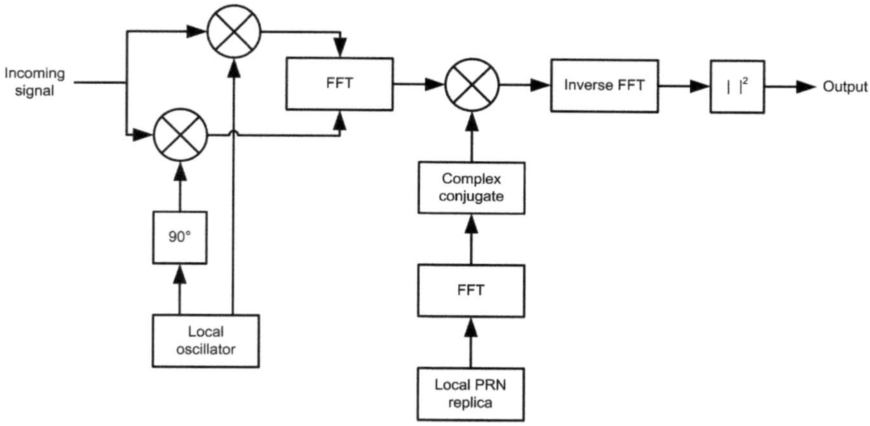

Fig. 6.2 Fundamentals of FFT acquisition, (van Nee and Coenen 1991)

available, nowadays the parallel acquisition implementations such as in Fig. 6.2 are most common.

Newer signals require longer registers in both matched filter and FFT acquisitions. This is caused by the increased spreading code length and BOC modulation present in Galileo (by comparison with GPS). Additionally the composite signals may add the complexity of acquisition engine, but also might improve the quality of the acquisition. The acquisition of composite signals is discussed with details in literature (Borio et al. 2009).

Tracking Is the Measuring Tool for Signals

Typical receivers consist of a number of parallel channels. Depending on the definition of a channel, such a channel is either a structure capable of receiving all signals from one satellite or it is a structure of correlators which are capable of receiving one signal from one satellite. The first definition was more aligned with legacy receivers where maximum of two GPS signals (C/A and P(Y)) were received, but as the number of signals transmitted per satellites has also been increasing, the latter definition is the one more frequently used today and it is adopted for the rest of this discussion.

The number of channels in receiver has an increasing trend: four-channel GPS receivers are in past and modern multi-system, and maybe multi-frequency receivers, should be capable for receiving tens of signals in parallel. Also some advanced algorithms require multiple channels, which again increase the requirements for the total amount of channels. The main functions of the tracking channel are to lock to the received signals and to fine tune the local replicas so that the timings of the received stream and local replica components are closely matching.

When this has being successfully done the receiver can demodulate the data bit stream to be forwarded to data decoding and frame synchronization. A block diagram of tracking hardware is given in Fig. 6.3.

The leftmost section of the channel removes the remaining carrier frequency component together with the Doppler frequency ($f_{IF} + f_{Doppler}$). This is illustrated also in Fig. 6.1 as the block called IF Processing. Numerically Controlled Oscillator (NCO) is used to create the desired frequency. The staircase function of the NCO (internal register value) is mapped to a few bit (digital representation) sine on cosine waves. These waves are multiplied with incoming sample stream, resulting a conversion to the baseband frequency. In this point the input stream can be either real-only or complex, depending on the type of the used radio front end and it is analog to digital converted. In both cases, the resulting baseband signal will be divided to in-phase (I) and quadrature (Q) phase branches. The outputs from I and Q branches are used to solve the unknown phase of the carrier ($\varphi_{carrier}$).

Baseband I and Q signals are fed to identical structures (in Fig. 6.1 labeled as baseband processing). There the signal is multiplied (correlated) with a numerous, differently delayed copies of local PRN replica. The details of code generation are discussed later in this chapter. The code generation rate is controlled by another NCO, which, if overflow occurs, it gives a pulse for the code generator.

The different correlator outputs are forwarded to the accumulators, where signal samples are typically accumulated over a period equivalent to the code epoch time. In state-of-the-art receivers, the accumulator results are typically read by the software, where also the discriminator computation and the tracking loop filtering takes place. The corrected values for NCOs (tracking loop controls) are then fed back. The distribution between the hardware and software components has not been always the same, and even in current situation there are variations or different approaches in here. These issues are elaborated a bit more in the last section of this chapter.

In legacy GPS receivers the number of tracking correlators used to be three per channel. Usually, the correlators were labeled early, prompt and late (E,P,L), also originating the discriminator names such as early-minus-late, etc. Since Galileo has a novel binary carrier modulation, the three correlator structure has become obsolete due its lack of ability to perform the detection of the false peaks. The false peak situation has already been illustrated in Chap. 5. An example of the autocorrelation function shapes for the current Galileo signals is given in Fig. 6.4. From the figure, it is easy to see how the CBOC modulated E1 signals have narrower main peak and additional side peaks (solid blue line) in comparison to the BPSK modulated signals.

An algorithm named bump-jumping was developed (Fine and Wilson 1999) to cope with the additional peaks of BOC modulated spreading codes. This algorithm deploys five correlators instead of three and can be considered as a minimum setup required for Galileo signal tracking. In practice, the tracking channels are likely to have even more correlators due the requirements from advanced algorithms, for example Multiple Gate Delay designed for better multipath mitigation (Hurskainen et al. 2008, 2009). More algorithms to deal with false peak tracking have been discussed in Chap. 5.

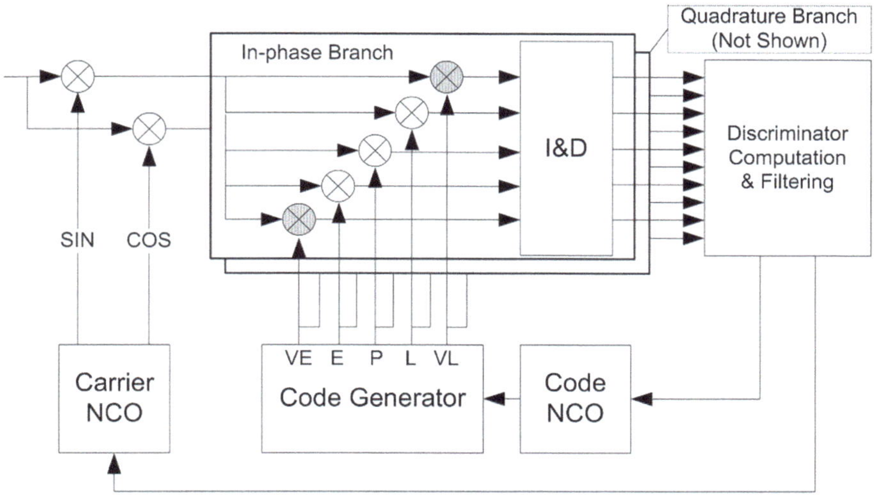

Fig. 6.3 A block diagram of typical GNSS tracking channel

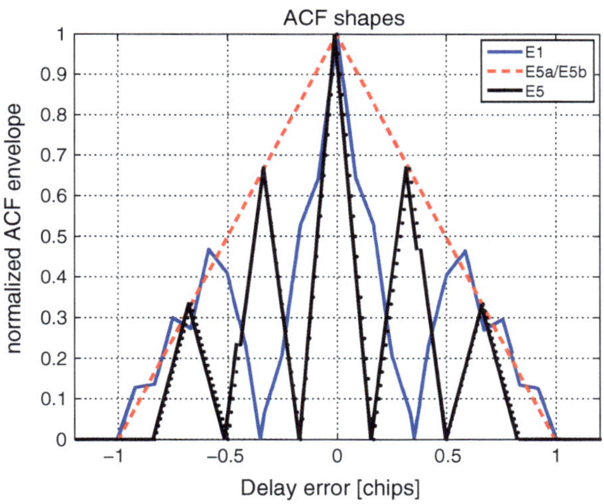

Fig. 6.4 ACF shapes for Galileo E1 (CBOC) in *solid blue*, E5a/E5b (BPSK) in *dashed red* and combined E5 (BOC(5,10)). The ACFs are scaled to the chips, in reality the ACF of E5 signals is much narrower than E1 due the faster chipping rate

The focus of this chapter is in hardware but the essential part of the tracking remains on the software where the tracking loops are closed. Tracking loops developed for GPS are usable also with Galileo signals and they can be studied for example from (Kaplan and Hegarty 2006). A good reference for dedicated Galileo tracking loops is also available in (Julien 2005).

GALILEO Specific Implementations

In this section we discuss the hardware implementation issues that are driven by the new signals introduced by the Galileo system (Galileo 2006, 2008, 2010). The examples given in this section are actual implementations on 16-channel, FPGA based prototype receiver (TUTGNSS). The full receiver is presented in detail in Chap. 9.

A Memory Resource Friendly Implementation of Galileo PRN Codes

From a receiver designer point of view, the main problem with the memory codes is the memory consumption. If all the channels are equipped with the memory large enough for all memory codes (a situation equivalent to having a code generator in each channel) the total amount of used memory will grow to be too expensive and energy hungry to be successful in handheld consumer electronics.

A solution to cope with this could be the usage of shared memory, where a single read-only memory (ROM) containing all usable, memory based spreading codes is talking to all channels simultaneously. This approach has been followed in our prototype, where new hardware entity, called Galileo Memory Unit (GMU) implements the memory codes for Galileo E1 OS signals. The receiver contains only one Galileo code memory where all the 16 tracking channels and acquisition unit fetches the codes on blocks of 32 bits. An arbiter is implemented to ensure the priorities between channels. The arbiter and 32-bit data blocks handle the time sharing between channels, which allows multiple channels to be served at the same time (i.e. channel communicates to GMU in every 32nd PRN chip and it keeps the communication free to other channels otherwise). Tracking channels have (arbitrary from 0 to 15) priorities over acquisition, based on the necessity to use tracking continuously whereas acquisition is performed only periodically.

GMU communicates with tracking and acquisition via 17 buses of sixteen bit-wide address. A single data bus (32 bits) is used to output the memory codes and the implemented single bit request/acknowledgement signals ensure that each channel reads only the data it should have been reading. A block diagram of GMU is given in Fig. 6.5.

The implemented single port ROM memory contains 32 PRN codes (first 32 codes from ICD) for both data (E1-B) and pilot (E1-C) component of the Galileo E1 signal. Four '0' bits are padded after each PRN code to align the starting point for a code better (i.e. for 4,092 chips 4,096 bits of memory space is used). The codes are stored without any modulation (i.e. BOC). Each channel contains a small entity that takes care of fetching the 32 bit memory blocks from the shared ROM. BOC(1,1) modulation is added in here to the spreading code channel, for example XOR-ing the code stream with the MSB of the code NCO register results 1 bit sine BOC(1,1) modulation.

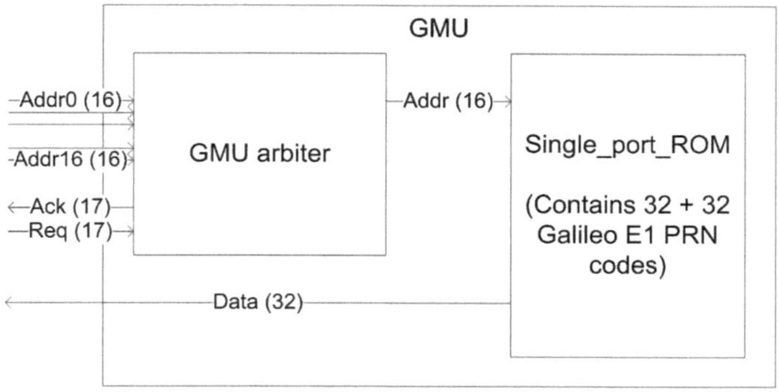

Fig. 6.5 GMU block diagram

It is noteworthy that the memory usage of Galileo E1 PRN codes is acceptable when compared to the amount of memories used elsewhere in the receiver prototype. For example, a parallel acquisition engine needs over five times the amount of memory that the plain Galileo PRN codes use. The 32 + 32 PRN codes (without BOC) consume approximately 262 kbits of memory which equals 32.7 kBytes. All the memory used in tracking unit comes from the needs of GMU. Additionally, the hardware increase per channel is also quite small. The entity handling the reading from shared code memory and adding the BOC(1,1) to the replica consumes approximately 25 % less hardware than combined Galileo/GPS code generator.

MBOC: to Track or Not to Track?

In Chap. 3, we have discussed MBOC modulation in transmission/specification point of view. To get the best out of the new modulation, it needs to be implemented in the receiver in such way that the performance is optimal but still the receiver resources, especially when implemented in hardware, are not over consumed. The latter has an important role when we are dealing with the mass-market segment, where receivers should be low in price and also in power consumption.

TMBOC, which is decided to be used with future GPS L1C signals, has only two levels and thus replica codes can be implemented with only one bit in the receiver. This results in a simple implementation where typically only a single XOR gate per correlator is needed. Such implementation is enough for executing the possible sign change for the received code stream samples. On the other hand, something needs to control/generate TMBOC carrier and the complexity of this, e.g. state machine, is depending of the TMBOC repetition period.

On the other hand, the CBOC modulation used in Galileo (see Chap. 3) has four levels which automatically increase the number of needed bits for the implementation of the replica code. Also the correlator itself needs to have functionality not only for sign change but also for magnitude change. Depending on the bit width of the incoming samples suitable implementation detail could range from combined XOR and shift register to Look-Up Table (LUT).

If the evolution of GNSS receivers follows the current trend and moves closer to ideal Software-Defined Radio (Mitola 1995; Akos 2003), then all the hardware issues are to be turned to software issues. In ideal SDR receiver, the CBOC implementation in the receiver is not anymore an issue since in software implementation the natural bit widths are the ones used by the selected processor, which in current efficient embedded processors are already 32 bits at least.

Further, a detailed analysis needs to be dedicated to the effect of the radio front end characteristics, sampling frequency and number of quantization bits in analog to digital conversion. These parameters have vital role in performance of the MBOC modulation in the receiver, especially in mass market receivers, where both are usually tuned to be as low as possible to lower the cost of the product.

Low cost hardware implementation can be achieved by using only the SinBOC (1,1) modulation component in receiver. This reduces the needed bandwidth from the radio and the maximum power loss of received signal is still only −0.9 dB (Hein et al. 2006).

Another method for the implementation of MBOC tracking in one-bit replica environment is to utilize two channels for the tracking. The problem in this approach is that it uses double amount of hardware and the benefit might not justify the resource consumption overhead. Especially if there are other algorithms that might deploy two channels, composite tracking for example, the analysis of benefits should be made. This comparison was made in (Lohan and Hurskainen 2011) and indeed there is no benefit of utilizing the channels two CBOC tracking over composite tracking in low bandwidths (below 12 MHz), but when the pre-detection bandwidth is high enough (above 12 MHz) the CBOC tracking actually gives better performance in multipath environment.

As a third alternate, one-bit CBOC tracking implementation presented by (Julien et al. 2007) seems to be promising from the hardware complexity point of view. There, the SinBOC(1,1)-modulated component is used for prompt replica and the SinBOC(6,1)-modulated component is used for the early and late phases of the discriminator function. This setup gives reasonable performance with adding almost no extra complexity to the hardware channel. The only change in comparison to usual structure is that the code generator needs to be able to produce both BOC(1,1) and BOC(1,6) modulated PRN codes simultaneously. The small complexity increase is seen on the clocking and handling memory accesses in case of memory mode implementation since these modulations have different requirements for sampling rates.

Pilot Signal to the Rescue

Naturally, the receiver should be able to cope with the new composite signals, and the research over this area has been active. The studies (Borio et al. 2009) on the

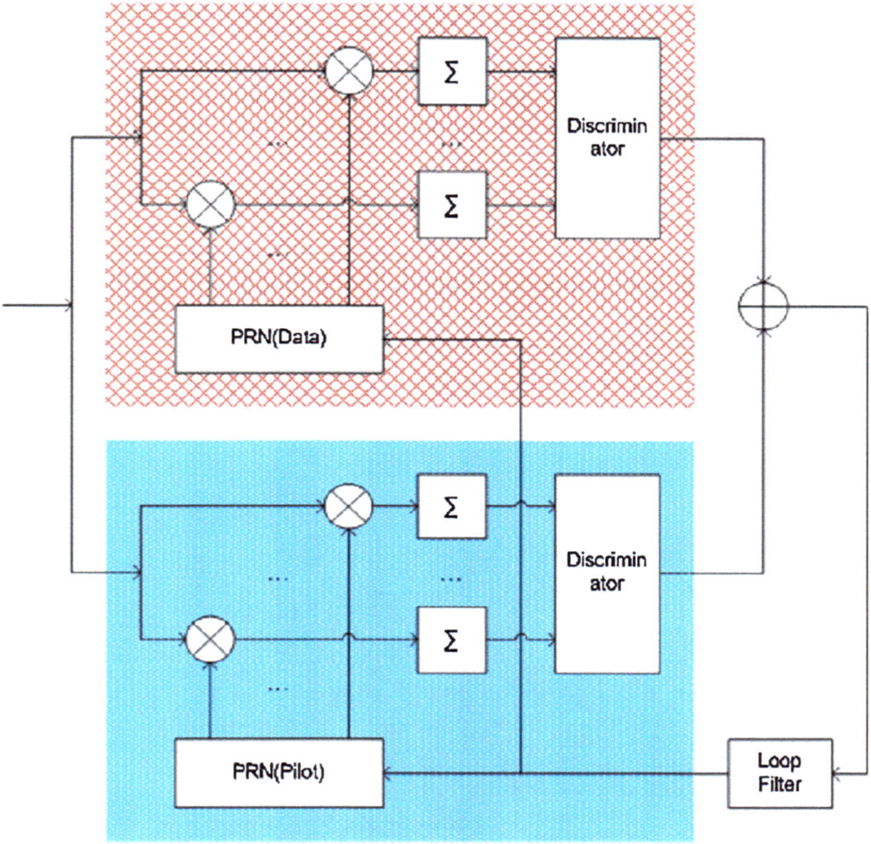

Fig. 6.6 Non-coherent collaborative tracking

collaborative tracking have divided the algorithms in two categories, non-coherent and coherent.

The principle for non-coherent collaborative code tracking is illustrated in the Fig. 6.6. The incoming signal is correlated with multiple correlators in two parallel channels. The correlation results are integrated and fed to discriminator computation. The number of correlators used per channel has a close relation to the tracking algorithm used, e.g. a typical discriminators for a single channel tracking consume three, e.g. a narrow early-minus-late correlator (NEML) (Dierendonck et al. 1992) to five e.g. high resolution correlator (HRC) (McGraw and Braasch 1999). Discriminator output S is filtered and used to control the frequency of replica code generation. The composite discriminator for non-coherent channel combining is given in Eq. (6.2):

$$S = \left(|E_D|^2 + |E_P|^2 + |L_D|^2 + |L_P|^2 \right) \tag{6.2}$$

where E denotes correlation result for early and L for late correlator, results achieved both from data X_D and pilot X_P (upper and lower channel in Fig. 6.6 respectively).

Figure 6.7 illustrates the coherent channel combining. In the coherent tracking, both channels use composite signal replicas, the difference being in the relative sign between the data and pilot components. There exists relative sign ambiguity due to the data bit modulation, thus both variants are tracked and the maximum function is used to select the correct discriminator result. The discriminator for coherent early minus late channel combining can be expressed as below:

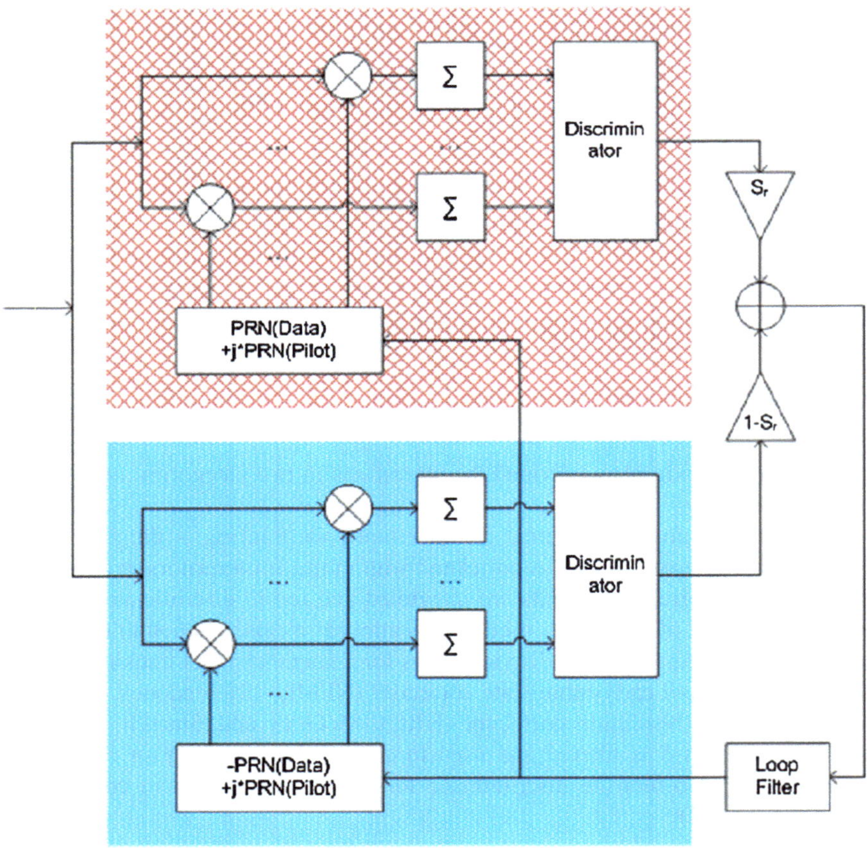

Fig. 6.7 Coherent collaborative tracking

$$
\begin{aligned}
S =& (E^{+} - L^{+})S_r + (E^{-} - L^{-})(1 - S_r) \\
=& \frac{1}{\sqrt{2}}[(E_D + jE_P) - (L_D + jL_P)]S_r \\
&+ \frac{1}{\sqrt{2}}[(-E_D + jE_P) - (-L_D + jL_P)](1 - S_r)
\end{aligned}
\tag{6.3}
$$

The analysis presented in (Borio et al. 2009) shows that coherent channel combining outperforms non-coherent algorithms.

When implementing the collaborative tracking algorithm, two channels need to be used. One channel is used to perform the tracking for E1-B (data) and another one for the E1-C (pilot). Additional care needs to be taken in the synchronization of the replica code generation and carrier removal in both channels. A method for this has been presented in (Hurskainen et al. 2010), where hardware multiplexers were presented in the TUTGNSS receiver to enable the copying of tracking channel controls from channel to another with guaranteed synchronization.

This approach is quite straightforward. Additional multiplexers are presented to the tracking channels structure enabling the channels to "steal" their controls from other channels. This "stealing" mechanism is implemented in hardware to guarantee the synchronization between data and pilot tracking channels. Software implementation would not guarantee the synchronization due to the unpredictable nature of the software interrupts and operating system priority assignments. The synthesis results showed the hardware overhead of 26.6 % in a 16 channel tracking unit when the structure was implemented.

For comparison, the coherent collaborative method could be implemented also by introducing more complex code generators, with multi-bit replicas (complex spreading codes). Still, further analysis is needed in here to justify whether the gain in performance is greater than the loss in receiver resource consumption and complexity.

Future of Hardware Implementations for Galileo Receivers

The main idea behind the hardware basebands is to ease the computational demands for the receiver processor by executing these highly parallel parts of computation in hardware accelerators. Still, as discussed so far, some of the features represented by the new Galileo signals might be more favorable for software implementation already in the baseband part of the receiver.

We could start from the question: "How computationally demanding are the baseband processes?" The literature gives some figures for this. In one example (Söderholm et al. 2008), a software GPS receiver was realized to ASUS Eee PC platform. The estimations for computational load for this full GPS receiver (including navigation) were reported to be 400 MIPS (million instructions per second). From this figure a major part is consumed by the acquisition and tracking processes. The same study reports a processing load of GPS acquisition to be 22–225 MIPS and similarly for tracking a computational power between 2.5–25 MIPS per channel is constantly needed.

In (Hurskainen et al. 2009), a comparison between GPS and Galileo baseband processing can be found. The figures presented in (Hurskainen et al. 2009) estimate that the parallel (in frequency domain) acquisition process is approximately 4.49 times more complex for Galileo signals in terms of required calculus. The presented study assumed that the acquisition execution time for both systems was equal and thus it is not surprising to get a multiplier over four due the code epoch increase from GPS's 1 ms to Galileo's 4 ms. Similar comparison was made for tracking and there the Galileo tracking was declared to be 1.83 times more complex in terms of required calculus than GPS. The increase can be roughly explained by the need of two extra correlators per tracking channel to solve the possible ambiguities caused by the used binary offset carrier (BOC) modulation.

Thus, indeed the computational load is highest in baseband and it is also increased by the advances in signal design when comparing the Galileo to the existing GPS. But on the other hand, also processor technologies are advancing, still meeting the roadmap set by the Moore's law (Moore 1965). Following the increase in processing power the trends in GNSS receiver architectures seem to pushing the implementation towards pure software receivers (Akos 2003). Already some of the functionality earlier executed in separate GPS engine has been moved to a host processor of the chip. The main drivers to this development are: reduction of hardware cost, increased computational power available in the processor platforms, flexibility needs of receiver due evolving new GNSS signal specifications.

The GNSS receiver implementations can be divided in four main groups; (i) standalone hardware GNSS receivers, (ii) measurement front-ends, where the hardware part provides measurements (pseudoranges), (iii) HW accelerated software receivers, where baseband hardware only accelerates correlation and loops are closed in software, and finally (iv) pure software receivers, where radio front end is directly connected to processor. Estimated trends of both requirements of host processors load (Söderholm et al. 2008) and the development of the relative unit price of receivers are given in Figs. 6.8 and 6.9. We remarks that the requirements

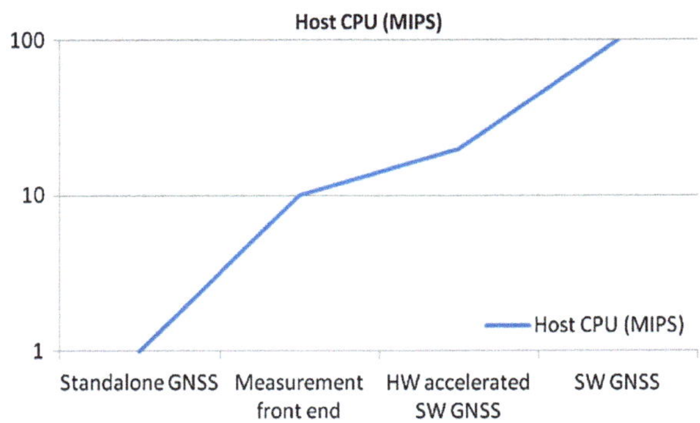

Fig. 6.8 Figure of the CPU requirement development in the evolving GNSS receivers (MIPS = million instructions per second) (data from Söderholm et al. 2008)

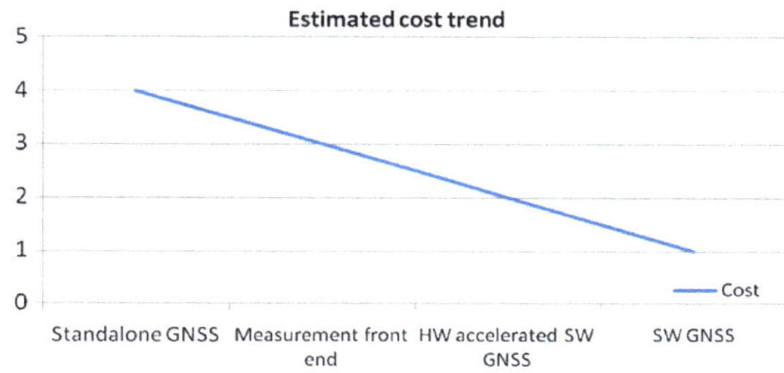

Fig. 6.9 Estimated relative receiver cost evolution in GNSS receiver technology

for processing the power in the receiver CPUs (central processing unit) is increasing in logarithmic scale, whereas the estimates of hardware costs are decreasing linearly.

A good real-life example of this trend towards software radios is the evolution from SirfStarIII to IV. SiRFstar III is one of the most used GPS chipsets in commercial receivers. The latest versions of it, GSC3e/LPx and GSC3f/LPx, were characterized by having up to 20 channels and by a power consumption of 46 mW in autonomous reception. The architecture was based on a GPS engine and ARM7 CPU (CSR 2009a). In the next generation, called SiRFstarIV, the architecture was changed dramatically. The fourth generation of chips was following the host-based architecture, where the navigation computation was moved from the chip to the host processor. This solution naturally lowers the bill-of-material (BOM) costs since less dedicated hardware is needed for GPS functionality. The updated 48-channel receiver has been reported to introduce 5 to 6 MIPS loads to the host processor. (CSR 2009b)

Summary

As this chapter discussed, the main functions of Galileo (or any GNSS) receiver baseband hardware are to find the signals of interest, to decode the navigation data when found and to measure signals transmission time from the tracking channels. These processes, namely acquisition and tracking, require a large amount of computational power in comparison of the rest of the signal reception.

Due to the combined hard deadline caused by real-time limits, continuous reception of new signal data and high computational requirement, the baseband functionality has been and is still mostly implemented in hardware. Quite many of the baseband solutions created for GPS signals are usable also for Galileo reception, due to the similar approach for signal coding. But still there exist a few distinctive features in Galileo signals, such as the memory based codes, pilot signals and usage of composite BOC subcarrier, that require special attention in hardware implementation.

Memory based code implementation is only feasible in a receiver when a single memory is used. A future analysis should be made whether it is more cost effective to implement other signals with code generators or should all spreading codes be placed on a single memory located in a receiver?

CBOC modulation guarantees the future interoperability and compatibility between GPS and Galileo signals but the benefit of it is not that straightforward in receiver designer point of view. The low bandwidths typical to cheap radio front ends remove efficiently all the potential benefits of the usage of CBOC in receiver and thus plain BOC(1,1) receivers are a competitive choice, especially in mass market segment.

Composite signals offer an optimal solution for both data reception (data channel) and high sensitivity due potentially longer integrations (pilot channels). From the hardware point of view the tradeoff is to utilize two channels instead of one per signals, and also to maintain the synchronization between data and pilot tracking. The real implementations of collaborative tracking methods pose still some unanswered questions (for example, the feasibility of coherent method in hardware) and there is room for future studies in this field.

All the main issues discussed in this chapter contain some elements which are more favorable to be implemented in software environment (i.e. memory codes, multi-bit replicas to name a few). Since the trend in navigation receiver development is indeed towards SDR, a concluding question comes to one's mind: Could the Galileo system be designed for software receivers?

References

Akos DM (2003) The role of global navigation satellite system (GNSS) software radios in embedded systems, GPS solutions

Braasch M, van Dierendonck AJ (1999) GPS receiver architectures and measurements. In: Proceedings of IEEE, pp 48–64

Borre K, Akos DM, Bertelsen N, Rinder P, Jensen SH (2007) A software-defined GPS and GALILEO receiver- a single frequency approach, Birkhäuser, Boston

Borio D, Mongredien C, Lachapelle G (2009) Collaborative code tracking of composite GNSS signals. IEEE J Sel Top Sign Proces 3(4)

CSR (2009a) CSR, SiRFstarIII GSC3e/LPx & GSC3f/LPx—flagship performance, lowest power flash GPS solution

CSR (2009b) CSR, SiRFstarIV GSD4t—host-based GPS receiver featuring SiRFaware technology

van Dierendonck AJ, Fenton P, Ford T (1992) Theory and performance of narrow correlator spacing in a GPS receiver. J Inst Navig 39(3):265–283

Fine P, Wilson W (1999) Tracking algorithms for GPS offset carrier signals. In: Proceedings national technical meeting of the institute of navigation

Galileo ICD (2006) Galileo open service, signal in space interface control document (OS SIS ICD). Galileo joint undertaking

Galileo ICD (2008) Galileo open service. Signal in space interface control document (OS SIS ICD). European space agency/European GNSS supervisory authority

Galileo ICD (2010) European GNSS (Galileo) open service signal in space interface control document, (OS SIS ICD), issue 1.1. European Union 2010

GPS ICD (2003) GPS space segment/user segment L5 interfaces, IS-GPS-705, U.S. Air Force, 24 Nov 2003

GPS ICD (2004) GPS interface control document (ICD-GPS-200D), IRN-200C-004, U.S. Air Force, 7 Dec 2004

GPS ICD (2006) GPS space segment/user segment L1C interfaces, Draft IS-GPS-800, U.S. Air Force, 19 April 2006

Hein G, Avila-Rodriguez J, Wallner S, Pratt A, Owen J, Issler J, Betz J, Hegarty C, Lenahan S, Rushanan J, Kraay A, Stansell T (2006) MBOC: the new optimized spreading modulation recommended for galileo L1 OS and GPS L1C. In: Proceedings of IEEE/ION PLANS, pp 883–892

Hurskainen H, Lohan ES, Hu X, Raasakka J, Nurmi J (2008) Multiple gate delay tracking structures for GNSS signals and their evaluation with Simulink, SystemC and VHDL. Int J Navig Obs 17. doi:10.1155/2008/785695

Hurskainen H, Raasakka J, Ahonen T, Nurmi J (2009) Multicore software-defined radio architecture for GNSS receiver signal processing. EURASIP J Embedded Syst 10 Article ID 543720. doi:10.1155/2009/543720

Hurskainen H, Raasakka J, Nurmi J (2010) Flexible hardware implementation of collaborative tracking channel. In: Proceedings of NORCHIP

Julien O (2005) Design of galileo L1F receiver tracking loops. Ph.D. dissertation. University of Calgary

Julien O, Macabiau C, Issler J, Ries L (2007) Two for one: tracking galileo CBOC signal with TMBOC, InsideGNSS (Spring 2007), pp 50–58

Kaplan ED, Hegarty C (2006) Understanding GPS—principles and applications

Lohan ES, Hurskainen H (2011) Galileo dual-channel CBOC receiver processing under limited hardware assumption. In: Proceedings of PSATS

McGraw G, Braasch M (1999) GNSS multipath mitigation using high resolution correlator concepts. In: Proceedings of ION national technical meeting

Mitola J (1995) The software radio architecture, IEEE communications magazine IEEE press, New York, pp 26–38

Moore GD (1965) Cramming more components onto integrated circuits, Electronics, pp 114–117

Söderholm S, Jokitalo T, Kaisti K, Kuusniemi H, Naukkarinen H (2008) Smart positioning with Fastrax's software receiver solution. In: Proceedings of ION GNSS

US-EU (2007) U.S., EU Announce final design for GPS-Galileo civil signal, http://useu.usmission.gov/Dossiers/Galileo GPS/Jul2607 Civil Signal Accord.asp, The United States Mission to the European Union, Jul. 26, 2007, visited 16 Aug 2009

van Nee D, Coenen A (1991) New fast GPS code acquisition technique using FFT. In: IEEE electonics letters

Chapter 7
PVT Computation Issues in Mixed Galileo/GPS Reception

Tommi Paakki, Francescantonio Della Rosa and Jari Nurmi

Abstract The navigation software of a GNSS receiver is responsible for producing the final Position, Velocity and Time (PVT) solution. The receiver needs to acquire and track enough satellites to solve the unknown parameters, namely the x-, y-, and z- coordinates and time biases. In this chapter, we address the basic navigation software tasks, with emphasis on the issues which arise in a multi-system GNSS receiver. In particular single-frequency GPS/Galileo multisystem is selected as an example. Since these GNSS systems both have a different reference time, the chapter includes a section about synchronization to either GPS or Galileo time. Also, this chapter covers topics such as the pseudorange construction, the calculation of the satellite locations, the atmospheric corrections, the navigation solution and navigation data decoding in dual system receiver, and, finally, few words about the extra availability of satellites.

Introduction

This chapter will give an overview of the issues that the navigation software must handle in a multisystem GNSS receiver. The chapter is written choosing GPS and Galileo systems to be the focus area for the topics considered. GPS—related topics have been included in order to provide a benchmark for the Galileo discussion. This chapter starts with an introduction to the basic differences that can be expected from combining two GNSS systems in the receiver, with respect to using a single GNSS. After the short explanation to system differences, a more detailed look is taken into the tasks which the navigation software processes. The details start with timing topic, synchronization to GNSS time. Since both GPS and Galileo use their own timing reference, the receiver must make the choice on which GNSS time to synchronize.

T. Paakki (✉) · F. Della Rosa · J. Nurmi
Tampere University of Technology, Tampere, Finland
e-mail: tommi.paakki@tut.fi

J. Nurmi et al. (eds.), *GALILEO Positioning Technology*,
Signals and Communication Technology 182, DOI 10.1007/978-94-007-1830-2_7

After successful timing synchronization, the receiver must make periodic measurement, named the pseudorange measurements. These pseudorange measurements need accurate timing information from the baseband in order to be of use. The method of constructing the pseudorange measurements for both systems will be introduced.

With multiple GNSS systems, the availability of satellites is notably increased. The amount of satellites might reach very high amount in an area with clear visibility to the sky. For such high availability scenario the chapter presents some mechanics for selecting a good set of satellites from the available satellites instead of using all of them. This will lessen the computational burden without reducing the accuracy of the final solution.

The algorithm of solving satellite location is included in the chapter. Also the Least-Squares (LS) solution for solving the final Position, Velocity and Time (PVT) is being looked at in this chapter. As before, for these algorithms main focus will be in the modifications caused by the multisystem approach.

Last section will be for conclusion where the main points of this chapter are collected together.

GNSS Time Synchronization

When the receiver is turned on for the first time, it has no previous knowledge of its current location or accurate timing information. In such situation no almanac data is available either. The first task of the navigation software is to recover the rough estimate of the current time, a process called synchronization to the GNSS time.

In a combined GPS/Galileo receiver the question is whether the receiver should synchronize to the GPS or to the Galileo time. In practice the choice does not matter much since both systems obviously provide excellent timing accuracy. Furthermore, after the first PVT solution when the clock bias is determined and GPS-to-Galileo time offset is solved, the receiver has essentially synchronized to both of the systems. Regardless of the choice of which system to synchronize to, we select one good candidate satellite from the list of tracked satellites. A good candidate here means usually the satellite with highest Carrier-to-Noise ratio (CNO). CNO values are estimated by the baseband, which provides this information to the navigation software. At this point the receiver does not have any information about the satellite location, so elevation angle is unknown. Generally high elevation angle would indicate good candidate for satellite, but since the receiver does not yet have this information, the selection of candidate satellite is based on the CNO values.

The GPS and Galileo satellites have well defined orbits, as shown in Table 7.1. This allows the receiver to use an accurate approximate of signal travel time for the timing synchronization. The approximate signal travel time is selected to be 72 ms. When multiplied by the speed of light, this time corresponds to a distance of 21,600 km between the user and satellite, which is an estimated range of a GPS/Galileo satellite at flight altitude.

Table 7.1 GPS/Galileo constellations

System	GPS	Galileo
Nominal satellites	24	27
Orbital planes	6	3
Inclination	55°	56°
Semi-major axis	26,550 km	29,600 km
Average flight altitude	20,180 km	23,222 km
Orbital period	11 h 56 min	14 h 7 min

In next part the equations for both GPS and Galileo synchronization are provided.

GPS Synchronization

$$T_{GPS} = T_{TOW} + T_{GPS\_code} + T_{GPS\_chip} + 0.072\ \text{s} \tag{7.1}$$

where T_{TOW} is the current Time Of Week (TOW) count decoded from the navigation data bits. In GPS case the TOW is found in start of each subframe, thus being updated every 6 s. The TOW reports how many seconds has elapsed since the start of the GPS week. By definition, both GPS and Galileo weeks start in the midnight between Saturday and Sunday. GPS week numbering has started on 6th of January 1980, whereas Galileo week numbering has started on 22nd August 1999 (Galileo ICD 2010; GPS ICD 2004).

T_{GPS_code} is the integer number of full GPS C/A—code epochs since the beginning of current subframe. In GPS case the C/A—codes are 1 ms long. The integer number of milliseconds is converted to seconds in the equation.

T_{GPS_chip} is composed of 2 parts. First part is the integer number of full code chips in the current GPS C/A—code (0-1022). The second part is the fraction of the current chip. The fraction information is provided by the baseband tracking loops. This combined timing information is then converted into seconds.

Galileo Synchronization

$$T_{Galileo} = T_{TOW} + T_{Gal\_code} + T_{Gal\_chip} + 0.072\ \text{s} \tag{7.2}$$

where T_{TOW} is the current TOW decoded from the navigation data. In Galileo I/NAV messages TOW is recovered at nominal page rate of 2 s. This is different from the GPS TOW rate, which was every 6 s.

T_{Gal_code} is the integer number of Galileo codes since the start of page. In E1-B frequency case code length is 4 ms.

T_{Gal_chip} is also composed of 2 parts. First part is the integer number of code chips (0-4091), and the second part is the fraction of the code chip, obtained from the baseband tracking loops.

Conclusion Concerning the Synchronization

As the previous synchronization Eqs. (7.1) and (7.2) suggest, the GNSS time is composed of 4 distinct components. Additionally a fifth parameter called Week Number (WN) is included in the GNSS time keeping. Next paragraph concludes the significance of each timing component.

- First part of the GNSS time is Week Number, which is, at minimum, needed to handle the cross-over of the week boundaries. The WN is decoded from the navigation message.
- Next major part of the composite time is the TOW, which is also decoded from the navigation data. The TOW counts full seconds, 0–604,799, since the beginning of a GNSS week. TOW is reset to zero at the end of each week.
- The second component is the integer number of C/A codes elapsed since the end of last subframe. This ranges between 0–6 s in GPS case, and between 0–2 s in Galileo I/NAV case.
- Next time component is the fraction of a code chip. This part ranges between 0–1 ms in GPS case, and between 0–4 ms in Galileo case. Now this part of composite time has already lot smaller impact in the total synchronization time.
- The fourth component of the time is the approximate signal travel time, for which 72 ms was selected. In practice the true travel time might be several milliseconds different from the chosen one. This explains why the code fraction component does not play major role in the coarse time synchronization.

The described GNSS synchronization technique should provide coarse time to GNSS time. It is expected to be within some millisecond range from the true time. Synchronization time errors of several milliseconds still provide sufficient accuracy to the PVT solution, which further allows fine tuning of the time by applying the clock bias correction. Once the clock bias is corrected, the receiver is accurately synchronized to the GNSS time.

Selecting Satellites for Navigation Solution

The modern GNSS receivers are able to track quite a few satellites simultaneously. When Galileo reaches full constellation of satellites, it is possible to have situation where the receiver is tracking around 10 GPS satellites and 10 Galileo satellites.

From the navigation solution point of view this abundance of satellite signals has its pros and cons.
Pros:

- Signal diversity
- Possibility to use selected set of best signals

Cons:

- Computational burden
- Increased chance erroneous measurements

This means that we have redundancy in the available signals, even to extent that we can deliberately choose not to use some satellites in the PVT solution. In urban canyon environment the amount of satellites is still going to remain as issue. The benefit from dropping some satellites would come in terms of reduced computational burden, and in reduced risk of including erroneous signal into the navigation solution.

Listing Accepted Satellites

Conventional single system receiver is able to operate with just having 4 satellites included into the navigation solution. This requirement will be one step stricter in dual system receiver, where we need 5 satellites for the solution. The requirement of 5th satellite comes from the time offset between the two satellite systems. The starting point for the dual system receiver navigation solution is to use 5 satellites, containing signals from both GPS and Galileo system. It is good to remember that dual system receiver can also operate as single system receiver with 4 satellites from either of the systems, in case no satellites are available from the other system.

Now as the minimum set of satellites for a dual system receiver has been defined, it is time to consider the effect of including more available satellite signals to the PVT solution. Adding more satellites might first sound like a very good idea, but in practice, the gain from extra satellites quickly becomes negligible. A rule of thumb would be to limit amount of selected satellites to 8, since adding more satellites beyond 8 hardly improves the accuracy of the PVT solution. If best satellite signals are selected to the set, newly added satellite signals will likely be of lower quality ones that are likely to lower the overall quality. Also the accuracy issue becomes apparent when investigating the Dilution of Precision (DOP) parameter calculated in the navigation solution. DOP value describes how good geometry the selected satellites provide, which is directly linked to the expected accuracy of the navigation solution. More discussion of this parameter will be provided later in this chapter.

Satellite Selection Criteria

A satellite is usable to the navigation solution once it meets certain criteria, which can be divided into **mandatory** criteria:

- Full set of valid ephemeris data must be obtained
- Satellite health status ok
- Accuracy (URA in GPS and SISA in Galileo) estimates from navigation data must be acceptable (Galileo ICD 2010; GPS ICD 2004)
- Data validity indicator must be ok
- Synchronization to the TOW is done
- Code phase measurements from the baseband are available

Let us start with elaborating the mandatory criteria for satellite selection. All of these conditions are obvious from the ICD point of view. First of all, the satellite signal must be tracked for certain time period to receive all the necessary ephemeris data needed to determine the satellite location at any given time. More of the ephemeris data details will be found in later chapter. GPS and Galileo navigation data rates and data layout are different. Low level details are found in the respective ICDs (Galileo ICD 2010; GPS ICD 2004).

In addition to the full set of ephemeris, the satellites also provide health status indicator, which indicate whether the control segment has determined the satellite signal to be of usable quality.

Another similar indicator is the User Range Accuracy (URA) value for GPS, and Signal in Space Accuracy (SISA) for Galileo. These values describe the estimated error component for which the space and control segment are responsible. For GPS, complete list of URA values is presented in GPS ICD (2004). For Galileo SISA, the current Galileo ICD does not yet provide the complete list of expected values (Galileo ICD 2010). The principle of this value for both systems is the same. The receiver will check that the indicated value is at acceptable levels. Rule of thumb is to avoid using satellites which indicate high expected error in the signal. Furthermore, both GPS and Galileo data streams provide an indicator which shows whether the transmitted navigation data is usable or not.

Synchronization to the TOW means that the receiver has been able to synchronize to the frame structure, and has been able to successfully retrieve the TOW value from the navigation data. The TOW value is needed for constructing the pseudoranges. Usually this TOW extraction step is closely tied to the decoding of ephemeris data.

Code phase measurements will be available from the baseband once the tracking loops have been converged, thus ensuring that the measurements will be stable. Code phase measurements are also needed for the pseudorange construction.

Once the mandatory criteria are met, further ranking of the satellite signals can be made with **optional** criteria:

- Elevation angle
- CNO estimate from baseband

- BER from baseband
- Pseudorange noise estimate
- DOP of the chosen constellation.

Elevation angle is one important value in determining the usability of a satellite. Elevation angle directly determines how long distance the satellite signal has to travel in Earth atmosphere. The lower the elevation angle, the longer distance the signal will take in the atmosphere. Long distance traveled in the atmosphere will induce greater variation to the signal travel time due to the effects of troposphere and ionosphere delays. Most receivers use freely chosen elevation mask, to discard satellites below certain elevation threshold from the navigation solution. Typical elevation mask values are around 10–15°. When a new satellite is being tracked in the receiver, the elevation angle is initially unknown to the receiver, since the satellite location is unknown. The receiver should have a mechanism to handle this first time calculation, for example just ignoring the elevation mask criteria for unknown satellite location.

CNO estimate from the baseband gives fairly reliable way of indicating the overall quality of the satellite signal that the receiver is able to receive. If some satellite signal indicates considerably lower CNO values, it is a good reason not to include that satellite to the navigation solution. In general, only in cases where the receiver is not able to track many satellites, then the satellites with low CNO values can be included in the navigation solution, in order to have some position fix—even if it might end up being of worse accuracy.

Bit error rate (BER) rates from the baseband are very similar indicator as the CNO values. BER describes how many navigation bits are not received correctly. In practice, false navigation bits make the CRC checks on the data frames to fail. This will lead to the rejection of the frame, causing the risk of not being able to notice critical updates in the ephemeris data/signal health status indicator. In conclusion, satellites with high BER are to be avoided, especially satellites which continuously fail to pass the CRC checks.

Pseudorange noise estimation can be done in the navigation software simply by observing the continuous pseudorange measurements over a time window. If significant variation is noticed in the measurements, it will indicate problem with the baseband tracking loops not being able to track the signal perfectly. Once again, high value in pseudorange noise estimation is a good reason to avoid using of such satellite signal.

Once the receiver has made initial PVT solution with available satellites, the position of satellites and user are known. With the location information it is possible to calculate the DOP value for different combination of satellites. The DOP value indicates how good accuracy the chosen satellites can provide. The DOP value can be solved from the geometry matrix which is introduced later in this chapter when navigation solution details are discussed. The DOP value can be used as one factor in reducing the amount of satellites without decreasing the accuracy of PVT solution.

Satellite Blacklist

Previous part listed many factors how to evaluate the quality of a satellite signal to the PVT solution. A term closely related to this evaluation is satellite blacklist. Such blacklist is useful to have in order to save resource consumption on the receiver. For example, if one particular satellite is constantly causing problems, and is not getting included to the PVT solution, it can be blacklisted. Information about the blacklisted satellites can then be given to the baseband. Baseband can then stop the tracking of such satellite for certain period of time, thus freeing the channel for possibly other satellites, or just to save power in the baseband. After some time, the "penalty" of a blacklisted satellite has expired, and it can be checked if it now behaves better than before.

Pseudorange Construction

The pseuodranges are the fundamental measurements of a standalone GNSS receiver. Pseudoranges represent the measured distance between the receiver and satellite. They are not exact ranges, due to the receiver clock bias and other errors that the measurements contain. The GNSS receivers usually have cheap crystal oscillator clocks which are not synchronized to the GNSS time, and they are not very stable. The receiver clock problem will be compensated when the clock bias is solved in the PVT solution. The other measurement errors, caused by, e.g. control segment or atmospheric errors, cannot be directly accounted for. In order to minimize the effect of these errors, the criteria presented in previous section should be used for selecting the satellites.

Pseudoranges are composite measurements, combined from different factors from the receiver. The receiver makes pseudorange measurements at certain point of time, meaning that the pseudorange measurements are made for all tracked satellites simultaneously. Let's denote via t the time of taking pseudorange measurement. Then pseudorange measurement will be as follows:

$$\rho_i(t) = c \cdot [t_u(t) - t_i(t - \tau)] \tag{7.3}$$

where

subscript i	denotes satellite i,
$\rho_i(t)$	pseudorange measurement for satellite i,
c	the speed of light,
$t_u(t)$	receiver generated time value

$$t_i(t - \tau) = t_i^{TOW} + t_i^{ms} + t_i^{chip} \tag{7.4}$$

t_i^{TOW} is latest TOW received from the navigation data,
t_i^{ms} is total amount of C/A—codes passed since the last frame, expressed as integer multiple of milliseconds,
t_i^{chip} is fraction of current C/A—code

Observing the last 3 components of the pseudorange measurement we notice few differences between GPS and Galileo pseudorange construction. First of all, the TOW count will be received every 6 s for GPS, and in Galileo I/NAV messages TOW is received once every 2 s.

Second component is the amount of C/A—codes received since the beginning of the frame. For GPS case, each GPS C/A—code adds 1 ms to the value, and for Galileo, each C/A—code adds 4 ms.

Third component is the fraction of the current C/A-code. For GPS these values will be between 0–1 ms, and for Galileo between 0–4 ms.

As conclusion, pseudoranges are generated for all channels, meaning both GPS and Galileo satellites, at the same internal time t. The receiver clock at time t is showing $t_u(t)$, the receiver time can be synchronized to either GPS, or Galileo time, the difference is negligible. At this time instant the real measurement information is given by the baseband in the three aforementioned parameters. The TOW parameter provides roughest estimate of the time; number of C/A—codes provide time component with more accuracy; and finally the last component, part of the C/A—code, provides time component of finest precision.

These three levels of timing accuracy have analogy to an analog clock, where the TOW count represents the hour hand of a clock. Number of C/A—codes represent the minute hand, and the fraction of C/A—code measurement represents the second hand. This is very useful analogy in understanding how the pseudorange timing is composed of these parameters (Kaplan and Hegarty 2006).

Satellite Location Calculation

At this point the receiver has been able to acquire and track enough of the GNSS signals. For those satellites, full set of ephemeris has been obtained from the navigation data. Although the navigation data frame structure is different for GPS and Galileo, it is very convenient that both systems support the same ephemeris data set for describing the satellite location on orbits at any given time. The ephemeris data is composed of parameters listed in Table 7.2.

The Table 7.2 shows the 6 basic Keplerian elements, 9 correction terms of time perturbations, one time epoch parameter plus 3 clock correction parameters. Additionally the navigation data is provided with a timestamp called Issue of Data Ephemeris (IODE), which indicates any changes in the ephemeris data. For individual satellite, all used ephemeris must be time-stamped with same IODE. The

Table 7.2 Ephemeris parameters

Parameter	Number of bits		Scale factor		Units
	GPS	GAL	GPS	GAL	
T_{GD}	8^*	10^*	2^{-31}	2^{-32}	Seconds
t_{0c}	16	14	2^4	60	Seconds
t_{0e}	16	14	2^4	60	Seconds
a_{f2}	8^*	6^*	2^{-55}	2^{-59}	Sec/sec^2
a_{f1}	16^*	21^*	2^{-43}	2^{-46}	Sec/sec
a_{f0}	22^*	31^*	2^{-31}	2^{-34}	Seconds
C_{rs}	16		2^{-5}		Meters
Δn	16^*		2^{-43}		Semicircles/sec
M_0	32^*		2^{-31}		Semicircles
C_{uc}	16^*		2^{-29}		Radians
e	32		2^{-33}		
C_{us}	16^*		2^{-29}		Radians
$\sqrt{A}$	32		2^{-19}		Meters$^{1/2}$
C_{ic}	16^*		2^{-29}		Radians
Ω_0	32^*		2^{-31}		Semicircles
C_{is}	16^*		2^{-29}		Radians
i_0	32^*		2^{-31}		Semicircles
C_{rc}	16^*		2^{-5}		Meters
ω	32^*		2^{31}		Semicircles
$\dot{\Omega}$	24^*		2^{-43}		Semicircles/sec
$\dot{i}$	14^*		2^{-43}		Semicircles/sec

Parameters indicated with * are two's complement numbers

algorithm for solving the satellite location at given time uses few constants, which are defined in Table 7.3

The complete procedure for solving satellite positions is described step-by-step in the next section. Parameters used in this procedure are those that can be found in previous Tables 7.2, 7.3. Some of the parameters slightly differ between GPS and Galileo, but the procedure still uses the same approach for both systems.

The following set of equations also contain Sagnac effect correction, which accounts for Earth rotation between the time when signal was transmitted from the satellite, and the time when it was received. Sagnac correction changes the satellite ECEF-coordinates (Earth Centered, Earth Fixed) into ECI-coordinate (Earth Centered Inertial) system. Sagnac effect is illustrated in Fig. 7.1.

One remaining issue in satellite location calculation is about the GPS and Galileo geodetic coordinate reference frames. Both systems use their own coordinate

Table 7.3 Parameter definition

Parameter	Value
t_u	User equipment time
	Generated in the receiver
ρ^s	Pseudorange measurement for satellite *s*
	Generated in the receiver
c	Speed of light
	299,792,458 m/s
μ	Earth's gravitational constant
	Note difference in the respective ICDs
	GPS: 3.986005e14 m^3/s^2
	Galileo: 3.986004418e14 m^3/s^2
$F = \frac{-2 \cdot \sqrt{\mu}}{c^2}$	Minor difference between systems due to Earth's gravitational constant
	GPS: −4.442807633e − 10 $s/m^{1/2}$
	Galileo: −4.442807309e − 10 $s/m^{1/2}$
$\dot{\Omega}_e$	Earth's rotation rate
	7.2921151467e − 5 rad/s

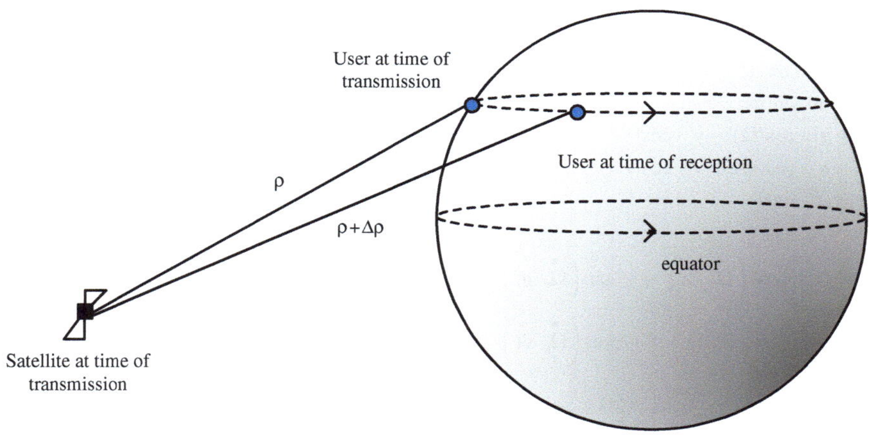

Fig. 7.1 Sagnac effect (magnified for illustrative purpose)

reference frames. For GPS the reference frame is WGS84, and for Galileo, Galileo Terrestrial Reference Frame (GTRF). The differences between the two models are expected to be within few centimeters, as both WGS84 and GTRF are supposed to follow closely the International Terrestrial Reference Frame (ITRF). Therefore, this error is hidden by the higher satellite orbital and clock estimation errors in the ephemerides and its impact in accuracy is negligible for a standard navigation servic. No further action is needed for correcting the coordinate differences (Table 7.4).

Table 7.4 Solution to satellite location

$t_{tr} = t_u - \rho^s/c$	Time of transmission*
$t_k = t_{tr} - t_{0e}$	Time difference between Time of transmission and Epoch time*
$A = \left(\sqrt{A}\right)^2$	Semi-major axis
$n = \sqrt{\frac{\mu}{A^3}} + \Delta n$	Corrected mean motion*
$M = M_0 + n \cdot t_k$	Mean anomaly
$E = M + e \cdot \sin(E)$	Eccentric anomaly—must be solved with iterative method
$v = \text{atan}\left(\frac{\sqrt{1-e^2} \cdot \sin(E)}{\cos(E) - e}\right)$	True anomaly
$\Phi = v + \omega$	Argument of latitude
$\delta u = C_{us} \cdot \sin(2\Phi) + C_{uc} \cdot \cos(2\Phi)$	Argument of latitude correction
$\delta r = C_{rs} \cdot \sin(2\Phi) + C_{rc} \cdot \cos(2\Phi)$	Radius correction
$\delta i = C_{is} \cdot \sin(2\Phi) + C_{ic} \cdot \cos(2\Phi)$	Inclination correction
$u = \Phi + \delta u$	Corrected argument of latitude
$r = A \cdot (1 - e \cdot \cos(E)) + \delta r$	Corrected radius
$i = i_0 + \delta i + \dot{i} \cdot t_k$	Corrected inclination
$\Omega = \Omega_0 + \left(\dot{\Omega} - \dot{\Omega}_e\right) \cdot t_k - \dot{\Omega}_e \cdot t_{0e}$	Corrected longitude of node
$x' = r \cdot \cos(u)$	In-plane x position
$y' = r \cdot \sin(u)$	In-plane y position
$x = x' \cdot \cos(\Omega) - y' \cdot \cos(i) \cdot \sin(\Omega)$	ECEF x-coordinate
$y = x' \cdot \cos(\Omega) + y' \cdot \cos(i) \cdot \cos(\Omega)$	ECEF y-coordinate
$z = y' \cdot \sin(i)$	ECEF z-coordinate
$dt = t_k - t_u$	Sagnac effect correction
$x_{SV} = x \cdot \cos\left(\dot{\Omega}_e \cdot dt\right) - y \cdot \sin\left(\dot{\Omega}_e \cdot dt\right)$	
$y_{SV} = x \cdot \sin\left(\dot{\Omega}_e \cdot dt\right) + y \cdot \cos\left(\dot{\Omega}_e \cdot dt\right)$	
$z_{SV} = z$	

Equations indicated with * have different parameters between GPS and Galileo

Atmospheric Corrections

The GNSS satellites are located on orbits above 26,000 km from the center of the Earth. Assuming that the typical GNSS receiver is located on the surface of the earth will lead to conclusion that satellites are typically sending signal 20,000 km from the receiver. These signal travel mostly in space, which can be considered as a vacuum. On the proximity of Earth, the signal enter Earth's atmosphere, which no longer can be considered as vacuum. In Earth's atmosphere signals are subject to atmospheric refraction which affects the signal propagation. The two most important layers of the atmosphere are called the ionosphere, and the troposphere. Next sections will discuss the effects of these layers, and of the methods how single frequency GNSS receiver can mitigate the effects.

Ionosphere

The first atmospheric layer the GNSS signals encounter is called the ionosphere. Within the ionosphere free electrons affect the signal propagation. The amount of free electrons varies with time, and the common goal in mitigating the ionosphere effect is to estimate the Total Electron Content (TEC) on the path between the user and the satellite. If not accounted for, ionosphere can cause pseudorange errors of tens of meters (Syrjärinne 2001). Ionosphere effects have greater magnitude than those of the troposphere.

The delay caused by the ionosphere is related to the TEC value via following equation:

$$I = \frac{40.3}{f^2} \cdot 10^4 TEC \tag{7.5}$$

where,

I is the delay caused by ionosphere (m),

f is the center frequency of the used GNSS band (Hz).

GPS and Galileo both have their own models for mitigating the ionospheric errors. Both systems rely on the broadcasted parameters within the navigation data. Next we are going to take closer look at both systems algorithms.

GPS Klobuchar Model

The GPS Klobuchar model is a simple model from the receiver design point of view. It solely relies on the inputs received from the satellites (Satellite Transmitted Terms) and Receiver Generated Terms. Complete procedure for computing the ionosphere delay is directly presented in GPS ICD 2004, and will not be repeated here. Using the algorithm of GPS ICD 2004 it is expected to reduce the ionosphere error with 50 % for single frequency GPS measurements.

Galileo NeQuick Model

The Galileo NeQuick model is another model for the ionospheric effects. Implementation of NeQuick capability on receivers needs bit more preparation, since the model assumes availability of so called CCIR maps, and dip latitudes grid file.

Table 7.5 Galileo Ionosphere Correction Terms

Parameter	Description
a_{i0}	Effective ionisation level 1st order parameter
a_{i1}	Effective ionisation level 2nd order parameter
a_{i2}	Effective ionisation level 3rd order parameter

It must be noted that the Galileo system itself does not provide this information on the navigation data. That is one major difference between the GPS Klobuchar model, which can operate solely on the data obtained from GPS system. On the other hand, the benefit of the NeQuick will be in improvement in the ionospheric corrections. A 70 % reduction of ionospheric effects is estimated for single frequency Galileo NeQuick model user (Galileo ICD 2010).

Galileo navigation data provides few transmitted terms to enable the ionospheric corrections. The three broadcast parameters are presented in Table 7.5.

Using the parameters in Table 7.5, the effective ionisation level is computed according to the following equation:

$$Az = a_{i0} + a_{i1}\mu + a_{i2}\mu^2 \tag{7.6}$$

where μ is the modified dip latitude MODIP.

The solved Az is used to replace the NeQuick model F10.7 value. Running NeQuick now provides the TEC value which eventually leads to the ionospheric delay through the use of Eq. (7.5).

For stand-alone GPS/Galileo receiver without availability to CCIR maps, and dip latitudes grid file, it is possible to apply GPS Klobuchar model to Galileo satellites. This procedure expects the receiver to retrieve the Klobuchar correction parameters from GPS satellite navigation data, and then applying Klobuchar corrections to both systems with the same parameters. This method avoids the requirement of extra files in Galileo NeQuick model, with the expense of losing some accuracy in the ionosphere correction for Galileo signals. The Klobuchar model does not work with full potential for Galileo, since Galileo orbits are different from GPS orbits, as indicated in Table 7.1.

Troposphere

The troposphere is the second important layer in Earth's atmosphere causing propagation delay to the received GNSS signals. Tropospheric delay is generally less significant than ionospheric delay. On low elevation satellites the tropospheric delay will be considerable, but as suggested earlier in the chapter, an elevation mask is used to exclude the low elevation satellite because troposphere effects.

From implementation point of view, the GNSS systems do not send the navigation data for correcting the tropospheric delays. In the literature there is wide variety of algorithms presented for correcting the tropospheric delay. For standard

Table 7.6 Saastamoinen model

$T_{z,d} = 0.002277 \cdot (1 + 0.0026 \cdot \cos(2\phi) + 0.00028 \cdot H) \cdot P0$	Dry zenith delay
$T_{z,w} = 0.002277 \cdot \left(\frac{1255}{T0} + 0.05\right) \cdot e0$	Wet zenith delay
$m_d(E) = \frac{1}{\sin(E) + \frac{0.00143}{\tan(E)+0.0445}}$	Mapping function for dry delay
$m_w(E) = \frac{1}{\sin(E) + \frac{0.00035}{\tan(E)+0.017}}$	Mapping function for wet delay
$t_{tropo} = T_{z,d} \cdot m_d(E) + T_{z,w} \cdot m_w(E)$	Total zenith delay for troposphere

GNSS user even the simplest models are usually enough to correct most of the delay. The simplest models work only as a function of elevation angle to the satellite, whereas the more sophisticated models can utilize information about local air pressure, temperature and other parameters to describe the atmospheric conditions.

Simple model for correcting tropospheric delay (Misra and Enge 2004):

$$t_{tropo} = \frac{2.47}{\sin(E) + 0.0121} \tag{7.7}$$

where the tropospheric delay will be in meters, and elevation angle E is given in radians.

An example of more sophisticated model based on Saastamoinen model is presented in Table 7.6 (Misra and Enge 2004):

In previous equations $\phi$is user latitude, H is orthometric height of the antenna in kilometers, P0 is total pressure in millibars, T0 is temperature in Kelvin, and e0is partial pressure due to water vapor in millibars.

Applying Clock Corrections

In addition to the delays caused by the ionosphere and troposphere, also relativistic effects cause errors to the measured signal. The relativistic correction term can be calculated based on the following equation. Another new clock correction term is for adjusting the bias and drift of the satellite onboard clock (Table 7.7).

The last step is to apply all the clock correction terms to the measured pseudorange. Tropospheric delay is given directly in meters; the rest of the correction terms must be multiplied with speed of light in order to convert them to meters.

$$\rho^s_{corr} = \rho^s + c \cdot (t_{clk} + t_r - t_{iono}) - t_{tropo} \quad \text{Corrected pseudorange} \tag{7.8}$$

Table 7.7 Relativistic correction term

$\Delta t_r = F \cdot e \cdot \sqrt{A} \cdot \sin(E)$	Clock correction due to relativistic effects
$t_{clk} = a_{f0} + a_{f1} \cdot (t_{tr} - t_{0c}) + a_{f2} \cdot (t_{tr} - t_{0c})^2 - T_{GD}$	Clock offset correction term

Note differences in parameters between GPS and Galileo, details found in Tables 7.2 and 7.3

Navigation Solution

The navigation solution will solve for the final PVT. With dual system operation having measurements from both systems, GPS and Galileo, there is one new issue to be taken care of. The new issue arises with the different system times of GPS and Galileo.

GPS satellites transmit their data based on GPS Time (GPST) and Galileo satellites transmit data based on Galileo System Time (GST). GPS and Galileo time systems have certain bias between them, and this bias is one more unknown which must be solved. For this problem two methods are described in the following section.

- Solution 1: Utilize broadcast Galileo/GPS Time Offset

Parameters to solve Galileo/GPS Time Offset (GGTO) value will be broadcast in the Galileo data message. These parameters are listed with Eq. 7.9. Once these parameters are obtained the Eq. 7.9 may be used to get the GGTO value for correcting all measurements to same time reference. The GGTO can be used for either fixing GPS time to Galileo or vice versa. While this approach is a simple way of fixing the time domain differences, the only remaining issue will be in the accuracy of the broadcasted parameters for the GGTO algorithm. The accuracy is nevertheless expected to provide sufficient accuracy for standard GNSS applications.

$$\begin{aligned} GGTO &= t_{Galileo} - t_{GPS} \\ GGTO &= A_{0G} + A_{1G}[TOW - t_{0G} + 604800 \cdot ((WN - WN_{0G}) \bmod 64)] \end{aligned} \tag{7.9}$$

where

$$\begin{aligned} A_{0G} &= \text{Constant term describing the offset to GGTO,} \\ A_{1G} &= \text{Rate of change of the offset GGTO,} \\ t_{0G} &= \text{reference time for GGTO data,} \\ WN &= \text{GSTWeek Number,} \\ WN_{0G} &= \text{Week Number of the GGTO reference} \end{aligned}$$

- Solution 2: Solve two time biases of both systems

The approach of solving the time bias for both systems comes with a drawback. Usual way of solving the PVT is to have 4 satellites for solving the 4 unknowns (x, y,z and bias). When having two biases to solve, a minimum of 5 satellites is needed. Naturally those 5 satellites must contain satellites from both systems, or otherwise the receiver can make a standard single system PVT solution with minimum of 4 satellites.

In the following part a short introduction to standard Least-Squares solution is derived. The standard method needs to be modified to be able to handle dual system navigation solution.

Least-Squares Solution

A receiver measuring pseudoranges can solve the PVT using the LS method. LS needs more measurements than there are unknowns. A basic observation equation for pseudorange is

$$P^s = \rho^s + c \cdot (\Delta t_u - \Delta t^s) + T^s + I^s + e^s, \tag{7.10}$$

where

$s =$ index denoting satellite s,
$\rho =$ geometrical range between satellite and receiver,
$\Delta t_u =$ receiver time bias,
$\Delta t^s =$ satellite clock offset,
$T^s =$ tropospheric delay,
$I^s =$ ionospheric delay,
$e^s =$ remaining measurement errors

The geometrical range between satellite and receiver is defined as:

$$\rho^s = \sqrt{(X^s - X_u)^2 + (Y^s - Y_u)^2 + (Z^s - Z_u)^2} \tag{7.11}$$

Out of the parameters present in the Eqs. 7.10 and 7.11, only $X_u, Y_u, Z_u, \Delta t_u$ and e^s are unknowns. The rest have been solved earlier in the navigation solution. In the end there are 4 unknowns which the LS will solve, and e^k, which is minimized in the process.

The Eq. 7.10 needs to be linearized before it can be used in LS. The linearization process needs initial information about user position. If an estimate is available, it is recommended to use it. In case no a priori information about location is available, it is sufficient to use center of Earth as initial guess (0, 0, 0).

$$f(X_u, Y_u, Z_u) = \sqrt{(X^s - X_u) + (Y^s - Y_u) + (Z^s - Z_u)} \tag{7.12}$$

The iteration increments $\Delta X, \Delta Y$ and ΔZ are defined as:

$$\begin{aligned} X_{u,i+1} &= X_{u,i} + \Delta X, \\ Y_{u,i+1} &= Y_{u,i} + \Delta Y, \\ Z_{u,i+1} &= Z_{u,i} + \Delta Z, \\ i &= \text{ iteration count} \end{aligned} \tag{7.13}$$

The Taylor expansion of (7.12) gives:

$$\begin{aligned} f\left(X_{u,i+1} + \Delta X, Y_{u,i+1} + \Delta Y, Z_{u,i+1} + \Delta Z\right) \approx f\left(X_{u,i}, Y_{u,i}, Z_{u,i}\right) - \frac{X^s - X_{u,i}}{\rho_i^s} \cdot \Delta X \\ - \frac{Y^s - Y_{u,i}}{\rho_i^s} \cdot \Delta Y - \frac{Z^s - Z_{u,i}}{\rho_i^s} \cdot \Delta Z \end{aligned} \tag{7.14}$$

For simplicity, (7.14) only contains first order terms. Thus the first order linearized observation equation becomes:

$$\begin{aligned} P^s = \rho_i^s - \frac{X^s - X_{u,i}}{\rho_i^s} \cdot \Delta X - \frac{Y^s - Y_{u,i}}{\rho_i^s} \cdot \Delta Y - \frac{Z^s - Z_{u,i}}{\rho_i^s} \cdot \Delta Z + c \cdot (\Delta t_u - \Delta t^s) + T^s \\ + I^s + e^s \end{aligned} \tag{7.15}$$

which can be rewritten in vector form

$$P^s = \rho_i^s + \left[-\frac{X^s - X_{u,i}}{\rho_i^s} \quad -\frac{Y^s - Y_{u,i}}{\rho_i^s} \quad -\frac{Z^s - Z_{u,i}}{\rho_i^s} \quad 1 \right] \cdot \begin{bmatrix} \Delta X \\ \Delta Y \\ \Delta Z \\ c \cdot \Delta t_u \end{bmatrix} - c \cdot \Delta t^s + T^s + I^s + e^s \tag{7.16}$$

A general form of the LS problem is $A \cdot x = b$. Modifying (7.16) to this form, it yields:

$$\left[-\frac{X^s - X_{u,i}}{\rho_i^s} \quad -\frac{Y^s - Y_{u,i}}{\rho_i^s} \quad -\frac{Z^s - Z_{u,i}}{\rho_i^s} \quad 1 \right] \cdot \begin{bmatrix} \Delta X \\ \Delta Y \\ \Delta Z \\ c \cdot \Delta t_u \end{bmatrix} = P^s - \rho_i^s + c \cdot \Delta t^s - T^s - I^s - e^s \tag{7.17}$$

A best choice $\hat{x}$ is selected such that $\|e\|^2 = (b - A \cdot x)^T \cdot (b - A \cdot x)$ gets minimized. (7.18) gives such solution.

$$\hat{x} = \left(A^T \cdot A\right)^{-1} \cdot A^T \cdot b \tag{7.18}$$

Finally, the LS form for m satellites is as follows:

$$A \cdot x = \begin{bmatrix} -\frac{X^1 - X_{u,i}}{\rho_i^1} & -\frac{Y^1 - Y_{u,i}}{\rho_i^1} & -\frac{Z^1 - Z_{u,i}}{\rho_i^s} & 1 \\ -\frac{X^2 - X_{u,i}}{\rho_i^2} & -\frac{Y^2 - Y_{u,i}}{\rho_i^2} & -\frac{Z^2 - Z_{u,i}}{\rho_i^2} & 1 \\ -\frac{X^3 - X_{u,i}}{\rho_i^3} & -\frac{Y^3 - Y_{u,i}}{\rho_i^3} & -\frac{Z^3 - Z_{u,i}}{\rho_i^3} & 1 \\ \vdots & \vdots & \vdots & \vdots \\ -\frac{X^m - X_{u,i}}{\rho_i^m} & -\frac{Y^m - Y_{u,i}}{\rho_i^m} & -\frac{Z^m - Z_{u,i}}{\rho_i^m} & 1 \end{bmatrix} \cdot \begin{bmatrix} \Delta X \\ \Delta Y \\ \Delta Z \\ c \cdot \Delta t_u \end{bmatrix} = b - e \tag{7.19}$$

After getting a solution from (7.19), it has to be updated to the user location estimate, in Eq. (7.13). After update, the new iteration will be performed with the newly obtained user location and bias estimates, until the solution has converged such that further iteration will not change estimates significantly. Usually LS will converge quickly to correct estimate, e.g. around 5 iterations.

Dual-System Least-Squares

Having both GPS and Galileo in the navigation solution, leads to the situation where 5th unknown for second time bias is needed. Following the Eq. (7.19) notation, a LS solution for case with 3 GPS satellites and 2 Galileo satellites is:

$$A \cdot x = \begin{bmatrix} -\frac{X^{1,GPS} - X_{u,i}}{\rho_i^{1,GPS}} & -\frac{Y^{1,GPS} - Y_{u,i}}{\rho_i^{1,GPS}} & -\frac{Z^{1,GPS} - Z_{u,i}}{\rho_i^{1,GPS}} & 1 & 0 \\ -\frac{X^{2,GPS} - X_{u,i}}{\rho_i^{2,GPS}} & -\frac{Y^{2,GPS} - Y_{u,i}}{\rho_i^{2,GPS}} & -\frac{Z^{2,GPS} - Z_{u,i}}{\rho_i^{2,GPS}} & 1 & 0 \\ -\frac{X^{3,GPS} - X_{u,i}}{\rho_i^{3,GPS}} & -\frac{Y^{3,GPS} - Y_{u,i}}{\rho_i^{3,GPS}} & -\frac{Z^{3,GPS} - Z_{u,i}}{\rho_i^{3,GPS}} & 1 & 0 \\ -\frac{X^{1,GAL} - X_{u,i}}{\rho_i^{1,GAL}} & -\frac{Y^{1,GAL} - Y_{u,i}}{\rho_i^{1,GAL}} & -\frac{Z^{1,GAL} - Z_{u,i}}{\rho_i^{1,GAL}} & 0 & 1 \\ -\frac{X^{2,GAL} - X_{u,i}}{\rho_i^{2,GAL}} & -\frac{Y^{2,GAL} - Y_{u,i}}{\rho_i^{2,GAL}} & -\frac{Z^{2,GAL} - Z_{u,i}}{\rho_i^{2,GAL}} & 0 & 1 \end{bmatrix} \cdot \begin{bmatrix} \Delta X \\ \Delta Y \\ \Delta Z \\ c \cdot \Delta t_u(GPS) \\ c \cdot \Delta t_u(GAL) \end{bmatrix} = b - e \tag{7.20}$$

Once a sufficient amount of iterations are performed on the LS method and the solution has converged, then the reliability of the solution is checked. If the solution is considered acceptable, the time bias adjustment can be applied to the user equipment clock. The first PVT solution should correct major part of the bias, which was caused in the rough synchronization to GPS/Galileo time. On the consecutive PVT solutions the remaining bias should be reduced into sub-millisecond range. Observing high bias values after PVT solution indicates some problems in the overall navigation process and appropriate action should be taken to fix the issue.

Similarly, the range residuals for individual satellites should be observed. If one satellite is having generally higher range residuals than other tracked satellites, this possibly indicates erroneous measurement for that particular satellite. A satellite with high range residuals should be considered to be excluded from the navigation solution if extra satellites are available. Alternatively weighted Least-Squares method can be used, where less weight is given for the lower quality satellites (Kaplan and Hegarty 2006).

With limited visibility to satellites, e.g. in urban canyons it might be harder to get 5 satellites for the position solution instead of the regular 4. Also due to increased size of the matrices, the computational requirement for calculating the matrix operations of the PVT is slightly increased.

Dilution of Precision

A DOP value indicates how good positioning accuracy can be expected from a given satellite combination. Geometrical Dilution of Precision (GDOP) is the most common DOP indicator, which describes the overall quality of the satellite geometry. The more the satellites are spread above the horizon, the smaller is the GDOP, and consequentially, the better is the expected positioning accuracy.

Using the earlier notation, GDOP is found from the geometry matrix **A** (introduced in Eq. 7.19)

$$GDOP = \sqrt{trace\left\{ \left[A^T \cdot A \right]^{-1} \right\}} \tag{7.21}$$

where the trace is the sum of diagonal elements in the matrix. In a combined GPS/Galileo receiver, the amount of diagonal elements in the GDOP calculation is increased. This is because the **A** matrix now contains the extra 5th column, containing the second time bias.

Navigation Data Decoding

A user receiver is expected to be able to recognize page types and to react properly and well controlled to page types unknown to its software as well as to variations in the flow of the received pages (Galileo ICD 2010). This is the reason why the process to accurately extract and correctly decode navigation data represents a fundamental task when developing GNSS receivers. The navigation data decoding is a process which cooperates with the baseband tracking. While the baseband tracking channels provide navigation data samples to the data decoding, data decoding provides information about data bit changes to tracking channel for allowing increasing the integration time. The aforementioned process is similar for both GPS and Galileo.

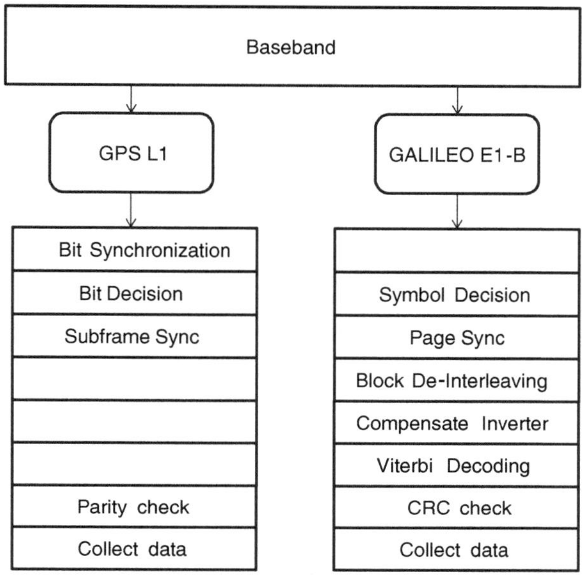

Fig. 7.2 Navigation data decoding tasks

In the rest of this chapter we describe the main steps needed for both systems (Fig. 7.2) like:

- GPS Bit Synchronization
- Bit/Symbol Decision
- Subframe/Page Synchronization
- Galileo Block De-Interleaving
- Galileo Compensate Inverter
- Galileo Viterbi Decoding
- Parity/CRC Check
- Collect Data

GPS Bit Synchronization

Finding the bit edges is a task mainly related to the GPS L1 signal. In GPS L1, the navigation data bit length is 20 ms, and the C/A-code length is 1 ms. In other words, the C/A-code repeats 20 times within a single navigation bit. In this scenario the task is to decide which of those 20 possible slots is the beginning of a data bit. With valuable signal strength conditions spotting the correct slot is a straightforward operation, since the sign of the samples only changes according to the data bit changes in multiples of 20 ms intervals. Different is the case of weak signal conditions, where the sign of the tracking sample may arbitrarily vary, making the decision more difficult. In such case we can observe many changes in the sign during the 20 ms period and a statistical method of detecting the correct bit edge is usually deployed.

A good example is provided in (Parkinson and Spilker 1996), where the overall process has 20 slots assigned. Each time a sign change is detected in the samples, the corresponding slot value gets incremented. Once time passes, the correct bit edge slot should have clearly the most increments. A threshold can be implemented in such a way, that once certain slot reaches the threshold, it is declared as the correct bit edge slot (Fig. 7.3).

Galileo E1-B signal does not have similar problem of determining the bit edges. In E1-B case the data bit length is 4 ms, and the code length is 4 ms as well resulting in an alignment of both navigation data bit and code, meaning that once code starts, a new data bit also starts. In this chapter we will assume that also for the Galileo implementation the tracking channel provides samples every 1 ms epochs. Thus, for E1-B data bit we get 4 samples. Further assumption is that every sample is now considered with a value of +1, or 0.

Bit/Symbol Decision

At this stage the receiver makes the decision of which data bit/symbol was received. The decision is based on the majority of samples. For GPS L1 we have 20 samples per bit. When adding these samples from the beginning of a data bit, the sum will be from 0 to 20. If the sum is greater or equal to 10, then the data bit is determined as +1. If the sum remains below 10, it is determined to be a 0 bit.

Galileo E1-B case is similar, with the only difference of having 4 samples per symbol. Hence, when these samples are added, the sum is from 0 to 4. If the sum is greater or equal to 2, then the data symbol is determined as +1. If the sum remains below 2, it is determined to be a 0 symbol. The determined data bits/symbols are then stored to memory for further processing. It is worth to mention that for Galileo

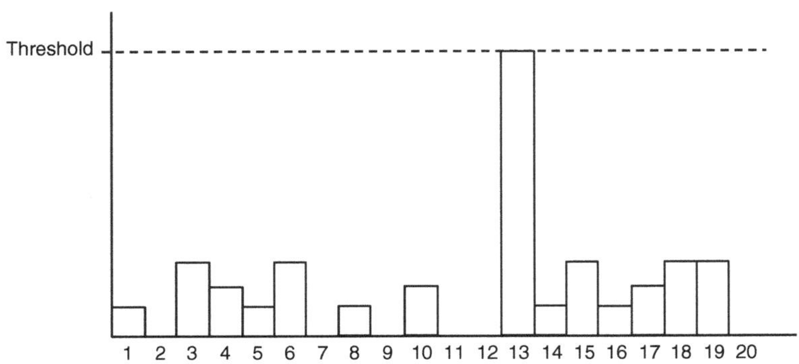

Fig. 7.3 Example of bit synchronization histogram

E1-B, the receiver may optionally utilize the tracking channels samples on a soft decision Viterbi decoder. In this case, also the Galileo E1-B samples need to be stored. Viterbi decoders are discussed later in this chapter.

Subframe/Page Synchronization

Once the received bits start flowing from the baseband, the next step represented by the synchronization of the frame structure of the specific GNSS system. The synchronization is based on finding a particular preamble word from the beginning of a subframe (GPS) or a page part (even or odd) (Galileo). This process has two problems:

1. Bit/symbol polarity is still uncertain. In addition to the normal preamble, we also need to search for an inverted preamble. Finding correct inverted preamble means that the data bits/symbols need to be inverted to their correct polarity.
2. Finding the correct preamble, or the inverted one, is not trivial. The preamble pattern might randomly appear within the data too, making the process bit more complicated. An easy solution is to check if the preamble pattern repeats with the subframe/page part rate. The subframe rate is 6 s for GPS L1, and page part rate is 1 s for Galileo E1-B. If the preamble appears to repeat with the subframe/ page part rate, it can be declared as correct preamble. In very unlucky condition the selection can still go wrong, but in these cases the error detection mechanisms will not pass, and the frame synchronization can be re-initiated.

The preambles for GPS L1 and Galileo E1-B are as it follows:
GPS L1: [1 0 0 0 1 0 1 1]
Galileo E1-B: [0 1 0 1 1 0 0 0 0 0]

Galileo Block De-interleaving

This section concerns Galileo E1-B data structure what happens once the whole page part has been received. For Galileo E1-B the page part it is represented by 250 symbols. A new feature introduced in the Galileo message structure is the presence of block interleaving. Block interleaving scrambles the message sent by the satellite. The message is then exposed to errors while propagating through the atmosphere to the receiver. These errors tend to appear in bursts. When the message is received, the receiver performs the opposite action, called de-interleaving, which reorders the original message, and scrambles the possible burst errors. The block interleaving, and its effect on error bursts is illustrated in Fig. 7.4. The Galileo block interleaver dimensions (n columns × k rows) are 30 × 8, for the 240 symbols of E1-B message (preamble is not included).

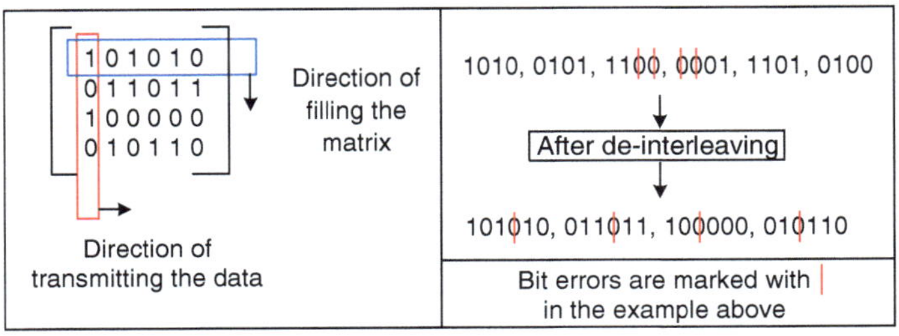

Fig. 7.4 The effect of block interleaving on bit error bursts

Galileo Compensate Inverter

Next step after Galileo de-interleaving is to compensate for the inverter effect on the G2 branch of the signal, as shown in Fig. 7.5. This is simply accomplished by inverting every second symbol of the received page part.

Galileo Viterbi Decoding

As mentioned in previous sections, Galileo E1-B utilizes block interleaving to reduce the amount of sequential errors in the message. This is beneficial since Galileo uses Forward Error Correction (FEC) encoding to the message. The convolutional coded message is decoded with a standard Viterbi decoder which is able to correct error bursts of length t.

$$t = \lfloor (d_{free} - 1)/2 \rfloor, \tag{7.22}$$

where d_{free} is maximum free distance of the code. For Galileo $d_{free} = 10$ (Proakis 1995). Based on the aforementioned equation, a Galileo receiver is able to correct error bursts of 4 bits. The implementation of Viterbi decoder has two options, either a hard decision or soft decision Viterbi decoder. A hard decision Viterbi decoder operates on the symbols which we determined earlier in this chapter. A soft decision Viterbi decode operates on the tracking loop samples. For mass market receivers the use of hard decision Viterbi decoder is considered sufficient. It is less demanding from the implementation point of view, and still provides excellent performance.

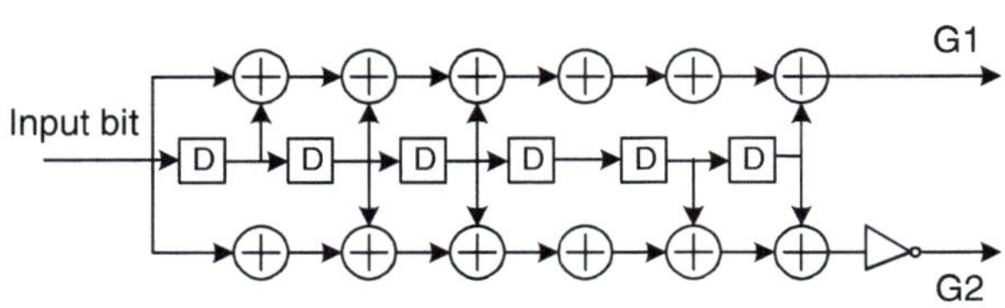

Fig. 7.5 Galileo convolutional coding scheme

A page part consists of 250 symbols, out of which 10 symbols are for the preamble pattern. The preamble pattern is not encoded, but the remaining 240 symbols must be decoded. From the practical point of view, it is not feasible to implement a Viterbi decoder directly for a 240 symbols. Instead, literature recommends using the Viterbi decoder in block mode that operates on lengths of 5x constraint length. Galileo constraint length is 7, so length of 35 would suffice. Since 240 is not a multiple of 35, a slight modification to the decoder specification is recommended. The closest options are to use depth of 30 or 40 for the decoder. In this way the decoder does not have to process any partial data blocks. Choosing depth of 30 is cheaper to implement, but might slightly underperform in weak signal conditions. On the other hand choosing depth of 40 is certainly sufficient, but also more expensive to implement. The Galileo Viterbi decoder always starts from zero state. This is also the reason why every Galileo page part has six zero bits at the end, called the tail bits. These tail bits enable the decoder to run continuously without explicitly resetting the states.

Parity/CRC Check

Galileo message structure includes block interleaving and FEC for correcting several bit errors. The benefits are obvious, but it also comes with a price: the receiver has to perform more computational work for these features. In the very low signal strength situation some errors can still remain. To handle the remaining errors, Galileo utilizes Cyclic Redundancy Check (CRC), and GPS utilizes parity bits, to notice these remaining errors.

- Galileo CRC

Galileo message structure has a CRC checksum included into the navigation data, which acts as a final check that no bit errors are present in the decoded data. The CRC checksum is calculated without the synchronization bit pattern and the tail bits. The 24 bit CRC for Galileo is specified by:

Generator polynomial: $G(X) = (1 + x) * P(X)$, where

$$P(X) = X^{23} + X^{17} + X^{13} + X^{12} + X^{11} + X^{9} + X^{8} + X^{7} + X^{5} + X^{3} + 1$$

R(X) is defined as the remainder of G(X) divided by $m(X) * X^{24}$, where $m(X) = m_k + m_{k-1}X + m_{k-2}X^2 + \ldots + m_1X^{k-1}$, and $m_1, m_2, \cdots, m_k$ represent the Galileo information bits.

The final CRC checksum will be composed of the coefficients of R(X). Implementation of the CRC algorithm is easily done utilizing the xor-operator. For a received bit sequence of m_i, the multiplication with X^{24} means in practice extending the sequence with extra zeros.

The first step is to start applying the xor-operation from the left bitwise with G (X) and m(X). After each round the G(X) is shifted one step to the right, until it

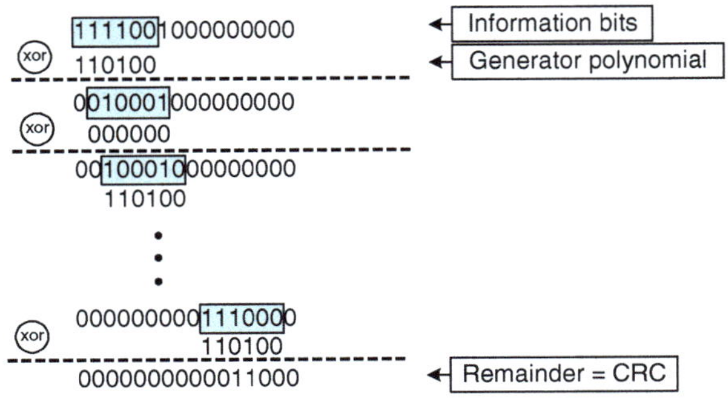

Fig. 7.6 Example of the CRC algorithm

reaches the end of m(X). If m(X) has a zero in the leftmost place being xor'ed with G(X), replace G(X) with zero sequence for that round. When the algorithm stops, all that is left will be the remainder R(X), which is the CRC checksum. This value will be compared to the received value and a decision is made whether CRC check pases or fails. Figure 7.6 illustrates the CRC algorithm with an arbitrary example (Ramabadran and Gaitonde 1988).

- GPS Parity Check

The GPS parity check enables the receiver to detect errors in the received message structure. Every GPS subframe is 300 bits long, and is divided into shorter words consisting of 30 bits, as shown in Fig. 7.7. Within each word, bits 1–24 are considered as data bits, while bit 25–30 are parity bits.

The actual parity check algorithm operates on complete words. It should also be noted that the algorithm requires two last bits, i.e. bits 29 and 30, from previous word. The algorithm itself is quite straightforward, and is implemented based on the equations presented in Table 7.8.

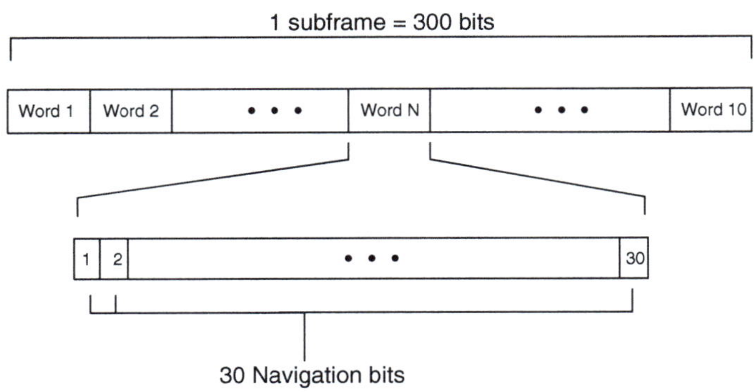

Fig. 7.7 The structure of a GPS subframe

Table 7.8 The GPS parity check equations

$D_1 = d_1 \oplus D_{30}$
$D_2 = d_2 \oplus D_{30}^*$
.
.
.
$D_{24} = d_{24} \oplus D_{30}^*$
$D_{25} = D_{29}^* \oplus d_1 \oplus d_2 \oplus d_3 \oplus d_5 \oplus d_6 \oplus d_{10} \oplus d_{11} \oplus d_{12} \oplus d_{13} \oplus d_{14} \oplus d_{17} \oplus d_{18} \oplus d_{20} \oplus d_{23}$
$D_{26} = D_{30}^* \oplus d_2 \oplus d_3 \oplus d_4 \oplus d_6 \oplus d_7 \oplus d_{11} \oplus d_{12} \oplus d_{13} \oplus d_{14} \oplus d_{15} \oplus d_{18} \oplus d_{19} \oplus d_{21} \oplus d_{24}$
$D_{27} = D_{29}^* \oplus d_1 \oplus d_3 \oplus d_4 \oplus d_5 \oplus d_7 \oplus d_8 \oplus d_{12} \oplus d_{13} \oplus d_{14} \oplus d_{15} \oplus d_{16} \oplus d_{19} \oplus d_{20} \oplus d_{22}$
$D_{28} = D_{30}^* \oplus d_2 \oplus d_4 \oplus d_5 \oplus d_6 \oplus d_8 \oplus d_9 \oplus d_{13} \oplus d_{14} \oplus d_{15} \oplus d_{16} \oplus d_{17} \oplus d_{20} \oplus d_{21} \oplus d_{23}$
$D_{29} = D_{30}^* \oplus d_1 \oplus d_3 \oplus d_5 \oplus d_6 \oplus d_7 \oplus d_9 \oplus d_{10} \oplus d_{14} \oplus d_{15} \oplus d_{16} \oplus d_{17} \oplus d_{18} \oplus d_{21} \oplus d_{22} \oplus d_{24}$
$D_{30} = D_{29}^* \oplus d_3 \oplus d_5 \oplus d_6 \oplus d_8 \oplus d_9 \oplus d_{10} \oplus d_{11} \oplus d_{13} \oplus d_{15} \oplus d_{19} \oplus d_{22} \oplus d_{23} \oplus d_{24}$
where,
$d_1, d_2, \ldots, d_{24}$ are source data bits;
the symbol * identifies the last 2 bits from previous subframe;
$D_{25}, D_{26}, \ldots, D_{30}$ are computed parity bits;
$D_1, D_2, \ldots, D_{30}$ are bits transmitted by the SV;
the symbol $\oplus$ is modulo-2 operator;

The last step of the algorithm is to check whether the calculated parity bits match to the received parity bits. When there is not any parity mismatch, the parity check passes and the word can be processes. If parity mismatches appear, the whole word will be discarded.

Data Collection

Once the navigation data message has passed through the error detection mechanisms, it can be safely adopted. For Galileo, one remaining issue is to combine even and odd page parts into complete messages. The receiver then processes pages for Galileo, and subframes for GPS. The layout of the page/subframe structures can be found in the corresponding GPS and Galileo ICDs, where the fundamental parameters can be listed as:

- Timing information, Time-Of-Week (TOW)
- Ephemeris data
- Clock/Ionosphere correction terms
- Health/status parameters
- Optional data, e.g. Almanac data

Such parameters usually remain constant for several hours (except time), so it is not necessary to update the information constantly. For this reason an Issue Of Data (IODE) parameter is provided. If the IODE value changes, it indicates that new ephemeris data is available and the receiver should proceed with updating the parameters. After successfully receiving all the necessary data for several satellites, the receiver can proceed with navigation solution calculation.

Conclusions

In this book chapter the procedure for obtaining navigation solution for GPS/Galileo dual system receiver was described. The procedure follows a similar approach that a normal single system GNSS would do. The dual system availability adds few details that must be taken into account when developing the dual system receiver software. The details found earlier in this chapter will be listed in the next paragraph.

The first decision with dual system navigation software was to make a decision whether the receiver wants to synchronize to GPS or Galileo time. The timing synchronization and pseudorange construction varied between GPS and Galileo with the C/A code lengths, navigation data bit length and with the measurements from the baseband. Once the receiver has synchronized to either GNSS system and first time bias correction is available from PVT, the receiver has access to both GPS/Galileo system times via the GGTO parameter.

Pseudorange measurements are performed with similar procedure for both systems. Due to the different code lengths and TOW structure, the pseudoranges must be measured with correct parameters.

GPS and Galileo satellites have slightly different orbits, but both system orbits are defined with the same set of Keplerian parameter, provided by the navigation data payload. Most of the parameters have exactly the same definition, but some parameters have differences in the amount of bits needed, and in scale factors. Also some constants defined in the respective ICDs have different values. These differences have been highlighted in the chapter describing the process of calculating satellite position.

Atmospheric corrections are provided differently for the ionosphere part, where GPS has it's Klobuchar parameters and Galileo has NeQuick parameters. Actually, if the parameters are available, the receiver can handle both systems with either model. Troposphere correction is handled with one common algorithm for both systems.

Navigation solution with the LS algorithm needs slight modification to normal single system operation. Due to both GPS and Galileo using their own time systems, time bias for these has to be solved. One can choose between two approaches, where first option is to solve the GGTO from parameters sent in the navigation data. With the GGTO, measurements from both systems can be represented in same time domain. The other option for time differences is to add 5th unknown to the LS algorithm, namely an additional time bias, which will be solved. This approach needs slightly more processing from the receiver due to the increased matrix sizes in the algorithm. Moreover, the required amount of satellites to produce a PVT is increased from 4 to 5.

Last concluding note is that a dual system receiver must include both time biases into the calculation of GDOP values.

References

Galileo ICD (2010) Galileo open service: signal In: Space interface control document (OS SIS ICD). European Space Agency/European GNSS Supervisory Authority

GPS ICD (2004) Navstar global positioning system. Interface specification, IS GPS-200, Revision D

Kaplan ED, Hegarty CJ (2006) Understanding GPS—principles and applications (2nd edn). Artech House, Norwood

Misra P, Enge P (2004) Global positioning system: signals, measurements, and performance. Ganga-Jamuna Press, Lincoln

Parkinson BW, Spilker JJ (eds) (1996) Global positioning system: theory and applications, vol 1. American Institute of Aeronautics and Astronautics (AIAA), New York

Proakis JG (1995) Digital communications (3rd edn). McGraw-Hill, New York. ISBN 0-07-051726-6

Ramabadran TV, Gaitonde SS (1988) A tutorial on CRC computations. IEEE Micro 8:62–75

Syrjärinne J (2001) Studies of modern techniques for personal positioning. Ph.D. dissertation, Tampere University of Technology

Chapter 8
Hybridization with Localization Information from Wireless Communications Systems

Stephan Sand

Abstract The fastest growing market segment for Galileo and other global navigation satellite system (GNSS) receivers are mobile phones. Users want to experience seamless location based services (LBS) no matter if they use them outdoors, in cities or indoors. To achieve this, hybridization of GNSS navigation signals with localization information from wireless communication systems is essential. Modern mobile phones are equipped today with several wireless communciation technologies that can provide localization information. In this chapter, we discuss localization information from wireless communication systems, e.g., proximity information, received signal strength, time of arrival, fingerprinting, and assisted GNSS. We further evaluate hybridization methods for GNSS and localization information from wireless communication systems, e.g., Gauss-Newton algorithm and extended Kalman filter. Additionally, we consider cooperative hybridization and error mitigation through hybridization. To conclude, hybridization of GNSS with localization information from wireless communication system is necessary for a seamless user experience of LBS.

Introduction

Market studies predict that in 2015 900 million GNSS devices will be shipped (GSA 2010). From these 900 million GNSS devices, 750 million will be GNSS-enabled mobile phones. Mobile users will request various LBS with these mobile phones. For instance, they may be interested in mobile and location aware advertisement of discounts or special offers, social networking to find recommendations about good restaurants or available entertainment, mobile gaming to spend spare or waiting time, or navigation and route planning.

Most LBS will be requested in city centers with under ground public transport, large shopping centers, or at airport terminals. These will be critical scenarios for

S. Sand (✉)
German Aerospace Center (DLR), Oberpfaffenhofen, Germany
e-mail: Stephan.Sand@dlr.de

J. Nurmi et al. (eds.), *GALILEO Positioning Technology*,
Signals and Communication Technology 182, DOI 10.1007/978-94-007-1830-2_8

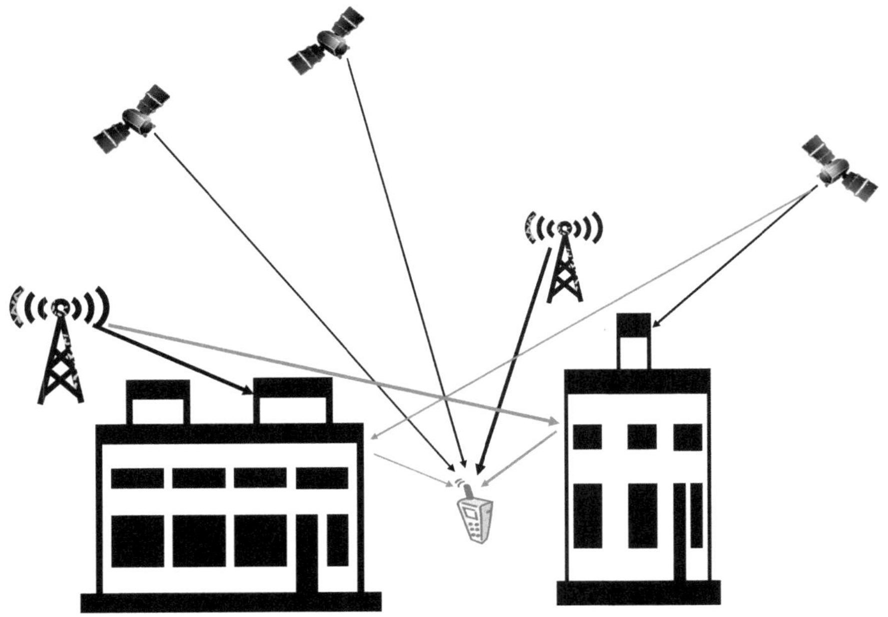

Fig. 8.1 Urban navigation scenario with LOS and NLOS satellites and base stations; received signal strength is indicated through thickness of lines

GNSS. Severe multipath and non-line-of-sight (NLOS) propagation, the low received signal power and the resulting poor satellite constellation will lead to large position errors or even to the unavailability of LBS, e.g., as can be seen in Fig. 8.1.

In contrast, communication systems are designed to provide good coverage for mobile communication in urban and indoor environments as most users exploit here voice and data services. Modern smartphones, which are expected to be the main enablers of LBS, are and will be equipped with a multitude of wireless communication systems. These include the Global System for Mobile Communication (GSM), Universal Mobile Telecommunication System (UMTS), and the 3rd Generation Partnership Project (3GPP) Long Term Evolution (LTE), Wireless Local Area Networks (WLANs) based on the IEEE 802.11 standard, Bluetooth, and Near Field Communication (NFC). Additionally, cellular communication systems such as GSM, UMTS, and 3GPP LTE support assisted GNSS (A-GNSS) for mobile phones to achieve a time-to-first-fix of the GNSS receiver within seconds or acquisition and tracking of GNSS signals at low received power.

In principle all of the above communication systems and technologies are suitable to provide location information. Location information may include proximity information, received signal strength (RSS), time of arrival (TOA), time-difference of arrival (TDOA), angle of arrival (AOA), or fingerprint measurements from access points (APs) or base stations (BSs). The location of these APs or BSs can be known precisely or partially and in some cases may be even unknown, which requires further processing such as simultaneous localization and mapping (SLAM) (Bailey and Durrant-Whyte 2006).

Although wireless communication systems can provide location information, their primary purpose is efficient communication over the air interface. Hence, these systems are optimized for communication and often made robust against impairments that will affect the accuracy of location information. For instance, pilot signals are not continuously transmitted to increase the data throughput. Further, in orthogonal frequency division multiplexing (OFDM) employed by 3GPP-LTE and WLANs, several hundred data bits or even thousands of them can be transmitted per OFDM symbol in 1 ms in contrast to Galileo, where a bit spreads over 4 ms. Also, most communication systems do not require the BS or AP transmitters to be synchronized in contrast to GNSS transmitters. Further, mobile phones employ non-ideal oscillators to save costs and receiver algorithms that can cope with these oscillators and coarse synchronization. In the case of OFDM, this can translate to synchronization inaccuracies of several samples, i.e., 100 m or more. The location of BS or AP transmitters as reference points is optimized to provide best possible coverage and data throughput often resulting in poor dilution of precision (DOP). As the modern third generation (3G) and upcoming fourth generation (4G) communication systems such as 3GPP LTE and 3GPP LTE-Advanced aim at a frequency reuse of one to achieve high spectral efficiency, 3G and 4G systems have to cope with inter-cell interference (ICI). ICI at the cell edge increases estimation errors and thus decreased the accuracy of location information (Mensing et al. 2009). On the other hand in the cell center, the frequency reuse of one significantly reduces the probability of detecting non serving BSs. Consequently, the location information of the mobile phone becomes very crude, e.g., the position of the BS or an arc spanning several hundreds meters. Besides the above impairments, cellular communication signals are affected in the same way by multipath and NLOS propagation as GNSS signals are (cf. Fig. 8.1).

Thus, hybridization of GNSS with localization information from communication system is necessary to overcome the impairments of each system and to achieve reliable and accurate position estimation. This in turn will increase the availability and quality of LBS enabling new emerging services with new requirements and revenue opportunities for smartphone manufacturers, application developers and mobile network operators.

Wireless Communication Measurements for Localization Information

In this section, we present different location dependent measurements of communication systems to obtain localization information. Generally, we can distinguish between five types of measurements:

1. Proximity information, e.g., cell identity (ID) for cellular communication systems or medium access control (MAC) address of WLAN APs
2. Distance dependent RSS, TOA, or TDOA measurements
3. Direction dependent AOA measurements
4. Fingerprinting of location dependent information such as the previous measurements
5. A-GNSS

These measurements will be explained in detail in the sequel.

Proximity Information

All wireless communication systems support signal acquisition and tracking through synchronization algorithms to establish a communication link. Thus, as soon as communication is established between a mobile station and a fixed station,[1] the mobile station has obtained the proximity information by identifying the fixed station through a unique number. For instance, the cell ID of BSs in cellular networks or the MAC address of a WLAN AP are such unique numbers. In order to translate proximity information into localization information, the mobile station needs to obtain at least the position of the fixed station. The position can be communicated by the fixed station or obtained through a fingerprint database. Then, the simplest method to determine the mobile station's position is to use the fixed station's position as the mobile station's position estimate. Alternatively, the mobile station's position is estimated at the center of gravity of the fixed station's coverage area (Figueira and Frattasi 2010). If the mobile station detects more than one fixed station, it can use the center of gravity of the fixed stations' joint coverage area as its position estimate as shown in Fig. 8.2. Further, if available from a data base, the mobile station may exploit statistics about the most likely position given the detected fixed stations.

The benefits of proximity information compared to the other types of measurements are that proximity information is obtained quickly, has the lowest computational complexity, requires the least amount of communication, consumes the smallest amount of battery power, and is simplest to implement in the fixed stations. However, this comes at the drawback of resulting in the least accurate localization information. The accuracy of proximity information strongly depends on the coverage area of the fixed stations. Whereas this information can be relatively accurate for NFC and radio-frequency identification (RFID) tags, in the order of meters, it can be very inaccurate for cellular communication systems, in the order

[1] The term fixed station is generally used to include BSs, APs or any other fixed transmitters.

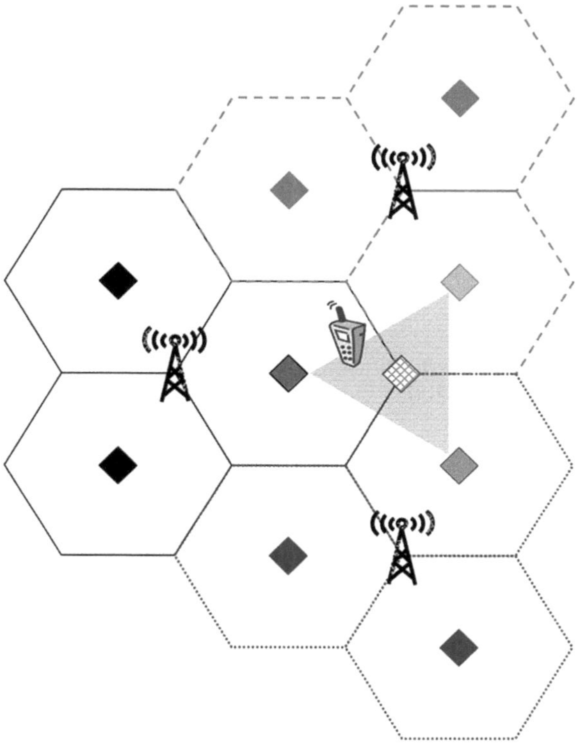

Fig. 8.2 Proximity based location information and center of gravities (*diamond markers*) for a cellular network with three hexagonal sectors per fixed station; mobile station identified three cells; position estimate of mobile station at *diamond marker* with crisscross pattern

of several hundred meters. Note that proximity based positioning is standardized in most cellular communication systems as a default positioning method, e.g., in GSM (3GPP TS 43.059 2011), UMTS (3GPP TS 25.305 2011), and LTE (3GPP TS 36.305 2011).

Received Signal Strength

Besides the proximity to a fixed station, many communication systems and standards support RSS measurements, e.g., Bluetooth (Bluetooth 4.0 2010), WLAN (IEEE 802.11n 2009), GSM (3GPP TS 45.008 2011), UMTS (3GPP TS 25.215 2011), and LTE (3GPP TS 36.214 2011). In communication systems, the RSS measurements are mainly used to adapt modulation and coding schemes to the link quality indicated by these measurements. The RSS measures the received signal power from a transmitter. The RSS depends on the distance between the receiver and transmitter as well as on the propagation environment (Rappaport 2002). Hence, RSS measurements are suitable location information from wireless communication systems.

The RSS $P_{\mathrm{R}}(r_i)$, which is depending on the distance r_i between a mobile station and a fixed station i, can be defined as

$$P_{\mathrm{R}}(r_i) = P_{\mathrm{T}} - PL(r_i). \tag{8.1}$$

Here, P_{T} and $PL(r_i)$ describes the transmit power and the distance and environment dependent path loss. We define the distance r_i between a mobile and fixed station as

$$r_i = \left\|\mathbf{x} - \mathbf{x}_{\mathrm{FS},i}\right\|_2 = \sqrt{\left(x - x_{\mathrm{FS},i}\right)^2 + \left(y - y_{\mathrm{FS},i}\right)^2}, \tag{8.2}$$

where $\mathbf{x} = (x, y)^{\mathrm{T}}$ and $\mathbf{x}_{\mathrm{FS},i} = \left(x_{\mathrm{FS},i}, y_{\mathrm{FS},i}\right)^{\mathrm{T}}$ denote the two-dimensional position of the mobile station and the fixed station i. Often, the path loss is modeled as (Rappaport 2002)

$$PL(r_i) = \overline{PL}(r_0) + 10n \log_{10}\left(\frac{r_i}{r_0}\right) + X_\sigma, \tag{8.3}$$

where $\overline{PL}(r_0)$ denotes the average path loss at a reference distance r_0 from the fixed station, n the path loss exponent, and X_σ is a zero-mean Gaussian distributed random variable (in dB) with standard deviation σ (also in dB). The path loss exponent n depends on the specific propagation environment and may vary between 1.6 and 6 (Rappaport 2002). For free space propagation, n is 2. X_σ models log-normal distributed shadow fading, i.e., shadowing effects that occur due to large obstacles in the surrounding environment of the receiver such as buildings, hills, etc. These shadowing effects cause large RSS variations for the same distance r_i between the fixed station i and the mobile station at different measurement locations.

With several RSS measurement, the mobile station can estimate the distance r_i to the fixed station given the path loss model in (8.3) as

$$\hat{r}_i = r_0 \cdot 10^{\frac{\overline{\widehat{PL}(r_i)} - \overline{PL}(r_0)}{10n}} = r_0 \cdot 10^{\frac{\frac{1}{N}\sum_{k=1}^{N} \widehat{P_R(r_i)} - P_T - \overline{PL}(r_0)}{10n}}. \tag{8.4}$$

Here, $\overline{\widehat{PL}(r_i)} = \frac{1}{N}\sum_{k=1}^{N} \widehat{P}_R(r_i) - P_T$ denotes the average estimated path loss. With averaged RSS measurements from three different fixed stations, the mobile station can use trilateration to estimate its two-dimensional position, i.e., the mobile station can solve the non-linear equation for its position $\mathbf{x}$

$$\mathbf{x} = f^{-1}(\hat{\mathbf{r}}) = f^{-1}\left(\begin{bmatrix} \hat{r}_1 \\ \hat{r}_2 \\ \hat{r}_3 \end{bmatrix}\right), \tag{8.5}$$

where the distance estimate $\hat{r}_i$ and the functional relation between $\hat{r}_i$ and $\mathbf{x}$ is given by (8.4) and (8.2). Note that the mobile station can estimate its three-dimensional position with averaged RSS measurements from four different fixed stations.

Theoretical analysis, e.g., Cramér-Rao lower bound (CRLB) evaluations (Qi et al. 2006), demonstrate that the accuracy of RSS based positioning methods can be very accurate for short range positioning and LOS propagation. However, RSS based positioning methods are not suitable for long range positioning as the CRLB depends on the path loss exponent n and the random variable X_σ of the shadow fading in (8.3). For instance, the influence of X_σ is neglected in (8.4).

As explained, RSS based positioning is supported by most wireless communication technologies although it is usually not included in the standards. An exception to this is the enhanced Cell-ID method standardized in 3GPP (3GPP TS 36.305 2011), which can use RSS measurements. In the remainder of this section, we focus on RSS based positioning in WLANs (GRAMMAR D1.4 2009). In applications targeted for mass market products, the availability of RSS measurements is more probable, as the mobile station can obtain them by passive scanning of WLAN beacon frames, which WLAN APs emit periodically. In addition, in many mobile stations, such as smartphones, tablet computers, and laptop computers, RSS measurements are easily available through application programming interfaces of their standard WLAN services.

In trilateration systems (cf. 8.4 and 8.5), path loss models of radio signals are used to translate RSS measurements to distances between the receiver and APs (GRAMMAR D1.4 2009). As in proximity based positioning methods, the mobile station needs prior information about the MAC addresses and locations of APs. In indoor environments, multipath and attenuation caused by walls, other structures, and even people complicate the modeling of signal propagation. This makes the simple path loss models (cf. 8.3) too inaccurate in many real life situations. To overcome this problem, the performance of triangulation can be enhanced using other models, such as pattern matching, probabilistic filtering approaches or fingerprinting.

Time of Arrival, Round Trip Delay, and Time Difference of Arrival

Compared to proximity or RSS based location information, TOA or TDOA based location information puts more stringent requirements on the communication system. For instance, both TOA and TDOA require that the fixed stations transmit synchronous signals, as is the case of GPS and Galileo, which are TOA-based systems. Alternatively, timing offsets between the transmissions of different fixed stations need to be approximately constant, measured and communicated throughout the network. Hence, TOA and TDOA based localization is often included in the standards, e.g., ultra-wideband (UWB) (IEEE 802.15.4a 2007), GSM (3GPP TS 43.059 2011), UMTS (3GPP TS 25.305 2011), or LTE (3GPP TS 36.305 2011). Nevertheless, some commercial products implement proprietary TOA or TDOA based localization information, e.g., (Nanotron Technologies 2011).

To measure TOAs or TDOAs, the mobile station synchronizes to the received signals of fixed stations mainly using reference signals that are known to the mobile station. For instance, LTE provides primary and secondary synchronization signals as well as cell-specific and positioning reference signals for time synchronization (3GPP TS 36.211 2011). The mobile station can use different time synchronization algorithms (cf. Chaps. 8 and 9, GRAMMAR D1.4 2009; WHERE D3.1 2009) to estimate the TOA. With the TOA estimates and a received system reference time, the mobile station can construct pseudo-ranges ρ_i similar to GNSS in Chap. 11.

In the following we assume that the fixed stations are all synchronized to a system time of the mobile radio network. Given the distance r_i between a mobile station and the fixed station i in (8.2), we can then define the pseudorange estimate $\hat{\rho}_i$ as

$$\hat{\rho}_i = ct_i + ct_{\text{MS}} + n_i = r_i + b + n_i = \sqrt{\left(x - x_{\text{FS},i}\right)^2 + \left(y - y_{\text{FS},i}\right)^2} + b + n_i, \quad (8.6)$$

where c, t_i, t_{MS}, n_i, and b denote the speed of light, the actual TOA of the i–th fixed station signal, the mobile station's clock offset relative to the system time of the fixed stations, the noise term combining all residual errors after correction, and the clock bias $b = ct_{\text{MS}}$. The position of the mobile station can be uniquely determined in two dimensions with four pseudorange estimates from four different fixed stations and the non-linear relation between the pseudorange estimates and the mobile station position in (8.6) as shown in Fig. 8.3. Although, there are only three unkowns, x, y, b, the equation system of (8.6) is non-linear.

Hence, the equation system requires four pseudorange estimates to obtain a unique and analytic solution in two dimensions without any additional assumptions

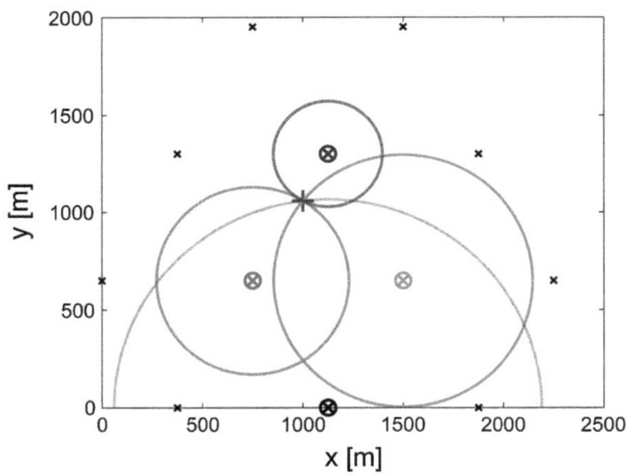

Fig. 8.3 Two-dimensional TOA-based pseudorange measurements and positioning: four measurements from four fixed stations (*encircled x markers*); mobile station at intersection of four circles; note that four measurements are necessary to obtain a unique and analytic solution of the position estimate (Bucher and Misra 2002)

(Bucher and Misra 2002). Note, when the mobile station knows that its position is within the area of a triangle spanned by three of the four fixed stations, three pseudorange estimates from these three different fixed stations suffice to determine its position uniquely.

Similar for three dimensions, five pseudorange estimates from five different fixed stations are needed to calculate the position of the mobile station (Bucher and Misra 2002). With the knowledge of the mobile station position being within a tetrahedron spanned by four of the fixed stations, four pseudorange estimates from these four fixed stations are enough to determine the position of the mobile station uniquely (cf. Fig. 8.4). An example of how the clock bias b creates uncertainty in the pseudorange measurement is shown in Fig. 8.4a. The radii of the two concentric spheres differ due to the clock bias by approximately 7 ns corresponding to 2 m uncertainty.

If the fixed stations in the mobile radio network are not synchronized location measurement units (LMUs) can estimate the clock offsets between different fixed stations and communicate this information over the mobile radio network, e.g., in

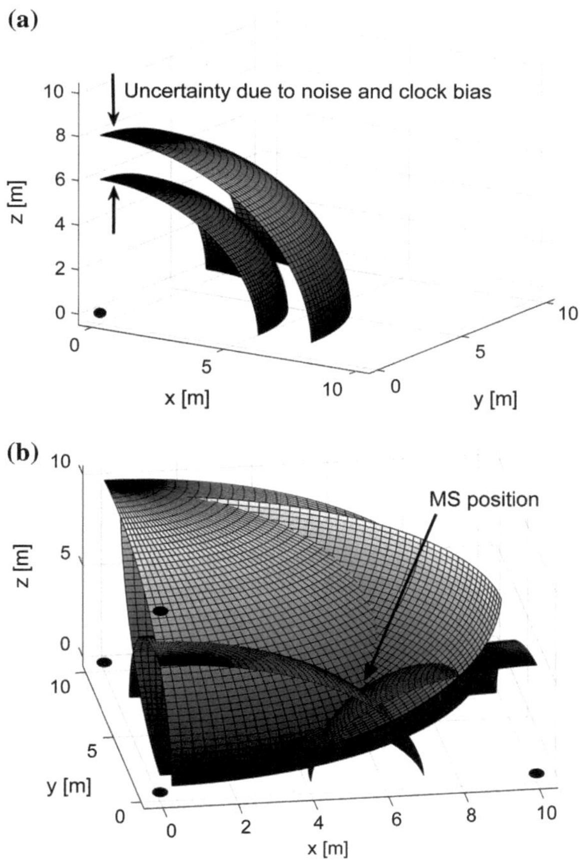

Fig. 8.4 Three-dimensional TOA-based pseudorange measurements and positioning: four fixed stations (*black points*); mobile station at intersection of spheres **a** uncertainty due to noise and clock bias of TOA measurement, **b** TOA-based pseudorange positioning

GSM (3GPP TS 43.059 2011), UMTS (3GPP TS 25.305 2011), or LTE (3GPP TS 36.305 2011).

TDOAs are another time based measurement, which do not require clock synchronization between the mobile station and the fixed stations. A TDOA between the fixed station i and the reference fixed station 0 measured at the mobile station is given by

$$
\begin{aligned}
\hat{d}_{i,0} &= \hat{\rho}_i - \hat{\rho}_0 = r_i + b + n_i - (r_0 + b + n_0) = r_i - r_0 + n_i - n_0 \\
&= \sqrt{\left(x - x_{\mathrm{FS},i}\right)^2 + \left(y - y_{\mathrm{FS},i}\right)^2} - \sqrt{\left(x_{\mathrm{FS},i} - x_{\mathrm{FS},0}\right)^2 + \left(y_{\mathrm{FS},i} - y_{\mathrm{FS},0}\right)^2} + n_{i,0} \\
&= d_{i,0} + n_{i,0}.
\end{aligned}
\tag{8.7}
$$

Here, r_i and r_0 denote the distance between the mobile station and fixed station according to (8.2). b, n_i, and n_0 are the clock bias between the mobile station and the system time of the fixed stations and the noise terms combining all residual errors after correction as in (8.6). To obtain a TDOA measurement, the mobile station measures two TOA-based pseudoranges and computes the delay difference between these two measurements (cf. 8.7). Compared to TOA-based pseudorange measurements according to (8.6), TDOA measurements have the benefit that the mobile station clock bias is eliminated from the measurement in (8.7). However, this is achieved by correlating the noise between all TDOAs with respect to the reference fixed station 0.

Whereas the possible locations of the mobile station given a TOA-based pseudorange result in circles (cf. Fig. 8.3), the possible locations of the mobile station given TDOAs result in hyperbolas (cf. Fig. 8.5). Comparing Figs. 8.3 and 8.5, we see that the TOA and TDOA intersection results in the same position estimates. However, removing the fixed station at (1100, 1400 m), we can see that the intersection of the remaining TDOAs occurs under a smaller angle compared to the remaining TOAs in Fig. 8.3. Thus, we would intuitively expect that TDOAs are more sensitive to noise in the y-direction for this example.

According to (Urruela et al. 2006), the CRLB of TDOA based positioning and TOA-based pseudo-range positioning is strictly the same. Hence, one can expect that the positioning accuracy for both methods is the same. Nevertheless, results may differ due to numerical accuracy and computational complexity.

TDOA measurements are mainly used in the cellular mobile communication systems such as GSM (3GPP TS 43.059 2011), UMTS (3GPP TS 25.305 2011), LTE (3GPP TS 36.305 2011), and Worldwide Interoperability for Microwave Access (WiMAX, IEEE 802.16 2009).

Instead of the above TOA or TDOA methods, the communication system can employ two-way ranging for non-synchronized fixed stations. For instance, UWB communication systems according to (IEEE 802.15.4a 2007) employ two-way ranging measuring the round-trip-delay (RTD) between the fixed station i and the mobile station. In this case, we define the RTD estimate $\widehat{RTD}_i$ as

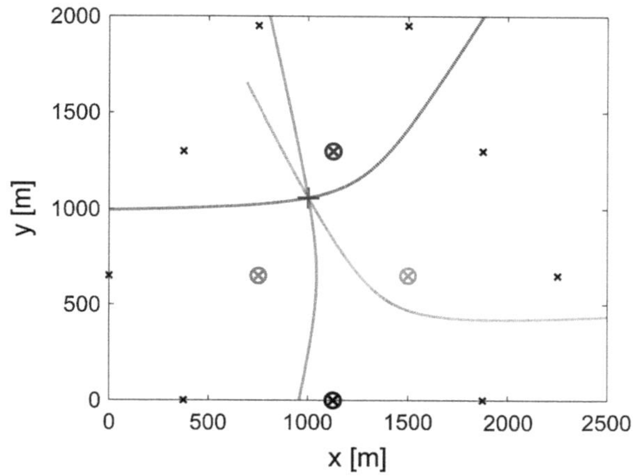

Fig. 8.5 Two-dimensional TDOA measurements and positioning: three measurements from four fixed stations (*encircled x markers*); mobile station at intersection of hyperbolas; note that three TDOA measurements are necessary to obtain a unique and analytic solution of the position estimate (Bucher and Misra 2002)

$$\begin{aligned}\widehat{RTD}_i &= \frac{ct_{\mathrm{RTD},i} - ct_{\mathrm{PROC}}}{2} + n_i = r_i - \frac{ct_{\mathrm{PROC}}}{2} + n_i \\ &= \sqrt{\left(x - x_{\mathrm{FS},i}\right)^2 + \left(y - y_{\mathrm{FS},i}\right)^2} - \frac{ct_{\mathrm{PROC}}}{2} + n_i,\end{aligned} \tag{8.8}$$

where $t_{\mathrm{RTD},i}$ and t_{PROC} are the RTD and the processing time. The mobile station needs t_{PROC} to process the initial, received ranging signal from the fixed station and to transmit an acknowledgement ranging signal upon the received ranging signal. The processing time t_{PROC} can be several milliseconds, whereas the estimated signal propagation time $\widehat{RTD}_i/c$ is only several nanoseconds (Nanotron Technologies 2011). Hence, it is critical that the processing time t_{PROC} is known very accurately. Another concern can be the clock drift of the fixed station i, which can be compensated by symmetric double-sided two-way ranging (Nanotron Technologies 2011). Given several RTD measurements, the position can be uniquely obtained with 3 RTD measurements in two-dimensions and 4 RTDs in three-dimensions. Thus, RTDs require one measurement less than TOA pseudorange to determine the position of the mobile station uniquely when there are no assumptions about the geometry of the mobile and fixed stations.

Angle of Arrival

All modern mobile high speed communication systems such as UMTS, LTE, LTE-Advanced, WiMAX, and WLAN will employ multiple-input multiple-output

(MIMO) transmission schemes to achieve the highest possible data rates. These techniques require transmit and receive antenna arrays for the fixed and mobile stations. For instance, LTE considers 4×4 MIMO and LTE-Advanced 8×8 MIMO transmission. Hence, it will become practical in the near future for mobile stations to estimate the AOA of transmit signals from fixed stations. Concerning UMTS, LTE, and LTE-Advanced, 3GPP has standardized AOA measurements as one technology to obtain location information from mobile cellular communication systems (3GPP TS 25.305 2011) and (3GPP TS 36.305 2011).

To measure AOAs, directional antennas need to be employed at the fixed stations. This subject has been extensively treated in the literature on radars and smart antennas, e.g., (Sarkar et. al. 2003). As for UMTS and LTE, we consider an up-link transmission, i.e., mobile station to fixed station communication. In this case, we can assume that the orientations of the fixed station antenna arrays are known. Further, as the fixed stations do not have any power or space limitations compared to the mobile stations, larger directional antennas can be employed that provide high horizontal and vertical angular resolution.

Given the distance r_i between mobile station and fixed station i, the estimated AOA ϕ_i in the horizontal direction and θ_i in the vertical direction at the directional antenna array of fixed station i, we have the following relation between the position of the mobile station and the estimated AOAs (cf. Fig. 8.6)

$$\begin{pmatrix} x - x_i \\ y - y_i \\ z - z_i \end{pmatrix} = r_i \begin{pmatrix} \cos(\phi_i)\cos(\theta_i) \\ \sin(\phi_i)\cos(\theta_i) \\ \sin(\theta_i) \end{pmatrix}. \tag{8.9}$$

To eliminate the radius r_i, we can reformulate (8.9), (8.10) to

$$\begin{aligned} (y - y_i) &= \tan(\phi_i)(x - x_i) \\ \cos(\phi_i)(z - z_i) &= \tan(\theta_i)(x - x_i). \end{aligned} \tag{8.10}$$

For each fixed station i, we can measure the horizontal and vertical AOAs ϕ_i and θ_i to obtain the two equations in (8.9) and (8.10). Note that (8.9) and (8.10) gives a linear set of equations for the mobile station position (x, y, z). Two horizontal AOA measurements as in Fig. 8.6 determine planes that intersect in a vertical line in the mobile station position. With one additional vertical AOA measurement the mobile station position can be determined analytically in three dimensions.

If the mobile station would determine its position based on AOA measurements, we would have to additionally estimate the unknown orientation of the mobile station, i.e., the angles ϕ and θ. In this case, (8.9) and (8.10) becomes

$$\begin{aligned} (y - y_i) &= \tan(\phi_i + \phi)(x - x_i) \\ \cos(\phi_i + \phi)(z - z_i) &= \tan(\theta_i + \theta)(x - x_i). \end{aligned} \tag{8.11}$$

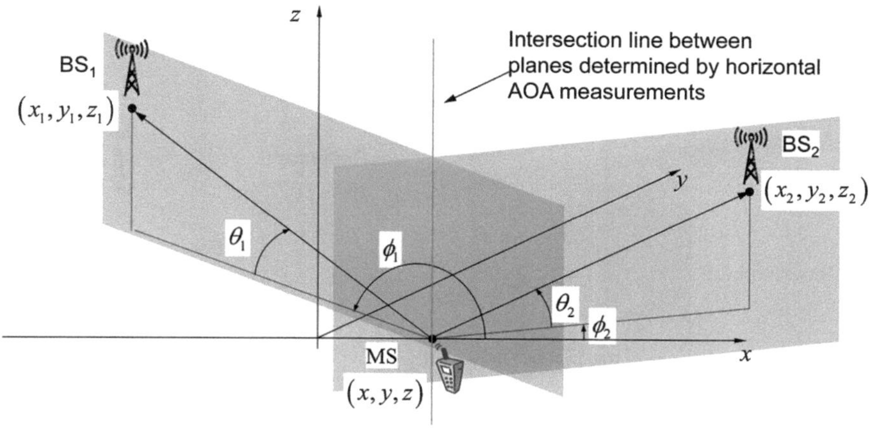

Fig. 8.6 Three-dimensional AOA positioning principle: the horizontal AOA measurements determine planes that intersect in a vertical line in the mobile station (MS) position; one additional vertical AOA measurement determines the mobile station position in three dimensions

Compared to (8.9) and (8.10) the mobile station now needs to estimate the unknowns (x, y, z) and (ϕ, θ) with the non-linear equations (8.11) and at least five independent AOA measurements. These can be obtained by measuring three horizontal AOAs ϕ_i and two vertical AOAs θ_i from at least three fixed stations.

Combining AOA measurements and RTD we obtain a radar, where a single fixed station with a directional antenna could suffice to estimate the mobile station position. Thus, in communication systems, the mobile station position can be determined with one or two signal transmissions in three dimensions whereas the RSS and timing based methods require more communication.

A drawback of AOA measurements is their need for directional antennas with good angular resolution, which are generally larger than omnidirectional antennas. Further, the angular resolution can vary with the AOA depending on the directional antenna. Especially for larger distances, the AOA position estimation error increases much faster than the one for RSS or timing based measurements (Qi et al. 2006). Last but not least, NLOS propagation results in totally wrong position estimates. For instance, in Fig. 8.7, the LOS signal from the MS to BS$_1$ is blocked by a building and so BS$_1$ only receives a reflected NLOS signal. This results in the biased position estimate at $(\hat{x}, \hat{y}, \hat{z})$.

Fingerprinting

In the previous sections, we have estimated the mobile station position through explicit functional relations of the position dependent measurements, e.g., for RSS,

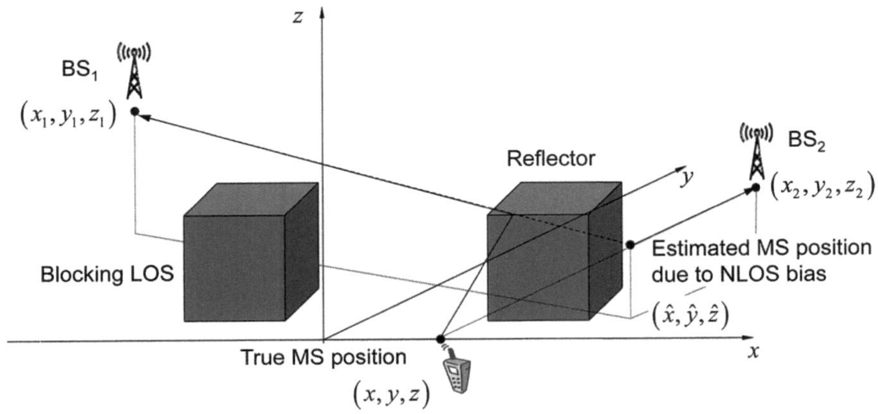

Fig. 8.7 AOA NLOS problem: blocking of LOS signal and reflection of NLOS signal results in false estimate of the mobile station position

TOA, RTD, TOA, and AOA. In this section, we generalize this idea to fingerprinting based positioning. A fingerprint in wireless communication systems is any measureable signal characteristic that depends on the location of transmission or reception. For fingerprinting, we do not require explicit functional relations between the location dependent measurement and the position estimate. Instead, the location dependent measurements are stored in a database, possibly for each location in the considered environment. If a mobile station now measures location dependent information, a service provider can compare the mobile station measurements with its database to determine the most likely position.

An example of a location dependent measurement, i.e., fingerprint, whose explicit functional relationship can be hardly determined, is the MIMO channel impulse response (CIR) in a multipath and NLOS propagation environment for modern multi-antenna mobile stations. Only if the complete propagation environment is known, the MIMO CIR can be determined, e.g., through wave propagation or raytracing calculations. This results in a very complex explicit functional relationship between the location dependent MIMO CIR and the mobile station position. Hence, it is computationally prohibitive even for small areas. Other examples of location dependent fingerprints are second order statistics of the channel such as power delay profiles, delay spread, angular spread, or the previously introduced proximity information, RSS, TOA, TDOA, RTD, AOA, etc.

Thus, if the determination of the functional relationship between the signal propagation is computationally prohibitive, e.g., in severe multipath and NLOS environments, fingerprinting is beneficial. However, fingerprinting requires all measurements to be communicated to database servers and regular updates of the database. With broadband cellular networks, e.g., UMTS, LTE, and LTE-Advanced, the amount of measurements communicated to the database servers can be handled nowadays.

In classical fingerprinting methods, it is assumed that in a first step calibration measurements are executed to build up the fingerprint database (GRAMMAR D1.4 2009). In a second step, position of the mobile station is determined by comparing the measured fingerprints with the fingerprints stored in the database. This can be done either deterministic with pattern recognition algorithms or probabilistic based on probability density functions, hidden Markov models or particle filters. Instead of using two phases, SLAM algorithms can be used to simultaneously estimate the mobile station position and to generate the fingerprint map or database (Bailey and Durrant-Whyte 2006, Bruno and Robertson 2011).

Further, it is possible to keep the database up-to-date through crowd sourcing. Although, the database will become large for large geographic regions, companies such as Apple, Google, or Skyhook Wireless employ fingerprinting for fast position determination mostly based on cell-IDs, MAC-IDs and RSS measurements by using distributed data centers and cloud computing. In some cases, these commercial fingerprinting services are more accurate than purely GNSS based positioning, e.g., in urban areas.

Assisted GNSS

Whereas the previous sections addressed localization information from wireless communication systems in the form of location dependent measurements, this section considers A-GNSS (GRAMMAR D1.4 2009). A-GNSS is localization information provided by the wireless communication network to a GNSS receiver equipped with a communication channel, e.g., as in a smartphone, to improve its performance on:

- Time to first fix.
- Weak signal reception in urban canyons or indoor especially for cellular-sized antennas.
- Reduced power consumption by reducing signal acquisition time.

The main idea of A-GNSS is to set-up a GNSS reference network (or equivalently a wide-area Differential GNSS (DGNSS) network) where the receivers have clear sky-views. This reference network is additionally connected with the cellular infrastructure, which continuously monitors the real-time constellation status, and provides data such as approximate mobile station position (or fixed station location), satellite visibility, ephemeris and clock correction, Doppler, and the code phases for each satellite, e.g., in LTE (3GPP TS 36.305 2011). On demand the assistance data is then transmitted to the mobile station or network nodes for fast start-up and increased sensitivity. When the available satellite signals are detected, the pseudorange measurements can be delivered to the network for position calculation or used internally in the mobile station to compute its position. Additional assistance data, such as real-time integrity, DGNSS corrections, satellite almanac, ionospheric delay, and Coordinated Universal Time offset can be transmitted.

There are two fundamental modes that are supported:

- The mobile-station-assisted solution shifts the majority of the GNSS receiver functions to the network processor. This method requires at least an antenna, RF section, and base-band processor in the mobile station for making measurements by generating replica codes and correlating them with the received GNSS signals. By A-GNSS an assistance message is sent to the mobile station, consisting of time, visible satellite list, satellite signal Doppler and code phase, as well as their search windows or, alternatively, approximate handset position and ephemeris. The assistance data of Doppler and code phase is usually valid for a few minutes, while ephemeris data for around two to four hours. From the mobile station the pseudorange data is returned to the network. Then, the location server estimates the position of the mobile station. Additionally, differential correction in terms of DGNSS can be applied to the pseudorange data or final result at the network side to improve the position accuracy.
- For mobile-station-based solutions a complete GNSS receiver is integrated in the handset. In the start-up phase, satellite orbital elements, i.e., ephemeris, must be provided to the mobile station. This data is valid for two to four hours and can be extended to cover the entire visible period of the GNSS satellite. For better positional accuracy or longer ephemeris life, differential correction, i.e., DGNSS, data can be transmitted to the mobile station. The final position of the mobile station is generated at the mobile station. Then, the estimated mobile station position can be sent to the network if required.

Comparing GSM, UMTS, and LTE, similar features are included in 3GPP/3GPP2. To reduce infrastructure investment, a shared location server could be implemented to support mobile stations operating in both GSM and UMTS or LTE networks.

However, the cellular time base is in general different for GSM and UMTS. In GSM, cellular network time can be expressed in terms of BCCH carrier, BSIC, frame number, time slot number, and bit number. For UMTS or LTE, it can be expressed in terms of universal terrestrial radio access network (UTRAN)-GPS or enhanced-UTRAN GNSS timing of cell frames, primary CPICH info, and SFN. The uncertainty between cellular and GNSS time is included in the GSM field "GPS Reference Time Uncertainty", in the UMTS field "SFN-TOW Uncertainty", and in the LTE field "GNSS-TOD-UNC".

Since GSM, UMTS, and LTE mobile stations do not have precise time information available internally, methods are included in the standards protocol to deliver precise time to the GSM/UMTS handset (3GPP TS 25.305 2011), (3GPP TS 36.305 2011). To accomplish precise time transfer in asynchronous GSM/UMTS/LTE networks, LMUs can be used. For A-GNSS, the LMU measures the relation between cellular frames from the serving fixed station with respect to GNSS time and sends this information periodically to the network. The network collects the time stamp information, and maintains a data base of the relationship between cell timing and GNSS time for every fixed station. This information is then sent to the mobile station. Nevertheless, installing LMUs in the network is rather expensive for a network operator. Therefore, the GSM/UMTS/LTE standards also allow the mobile station to perform the LMU function (3GPP TS 25.305 2011), (3GPP TS 36.305 2011). In that sense, after a position request the mobile station reports the difference between cellular time and GNSS time to the network. The network then can use this information for assistance of other mobile stations. Hence, also in GSM/UMTS/LTE it is possible to substantially reduce the contribution of time error.

Hybridization of Location Information from GNSS and Wireless Communication Systems

This section presents hybridization methods of location information from GNSS and wireless communication systems. In the following, we address four aspects of hybridization:

1. Static hybridization methods can be used for fast initial or discontinuous localization without requiring tracking and movement information.
2. Dynamic hybridization methods exploit the tracking and movement information to improve availability and accuracy of localization services.
3. Cooperative hybridization methods extend the concept of hybridization from a single mobile station to several heterogeneous mobile stations that cooperate to obtain robust and accurate position estimates.
4. Error mitigation techniques through hybridization overcome problems such as interference and NLOS propagation.

Static Hybridization Methods

Mobile stations use static hybridization methods when the mobile is switched on to obtain quickly a first position estimate without movement history. Alternatively, the mobile station may use the static methods for irregular and discontinuous localization.

Usually, there is a non-linear dependency between the position of the mobile station and the measurements, e.g., pseudoranges, which results in the fact that the navigation equation may have no analytical or unique solution. Hence, approximate non-iterative or iterative procedures are necessary to get an accurate estimate of the position.

Gauss-Newton based approaches for the solutions of the navigation equation for TOA based GNSS measurements are commonly used (Misra and Enge 2004). Extensions to hybridization concepts are possible, e.g., TDOA measurements from cellular communication systems or RSS measurements from WLAN, with the corresponding covariance matrices can be included. Besides the Gauss-Newton algorithm (Foy 1976), also other algorithms, e.g., based on Steepest Descent (Kay 1993) or Levenberg-Marquardt (Levenberg 1944; Marquardt 1963) are used.

In the previous sections, we introduced localization information from wireless communication systems that can be used in a hybrid data fusion (HDF) entity together with measurements from GNSS. In the HDF, all available information is used to calculate the position of the mobile station, i.e., to solve the navigation equation.

Without loss of generality, we use the earth-centered earth-fixed coordinate system and consider in the remainder of this section HDF of GNSS based pseudorange measurements and TDOA measurements from a wireless cellular communication system. We assume $N = N_{\text{GNSS}}$ pseudorange measurements from GNSS and $M = N_{\text{FS}} - 1$ TDOA measurements from fixed stations. According to (8.6) and (8.7), we have

$$\begin{aligned}\hat{\rho}_i &= \sqrt{\left(x - x_{\text{GNSS},i}\right)^2 + \left(y - y_{\text{GNSS},i}\right)^2 + \left(z - z_{\text{GNSS},i}\right)^2} + b + n_i, \\ &\qquad\qquad i = 1, \ldots, N, \\ \hat{d}_{k,0} &= r_k - r_0 + n_k - n_0 = d_{k,0} + n_{k0},\ \mathrm{k} = 1, \ldots, M.\end{aligned} \tag{8.12}$$

We collect all pseudorange measurements and all TDOA measurements in the vectors $\boldsymbol{\rho} = (\rho_1, \ldots, \rho_N)^{\mathrm{T}}$ and $\mathbf{d} = \left(d_{1,0}, \ldots, d_{M,0}\right)^{\mathrm{T}}$. We define the noise vector for GNSS as $\mathbf{n}_{\text{GNSS}} = (n_1, \ldots, n_N)^{\mathrm{T}}$, where we assume each element n_i is Gaussian distributed with variance $\sigma^2_{\mathrm{n}_i}$. For the wireless communication system, we define the noise vector $\mathbf{n}_{\text{TDOA}} = \left(n_{1,0}, \ldots, n_{M,0}\right)^{\mathrm{T}}$, where $n_{k,0} = n_k - n_0$ and we assume each element n_k is Gaussian distributed with variance $\sigma^2_{\mathrm{n}_k}$. Thus, the covariance matrices of the noise vectors are

$$\boldsymbol{\Sigma}_{\mathbf{n},\text{GNSS}} = E\left\{\mathbf{n}_{\text{GNSS}}\mathbf{n}^{\mathrm{T}}_{\text{GNSS}}\right\} \in \mathbb{R}^{N_{\text{GNSS}} \times N_{\text{GNSS}}},$$

$$\boldsymbol{\Sigma}_{\mathbf{n},\text{TDOA}} = E\left\{\mathbf{n}_{\text{TDOA}}\mathbf{n}^{\mathrm{T}}_{\text{TDOA}}\right\} = \begin{bmatrix} \sigma^2_{\mathrm{n}_0} + \sigma^2_{\mathrm{n}_1} & \sigma^2_{\mathrm{n}_0} & \cdots & \sigma^2_{\mathrm{n}_0} \\ \sigma^2_{\mathrm{n}_0} & \sigma^2_{\mathrm{n}_0} + \sigma^2_{\mathrm{n}_2} & \cdots & \sigma^2_{\mathrm{n}_0} \\ \vdots & \vdots & \ddots & \vdots \\ \sigma^2_{\mathrm{n}_0} & \sigma^2_{\mathrm{n}_0} & \cdots & \sigma^2_{\mathrm{n}_0} + \sigma^2_{\mathrm{n}_M} \end{bmatrix} \in \mathbb{R}^{M \times M},$$

and

$$\boldsymbol{\Sigma}_{n,\text{HDF}} = \begin{bmatrix} \boldsymbol{\Sigma}_{n,\text{GNSS}} & \mathbf{0}_{\mathrm{N}\times\mathrm{M}} \\ \mathbf{0}_{\mathrm{N}\times\mathrm{M}} & \boldsymbol{\Sigma}_{n,\text{TDOA}} \end{bmatrix}. \tag{8.13}$$

Then, (8.12) can be written as

$$\mathbf{g}_{\mathrm{HDF}} = \begin{pmatrix} \boldsymbol{\rho} \\ \mathbf{d} \end{pmatrix} = \begin{pmatrix} \mathbf{r}(\mathbf{x}) + b\mathbf{1}_N + \mathbf{n}_{\mathrm{GNSS}} \\ \mathbf{d}(\mathbf{x}) + \mathbf{n}_{\mathrm{TDOA}} \end{pmatrix} \tag{8.14}$$

Using Eqs. (8.13) and (8.14) for the solution of the estimation problem for the mobile station location it is a common way to follow the weighted non-linear least squares approach which minimizes the cost function

$$\varepsilon(\mathbf{x}, b) = \mathbf{g}_{\mathrm{HDF}}^{\mathrm{T}} \boldsymbol{\Sigma}_{n,\mathrm{HDF}}^{-1} \mathbf{g}_{\mathrm{HDF}} \tag{8.15}$$

w.r.t. the unknown mobile station position and the unknown time offset. Note that due to the included covariance matrix of the noise, we address a weighted approach considering the quality of the measurements for the different links. This procedure results in the estimate

$$\begin{bmatrix} \hat{\mathbf{x}} \\ \hat{b} \end{bmatrix} = \underset{\mathbf{x},b}{\operatorname{argmin}}\, \varepsilon(\mathbf{x}, b). \tag{8.16}$$

In the general case, there exists no closed-form solution to this non-linear four-dimensional optimization problem, and hence, iterative approaches are necessary.

Gauss-Newton Algorithm

A standard approach to deal with the resulting minimization problem in (8.15) is based on the Gauss-Newton (GN) algorithm (Foy 1976). The GN algorithm linearizes the system model about some initial value $\mathbf{x}^{(0)}$ yielding

$$\begin{aligned} \mathbf{r}(\mathbf{x}) &\approx \mathbf{r}\left(\mathbf{x}^{(0)}\right) + \boldsymbol{\Phi}(\mathbf{x})\Big|_{\mathbf{x}=\mathbf{x}^{(0)}} \left(\mathbf{x} - \mathbf{x}^{(0)}\right), \\ \mathbf{d}(\mathbf{x}) &\approx \mathbf{d}\left(\mathbf{x}^{(0)}\right) + \boldsymbol{\Psi}(\mathbf{x})\Big|_{\mathbf{x}=\mathbf{x}^{(0)}} \left(\mathbf{x} - \mathbf{x}^{(0)}\right) \end{aligned} \tag{8.17}$$

with the elements of the $N \times 4$ Jacobian matrix

$$\boldsymbol{\Phi}(\mathbf{x}) = \nabla_{\mathbf{x},b}^{\mathrm{T}} \otimes \mathbf{r}(\mathbf{x}) = \begin{pmatrix} \frac{x-x_1}{r_1} & \frac{y-y_1}{r_1} & \frac{z-z_1}{r_1} & 1 \\ \frac{x-x_2}{r_2} & \frac{y-y_2}{r_2} & \frac{z-z_2}{r_2} & 1 \\ \vdots & \vdots & \vdots & \vdots \\ \frac{x-x_N}{r_N} & \frac{y-y_N}{r_N} & \frac{z-z_N}{r_N} & 1 \end{pmatrix} \tag{8.18}$$

and the $M \times 3$-dimensional Jacobian matrix

$$\boldsymbol{\Psi}(\mathbf{x}) = \nabla_{\mathbf{x}}^{\mathrm{T}} \otimes \mathbf{d}(\mathbf{x}) = \begin{pmatrix} \frac{x-x_1}{r_1} - \frac{x-x_0}{r_0} & \frac{y-y_1}{r_1} - \frac{y-y_0}{r_0} & \frac{z-z_1}{r_1} - \frac{z-z_0}{r_0} \\ \frac{x-x_2}{r_2} - \frac{x-x_0}{r_0} & \frac{y-y_2}{r_2} - \frac{y-y_0}{r_0} & \frac{z-z_2}{r_2} - \frac{z-z_0}{r_0} \\ \vdots & \vdots & \vdots \\ \frac{x-x_M}{r_M} - \frac{x-x_0}{r_0} & \frac{y-y_M}{r_M} - \frac{y-y_0}{r_0} & \frac{z-z_M}{r_M} - \frac{z-z_0}{r_0} \end{pmatrix}, \tag{8.19}$$

where $\nabla_{\mathbf{x},b} = \left[\frac{\partial}{\partial x}, \frac{\partial}{\partial y}, \frac{\partial}{\partial z}, \frac{\partial}{\partial b}\right]^{\mathrm{T}}$, $\nabla_{\mathbf{x}} = \left[\frac{\partial}{\partial x}, \frac{\partial}{\partial y}, \frac{\partial}{\partial z}\right]^{\mathrm{T}}$ and "$\otimes$" denotes the Kronecker product.

For HDF solutions, the Jacobian matrices for GNSS and TDOA can be combined as

$$\boldsymbol{\Pi}(\mathbf{x}) = \begin{bmatrix} \boldsymbol{\Phi}(\mathbf{x}) & \\ \boldsymbol{\Psi}(\mathbf{x}) & \mathbf{0}_M \end{bmatrix}. \tag{8.20}$$

The matrix $\boldsymbol{\Pi}(\mathbf{x})$ is also known as the observation or geometry matrix as it takes into account the geometric relations between the MS, the satellites, and the fixed stations.Afterwards, the standard linear least squares procedure (Kay 1993) is applied resulting in the iterated solution

$$\begin{aligned} \begin{bmatrix} \mathbf{x}^{(k+1)} \\ b^{(k+1)} \end{bmatrix} &= \begin{bmatrix} \mathbf{x}^{(k)} \\ b^{(k)} \end{bmatrix} + \underbrace{\left(\boldsymbol{\Pi}^{\mathrm{T}}\left(\mathbf{x}^{(k)}\right)\boldsymbol{\Sigma}_{\mathbf{n},\mathrm{HDF}}^{-1}\boldsymbol{\Pi}\left(\mathbf{x}^{(k)}\right)\right)^{-1}}_{\triangleq \mathbf{A}_{\mathrm{HDF}}^{(k),-1}} \\ &\quad \underbrace{\boldsymbol{\Pi}^{\mathrm{T}}\left(\mathbf{x}^{(k)}\right)\boldsymbol{\Sigma}_{\mathbf{n},\mathrm{HDF}}^{-1}\begin{pmatrix} \boldsymbol{\rho} - \boldsymbol{\rho}\left(\mathbf{x}^{(k)}\right) - b^{(k)}\mathbf{1}_N \\ \mathbf{d} - \mathbf{d}\left(\mathbf{x}^{(k)}\right) \end{pmatrix}}_{\triangleq \mathbf{g}_{\mathrm{HDF}}^{(k)}} \\ &= \begin{bmatrix} \mathbf{x}^{(k)} \\ b^{(k)} \end{bmatrix} + \mathbf{A}_{\mathrm{HDF}}^{(k),-1}\mathbf{g}_{\mathrm{HDF}}^{(k)} \end{aligned} \tag{8.21}$$

Normally, for pure GNSS solutions the GN algorithm achieves a sufficient performance. Nevertheless, especially for pure TDOA based solutions this type of algorithm can have convergence problems due to an ill-conditioned matrix $\mathbf{A}_{\mathrm{HDF}}^{(k)}$ that can become singular for bad geometric conditions or bad initial values. For these scenarios, the hyperbolic character of the TDOA measurements needs more robust approaches. Two of them are presented below. They generally hold both for GNSS only, TDOA based, and hybrid solutions.

Steepest Descent Algorithm

Contrary to GN, the steepest descent (SD) algorithm (Kay 1993) is a gradient based procedure with search direction $\nabla_{\mathbf{x}} = \varepsilon(\mathbf{x})$ and step size μ yielding

$$\mathbf{x}^{(k+1)} = \mathbf{x}^{(k)} + \mu^{(k)}\mathbf{g}_{\mathrm{HDF}}^{(\mathrm{k})}. \quad (8.22)$$

The easiest way to find a step size is to choose a constant $\mu^{(k)} = \mu$ for all iteration steps. An alternative solution is to perform a line-search procedure to find the optimum step-size in each iteration step. Main drawbacks of the SD method are the possibility for running in local minima and slow convergence in the final iteration steps.

Levenberg-Marquardt Algorithm

To cope with the problems of GN and SD (robustness and slow convergence), a method introduced by Levenberg and Marquardt (Levenberg 1944; Marquardt 1963) is adapted to the positioning problem (Mensing and Plass 2006). It is based on a damped GN procedure given by

$$\mathbf{x}^{(k+1)} = \mathbf{x}^{(k)} + \left(\mathbf{A}_{\mathrm{HDF}}^{(\mathrm{k})} + \lambda^{(k)}\mathbf{I}_2\right)^{-1}\mathbf{g}_{\mathrm{HDF}}^{(\mathrm{k})}. \quad (8.23)$$

The damping parameter $\lambda^{(k)}$ makes sure that the appropriate matrix—in comparison to GN—can always be inverted yielding a much more robust implementation. The damping parameter can be calculated using a computational efficient algorithm which is based on a suboptimum line search procedure, e.g., (Mensing and Plass 2006). Note that for $\lambda^{(k)} = 0$ we obtain the GN solution whereas for $\left|\lambda^{(k)}\right| \gg 1$ we have a SD behavior. Levenberg-Marquardt algorithm provides fast convergence and is very robust against inaccurate initial values for TDOA positioning.

Dynamic Hybridization Methods

The algorithms presented in the previous section just consider static, i.e., snapshot solutions. These solutions do not exploit any information between different snapshots such as models of the user dynamics. Clearly, in the usual case the mobile station will move with a certain velocity. Nevertheless, the solutions presented before can be seen on the one hand as performance bound in the static case. On the other hand, they also can be used as input for tracking algorithms that make use of apriori information and can improve the positioning accuracy and coverage a lot.

Especially in critical scenarios – where the performance is restricted due to only few satellites, bad DOP, multipath or NLOS conditions—the accuracy can be increased and the coverage extended. For instance, (Chen and Feng 2005) proposes to apply two Kalman filters (one for GNSS, and one for a cellular position solution) which are fused in a HDF step every time-step. Ma (2003) integrates GNSS and TDOA measurements in a linearized Kalman filter environment. Heinrichs et al. (2004) compares the performance of recursive least mean square methods with Kalman filter solutions based on GNSS and TDOA measurements.

In the next subsection, we describe how to apply Kalman filter based approaches for positioning problems in the HDF context with focus on GNSS and TDOA measurements. After a short introduction in the background of Kalman filtering, we will show the necessary extensions for positioning and propose two approaches that combine tracking with HDF.

Kalman Filter

Fundamentals

The Kalman filter (KF) is a generalization of the Wiener filter, where the restriction of the Wiener filter that signal and noise are stationary is no longer necessary. It is a sequential minimum mean square error (MMSE) estimator of a signal embedded in noise, where the signal is characterized by a dynamical or state space model which—in case of positioning—is the mobility model.

If signal and noise are jointly Gaussian, the KF is the optimum MMSE estimator. If this assumption does not hold, it is the optimum linear MMSE estimator. A deeper analysis of the theory and a good overview of related topics are given in (Kay 1993).

Vector Kalman Filter

First we present the standard vector KF. In this case, the state space and observation equations can be written as

$$\mathbf{s}[n] = \mathbf{A}\mathbf{s}[n-1] + \mathbf{B}\mathbf{u}[n] \in \mathbb{R}^p \tag{8.24}$$

and

$$\mathbf{x}[n] = \mathbf{H}[n]\mathbf{s}[n] + \mathbf{w}[n] \in \mathbb{R}^q. \tag{8.25}$$

Note that $\mathbf{s}[n]$ and $\mathbf{x}[n]$ denote the state space vector and the measurement vector. We will use this notation throught this section although the measurement vector may refer to position measurements in the case of the position KF and to pseudorange or TDOA measurements in the case of the HDF-extended KF (EKF). The noise vectors for the state space equation $\mathbf{u}[n] \in \mathbb{R}^r$ and for the observation equation $\mathbf{w}[n] \in \mathbb{R}^q$ have the covariance matrices $\mathbf{Q} \in \mathbb{R}^{r\times r}$ and $\mathbf{C}[n] \in \mathbb{R}^{q\times q}$. The transition matrix $\mathbf{A} \in \mathbb{R}^{p\times p}$ describes the relation between the states of the different time-steps, and the matrix $\mathbf{B} \in \mathbb{R}^{p\times r}$ models the influence of the state space noise.

Finally, the matrix $\mathbf{H}[n] \in \mathbb{R}^{q \times p}$ is the so-called observation matrix. Note that in the general case also the matrices $\mathbf{A}$, $\mathbf{B}$, and $\mathbf{Q}$ may have time-varying character. However, for the considered positioning application this is not necessary in the performed simulations.

Using this model, the corresponding calculations for the vector KF are:

1. Prediction

$$\hat{\mathbf{s}}[n|n-1] = \mathbf{A}[n-1]\hat{\mathbf{s}}[n-1|n-1] \in \mathbb{R}^{p}; \tag{8.26}$$

2. Minimum prediction MSE matrix

$$\mathbf{M}[n|n-1] = \mathbf{A}[n-1]\mathbf{M}[n-1|n-1]\mathbf{A}^{\mathrm{T}}[n-1] + \mathbf{B}\mathbf{Q}\mathbf{B}^{\mathrm{T}} \in \mathbb{R}^{p \times p}; \tag{8.27}$$

3. Kalman gain matrix

$$\mathbf{K}[n] = \mathbf{M}[n|n-1]\mathbf{H}^{T}[n]\left(\mathbf{C}[n] + \mathbf{H}[n]\mathbf{M}[n|n-1]\mathbf{H}^{T}[n]\right)^{-1} \in \mathbb{R}^{p \times q}; \tag{8.28}$$

4. Correction

$$\hat{\mathbf{s}}[n|n] = \hat{\mathbf{s}}[n|n-1] + \mathbf{K}[n](\mathbf{x}[n] - \mathbf{H}[n]\hat{\mathbf{s}}[n|n-1]) \in \mathbb{R}^{p}; \tag{8.29}$$

5. MMSE matrix

$$\mathbf{M}[n|n] = \left(\mathbf{I}_p - \mathbf{K}[n]\mathbf{H}[n]\right)\mathbf{M}[n|n-1] \in \mathbb{R}^{p \times p}. \tag{8.30}$$

Hence, after each time-step the current estimate of the state space variables is included in the vector $\hat{\mathbf{s}}[n|n]$. The MMSE matrix $\mathbf{M}[n|n]$ is composed of the theoretical MMSEs for each time-step. The vector KF is initialized with

$$\begin{aligned} \hat{\mathbf{s}}[-1|-1] &= \mathrm{E}\{\mathbf{s}[-1]\} = \boldsymbol{\mu}_s \\ \mathbf{M}[-1|-1] &= \mathbf{C}_s, \end{aligned} \tag{8.31}$$

based on a distribution of the initial state vector according to $\mathbf{s}[-1] \sim \mathrm{N}(\boldsymbol{\mu}_s, \mathbf{C}_s)$.

Hybrid Data Fusion Position Kalman Filter

Assuming that a static solution of the navigation equation is available in each time-step, we can apply a HDF position Kalman filter (PKF). Generally, in each time-step the navigation equation is solved with all available sources, e.g., hybrid GNSS + TDOA. To do so, we apply the static solutions described in this Chapter. Afterwards, a mobility model is assumed and the position estimate for each time-step is processed with a linear vector KF for smoothing the estimated values. According to the previous section, we now describe the mobility model and the observation model for this procedure in detail.

As components of the state space vector, i.e., the parameters that have to be estimated, we use the mobile station position and its velocity in the 3-dimensional space. Hence, we obtain

$$\mathbf{s}[n] = \left[x[n], y[n], z[n], v_x[n], v_y[n], v_z[n]\right]^{\mathrm{T}}. \tag{8.32}$$

Similar as shown in (Ma 2003), we assume a mobility model that bases on random-walk behavior. The corresponding transition matrix is given as

$$\mathbf{A} = \begin{bmatrix} 1 & 0 & 0 & \Delta n & 0 & 0 \\ 0 & 1 & 0 & 0 & \Delta n & 0 \\ 0 & 0 & 1 & 0 & 0 & \Delta n \\ 0 & 0 & 0 & 1 & 0 & 0 \\ 0 & 0 & 0 & 0 & 1 & 0 \\ 0 & 0 & 0 & 0 & 0 & 1 \end{bmatrix}, \tag{8.33}$$

where Δn is the update period, e.g., one position is estimated each second. The state space noise is

$$\mathbf{u}[n] = \left[u_x[n], u_y[n], u_z[n]\right]^{\mathrm{T}}, \tag{8.34}$$

which is mapped to the state space vector by the matrix

$$\mathbf{B} = \begin{bmatrix} 1 & 0 & 0 & 1 & 0 & 0 \\ 0 & 1 & 0 & 0 & 1 & 0 \\ 0 & 0 & 1 & 0 & 0 & 1 \end{bmatrix}^{\mathrm{T}}. \tag{8.35}$$

This model implies that the change of the mobile station position is controlled by a process driving noise of a certain variance. This affects also the velocity of the mobile station and can be controlled by the covariance matrix of the state space noise vector given as

$$\mathbf{Q} = \begin{bmatrix} \sigma_{u_x}^2 & 0 & 0 \\ 0 & \sigma_{u_y}^2 & 0 \\ 0 & 0 & \sigma_{u_z}^2 \end{bmatrix}. \tag{8.36}$$

For the observation equation the measurements produced by the "snap-shot" static solution are required, i.e., for each time-step the static navigation equation is solved. This can be done for only GNSS, only TDOA, and hybrid GNSS + TDOA. However, the focus is of course on general hybrid solutions. Nevertheless, the HDF-PKF can handle the change of sources "online", i.e., for different time-steps the number and kind of sources can vary.

As a first approximation we do not get any velocity information from the GNSS and cellular entities. Thus, the velocity states in the HDF-PKF are assumed to be "hidden", i.e., for the delivered observations

$$\mathbf{x}[n] = [x[n], y[n], z[n]]^{\mathrm{T}} \tag{8.37}$$

we assume the observation matrix

$$\mathbf{H}[n] = \begin{bmatrix} 1 & 0 & 0 & 0 & 0 & 0 \\ 0 & 1 & 0 & 0 & 0 & 0 \\ 0 & 0 & 1 & 0 & 0 & 0 \end{bmatrix}. \tag{8.38}$$

Note that even though there is no velocity information delivered to the observation equation, the velocity is estimated by the HDF-PKF implicitly by making use of the mobility model. The observation noise vector is given as

$$\mathbf{w}[n] = \left[w_x[n], w_y[n], w_z[n]\right]^{\mathrm{T}}, \tag{8.39}$$

where the corresponding covariance matrix can be calculated by using the Jacobian matrix required for the solution of the static navigation equation. Hence, we obtain

$$\mathbf{C}[n] = \mathbf{\Pi}^{\mathrm{T}}(\mathbf{x})\mathbf{\Sigma}^{-1}_{\mathbf{n},\mathrm{HDF}}\mathbf{\Pi}(\mathbf{x}), \tag{8.40}$$

e.g., for the HDF solution according to (8.13) and (8.20). The matrix $\mathbf{C}[n]$ has time-varying character due to the fact that the geometric constellations or number of visible sources may change over time, and thus, also the covariances of the errors can change.

The HDF-PKF is initialized with the parameters produced by the static solution of the navigation equation in the first time-step.

Extended Kalman Filter

The main problem of the Kalman filter is that it requires a linear state equation and a linear observation equation (and zero-mean Gaussian noise processes) to be optimum. Clearly, to track just the position of the terminal—based on recent position estimates and the mobility model—would result in such a linear relation. However, if we want to include measurements of all kinds (pseudoranges, TDOAs, …) that have a strongly non-linear character w.r.t. the current position, the standard linear vector KF is not a suitable approach to deal with this problem. Therefore, we propose to use the so-called EKF (Kay 1993), which is an extension of the KF for non-linear problems. Generally, state space and observation equations can have a non-linear character. However, we just consider here a non-linear observation equation

$$\mathbf{x}[n] = \mathbf{h}(\mathbf{s}[n]) + \mathbf{w}[n] \tag{8.41}$$

that includes the non-linear character of the observations or measurements. Hence, $\mathbf{h}(\mathbf{s}[n]) \in \mathbb{R}^q$ is the non-linear dependency between the state-space variables and the observations. Adapted for the positioning problem, this could be, e.g., the pseudorange and TDOA equations. In comparison to the linear vector KF, modifications of the standard vector KF equations are necessary. Therefore, we have to replace $\mathbf{h}(\mathbf{s}[n])$ in (8.41) by a linearization about the estimate $\hat{\mathbf{s}}[n|n-1]$ of $\mathbf{s}[n]$ in the current time-step, i.e.,

$$\mathbf{h}(\mathbf{s}[n]) \approx \mathbf{h}(\hat{\mathbf{s}}[n|n-1]) + \mathbf{H}[n](\mathbf{s}[n] - \hat{\mathbf{s}}[n|n-1]), \tag{8.42}$$

where the Jacobian observation matrix is calculated as

$$\mathbf{H}[n] = \left.\frac{\partial \mathbf{h}}{\partial \mathbf{s}[n]}\right|_{\mathbf{s}[n]=\hat{\mathbf{s}}[n|n-1]}, \tag{8.43}$$

i.e., it includes the derivation of the measurement model w.r.t. the variables of the state space vector. Then, the linearized observation model according to (8.41) can be rewritten as

$$\mathbf{x}[n] \approx \mathbf{h}(\hat{\mathbf{s}}[n|n-1]) + \mathbf{H}[n](\mathbf{s}[n] - \hat{\mathbf{s}}[n|n-1]) + \mathbf{w}[n]. \tag{8.44}$$

If we apply the standard vector KF equations to the linearized model in (8.44), we arrive at the EKF equations given as follows in matrix-vector form:

1. Prediction

$$\hat{\mathbf{s}}[n|n-1] = \mathbf{A}[n-1]\hat{\mathbf{s}}[n-1|n-1] \in \mathbb{R}^p; \tag{8.45}$$

2. Minimum prediction MSE matrix

$$\mathbf{M}[n|n-1] = \mathbf{A}[n-1]\mathbf{M}[n-1|n-1]\mathbf{A}^{\mathrm{T}}[n-1] + \mathbf{B}\mathbf{Q}\mathbf{B}^{\mathrm{T}} \in \mathbb{R}^{p\times p}; \tag{8.46}$$

3. Kalman gain matrix

$$\mathbf{K}[n] = \mathbf{M}[n|n-1]\mathbf{H}^T[n]\left(\mathbf{C}[n] + \mathbf{H}[n]\mathbf{M}[n|n-1]\mathbf{H}^T[n]\right)^{-1} \in \mathbb{R}^{p\times q}; \tag{8.47}$$

4. Correction

$$\hat{\mathbf{s}}[n|n] = \hat{\mathbf{s}}[n|n-1] + \mathbf{K}[n](\mathbf{x}[n] - \mathbf{h}(\hat{\mathbf{s}}[n|n-1])) \in \mathbb{R}^p; \tag{8.48}$$

5. MMSE matrix

$$\mathbf{M}[n|n] = \left(\mathbf{I}_p - \mathbf{K}[n]\mathbf{H}[n]\right)\mathbf{M}[n|n-1] \in \mathbb{R}^{p\times p}. \tag{8.49}$$

The structure of the equation is similar compared to the vector KF equations. Merely the observation matrix is replaced by the Jacobian matrix. In addition, for the correction step in (8.48) the non-linear mapping from the state space vector to the measurement vector is applied. This is directly a result of the linearized observation model in (8.44) in the context of the derived filter equations. For more details we refer to (Kay 1993). Note that compared to the vector KF the EKF has no more the MMSE optimum behavior as it is a property of the standard KF. Furthermore, Kalman gain and MMSE matrix can no longer be calculated “offline”, i.e., apriori and without any knowledge of the observations. Hence, for the EKF these matrices have to be recalculated in every time-step depending on the current observations. The initialization of the EKF is the same as it was shown for the vector KF in (8.31).

Hybrid Data Fusion Extended Kalman Filter

Where the HDF-PKF uses already estimated positions as input, the HDF-EKF directly uses all available measurements (pseudoranges and/or TDOAs) as input, i.e., observed signals. This may be beneficial in environments where only a few sources are available and the HDF-PKF solution has limited performance.

The assumed mobility model and the state space model are nearly the same as described for the HDF-PKF. Merely the bias of the mobile station clock $b[n]$ has to be estimated as an additional parameter. Hence, the state space equations are based on the state space vector

$$\mathbf{s}[n] = \left[x[n], y[n], z[n], v_x[n], v_y[n], v_z[n], b[n]\right]^{\mathrm{T}}. \tag{8.50}$$

As observables in the HDF-EKF, we use the pseudoranges from the GNSS unit and the TDOAs from the cellular unit. Hence, the observation equation for the HDF-EKF is generally given by

$$\mathbf{x}[n] = \mathbf{h}(\mathbf{s}[n]) + \mathbf{w}[n] \tag{8.51}$$

as shown above in (8.41). However, the observation vector is defined as

$$\mathbf{x}[n] = \left[\rho_1[n], \rho_2[n], \ldots, \rho_N[n], d_{1,0}[n], d_{2,0}[n], \ldots, d_{M,0}[n]\right]^{\mathrm{T}} \tag{8.52}$$

which results in an $(N + M)$-dimensional vector, i.e., N pseudoranges are available from GNSS measurements, and M differential TDOA measurements performed with M fixed stations are used. The definition of the HDF-EKF is very general, thus, it is possible that the number of observables change “online”, i.e., in each time-step. For instance, for different time-steps the number of satellites can vary due to, e.g., an occurring urban canyon situation. Therefore, besides the hybrid case also only GNSS or only cellular environments can be handled easily. Even the situation that no measurements are available for a certain time can be handled. In this case, just the prediction property of the filter is used.

Where the observation vector $\mathbf{x}[n]$ is composed of the measurements, the vector $\mathbf{h}(\mathbf{s}[n])$ includes the according measurement models for the pseudoranges and TDOAs. Hence, it can directly be obtained from the models defined previously, so the mapping function for the observable equation is

$$\mathbf{h}(\mathbf{s}[n]) = \begin{bmatrix} r_1(\mathbf{x})[n] & \cdots & r_N(\mathbf{x})[n] \, d_{1,0}(\mathbf{x})[n] & \cdots & d_{M,0}(\mathbf{x})[n] \end{bmatrix}^{\mathrm{T}}. \tag{8.53}$$

Correspondingly, the noise vector is defined as

$$\mathbf{w}[n] = \left[n_1[n], n_2[n], \ldots, n_N[n], n_{1,0}[n], n_{2,0}[n], \ldots, n_{M,0}[n]\right]^{\mathrm{T}}, \tag{8.54}$$

with the corresponding covariance matrix

$$\mathbf{C}[n] = \mathbf{\Sigma}_{n,\mathrm{HDF}} \tag{8.55}$$

defined in (8.13). Clearly, also this covariance matrix may be time-varying. Finally, for the calculation of the HDF-EKF equations the Jacobian matrix of the observations is required, which includes the derivation of the observation function in (8.53) w.r.t. the state space variables defined in (8.50) (cf. 8.43). Hence, for the considered HDF case this matrix is given as

$$\mathbf{H}[n] = \begin{bmatrix} \nabla_{\mathbf{x}}^{\mathrm{T}} \otimes \mathbf{r}(\mathbf{x})[n] & \mathbf{0}_{N\times 3} & 1_N \\ \nabla_{\mathbf{x}}^{\mathrm{T}} \otimes \mathbf{d}(\mathbf{x})[n] & \mathbf{0}_{M\times 3} & \mathbf{0}_M \end{bmatrix}, \tag{8.56}$$

where the N equations of the upper part correspond to the N pseudorange equations from GNSS and the M equations of the lower part belong to the TDOA equations provided by $M+1$ fixed stations. Nevertheless, the structure of this matrix can change for varying kind and number of observables over time.To summarize, the HDF-EKF algorithm consists of the five Eqs. (8.45)–(8.49), where $\mathbf{x}[n]$ is given by (8.52), $\mathbf{h}(\mathbf{s}[n])$ by (8.53), $\mathbf{C}[n]$ by (8.55), $\mathbf{H}[n]$ by (8.56). The transition matrix $\mathbf{A}[n]$, the state space noise matrix $\mathbf{B}$, the state space noise vector $\mathbf{u}[n]$ and its covariance matrix $\mathbf{Q}$ are given by

$$\mathbf{A}[n] = \begin{bmatrix} 1 & 0 & 0 & \Delta n & 0 & 0 & 0 \\ 0 & 1 & 0 & 0 & \Delta n & 0 & 0 \\ 0 & 0 & 1 & 0 & 0 & \Delta n & 0 \\ 0 & 0 & 0 & 1 & 0 & 0 & 0 \\ 0 & 0 & 0 & 0 & 1 & 0 & 0 \\ 0 & 0 & 0 & 0 & 0 & 1 & 0 \\ 0 & 0 & 0 & 0 & 0 & 0 & 1 \end{bmatrix}, \mathbf{B} = \begin{bmatrix} 1 & 0 & 0 & 0 \\ 0 & 1 & 0 & 0 \\ 0 & 0 & 1 & 0 \\ 1 & 0 & 0 & 0 \\ 0 & 1 & 0 & 0 \\ 0 & 0 & 1 & 0 \\ 0 & 0 & 0 & 1 \end{bmatrix},$$

$$\mathbf{u}[n] = \begin{bmatrix} u_x[n] \\ u_y[n] \\ u_z[n] \\ u_b[n] \end{bmatrix}, \text{ and } \mathbf{Q} = \begin{bmatrix} \sigma_{u_x}^2 & 0 & 0 & 0 \\ 0 & \sigma_{u_y}^2 & 0 & 0 \\ 0 & 0 & \sigma_{u_z}^2 & 0 \\ 0 & 0 & 0 & \sigma_{u_b}^2 \end{bmatrix}. \tag{8.57}$$

Similar as for the HDF-PKF, the HDF-EKF is initialized with the parameters produced by the static solution of the navigation equation in the first time-step.

Unscented Kalman Filter

For non-linear observation models we have seen that the EKF extension of the KF linearizes all non-linear models (in the considered case the observation model). Thus, the traditional KF can be applied to the problem. Unfortunately, the EKF has two important drawbacks (Wan and van der Merwe 2000). On the one hand, the derivation of the Jacobian matrices, i.e., the linear approximations to the non-linear functions, may be complex and can cause implementation difficulties. On the other hand, these linearizations can lead to filter instabilities if the time-step intervals are not sufficiently small.

To address these limitations the unscented KF (UKF) can be an alternative, especially in highly non-linear environments or for users that change their direction rapidly (Julier and Uhlmann 1997). The UKF uses the premise that it is easier to approximate a Gaussian distribution than it is to approximate an arbitrary non-linear function. Instead of linearizing the system by Jacobian matrices, the UKF uses a deterministic sampling approach to capture the mean and covariance estimates with a minimal set of sample points. Another interpretation of the UKF is that rather than tracking the user directly it tries to track the estimation error in each time-step.

Clearly, the UKF can directly be integrated in the considered HDF context where GNSS and TDOA measurements have to be processed in parallel.

Particle Filter

As mentioned above, if the assumption of Gaussian noise processes is not valid for the state and observation noise, the performance of the standard tracking filters can be bad. Especially in situations where a strongly non-linear relation between the measurements and the state or observation model exists. In those scenarios the necessary assumptions, e.g., for the EKF, are no longer valid. A possible solution for that kind of situation is the particle filter (PF) approach (Gustafsson et al. 2002), (Ristic et al. 2004). It is a solution for tracking problems within highly non-linear functions disturbed by non-Gaussian noise and can be seen as an important alternative to KFs for an environment with the mentioned properties.

Like KFs, also PFs solve the Bayesian filtering problem. PFs are based on a sequential Monte-Carlo methodology and calculate recursively the relevant probability distributions by importance sampling and approximation of probability distributions with discrete random measures. The basic principle of particle filtering is the representation of the state-PDF by a defined number of hypotheses rather than an analytical function. Hence, the PF approximates the optimal solution numerically based on the physical state space model, rather than applying an optimal filter to an approximate model as it can be seen for the KF. Compared to KFs the PFs have usually a higher complexity depending on the number of particles that are generated to model the PDF. In addition, they can suffer from the phenomena of sample degeneracy and sample impoverishment which could lead to unstable behavior (Ristic et al. 2004).

Similar to the UKF, the PFs can be fully integrated in the HDF environment. An example of a HDF-PF implementation for 3GPP-LTE and GPS can be found in (Gentner et al. 2012), which allows real-time processing.

Simulation Results for Urban Canyon Scenario

In the following, we present simulation results for an urban environment to compare the different hybridization algorithms and to demonstrate their benefits. For satellite navigation, we consider a fixed Galileo constellation and for navigation with wireless communication systems, we consider the cellular network structure depicted in Fig. 8.8 for a 3GPP-LTE communication system with an inter-BS distance of 750 m. Galileo has been selected as an example of using a GNSS. It is understood that most mobile users would use it in combination with GPS. For more details on the system parameter and simulation setup please refer to (Mensing et al. 2010a).

The simulated urban environment (cf. top view in Fig. 8.9) is located at the cell edge in Munich, Germany, where the average building height is about 26 m. In addition to the buildings, we also plotted one pedestrian track, which was generated with a gas diffusion model (Mensing et al. 2010a). The track also displays the LOS visible Galileo satellites. Only at the starting point at ($x \approx 50$ m,$y \approx 100$ m) and at the end of the track ($x > 210$ m,$y > 200$ m) more than four Galileo satellites are visible and LOS (cf. Fig. 8.10). Thus, we expect large positioning errors when employing a stand-alone Galileo navigation receiver. On the other hand, we see in Fig. 8.10 that most of the times we can measure more than two TDOAs from the LTE BSs, which allows stand-alone 2-D LTE-based positioning. So we expect that with the previously introduced tracking algorithms and hybridization of Galileo and LTE-based positioning reliable positioning is possible.

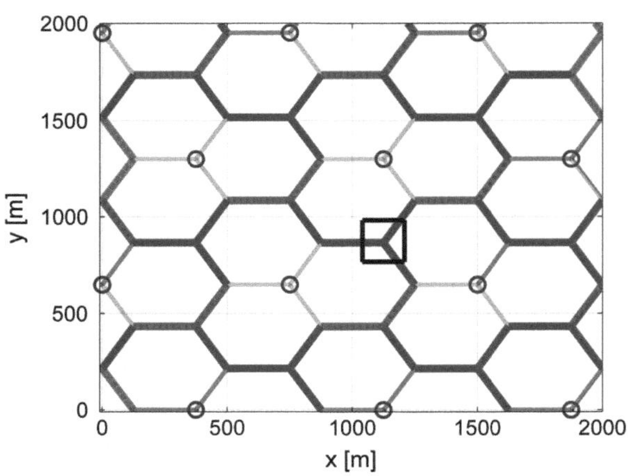

Fig. 8.8 Cellular network for urban environment: hexagonal grid with three sectors per BSs (*circular markers*), urban simulation environment at cell edge *inside rectangle* (for details see Fig. 8.9)

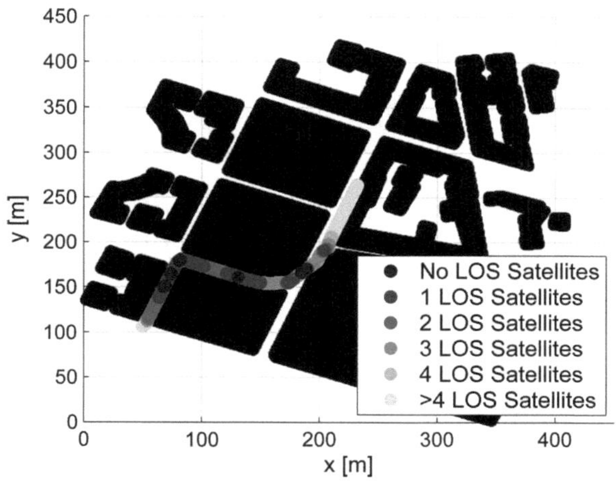

Fig. 8.9 Track through an urban canyon: LOS visibility of Galileo satellites

The root mean square error (RMSE) simulations with an EKF tracking algorithm for the track shown in Fig. 8.9 confirm this in Fig. 8.11. Whereas a stand-alone Galileo navigation receiver shows RMSEs above 40 m and 50 m, a stand-alone LTE navigation receiver achieves RMSEs below 15 m after initialization. The hybrid Galileo and LTE navigation receiver further reduces the RMSEs below 11 m after initialization (first 25 steps). Further, the hybrid receiver also has a smaller RMSE ($\approx$3 m) in the last 50 steps of the track compared to the stand-alone Galileo receiver ($\approx$8 m) where more than four Galileo satellites are in LOS.

In Figs. 8.12, 8.13 and 8.14, we compare different positioning algorithms for Galileo only, LTE only and hybrid Galileo and LTE-based positioning: Static, snapshot based GN algorithm, PKF, EKF, and UKF. As performance metric, we use the cumulative distribution function (CDF) of the positioning error x [m] on the

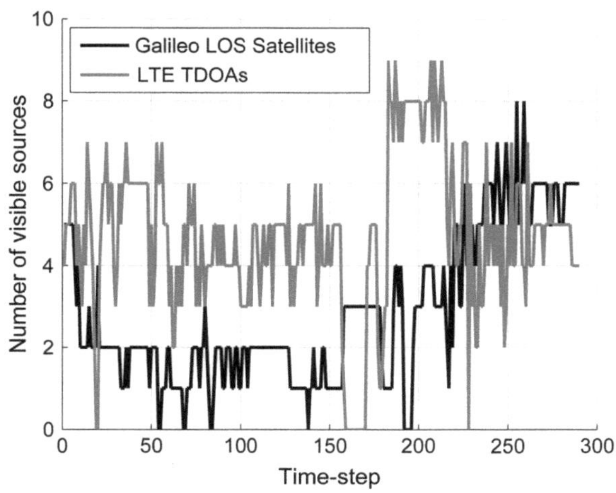

Fig. 8.10 Track through an urban canyon: number of LOS visible Galileo satellites and measured TDOAs from LTE BSs versus time-step

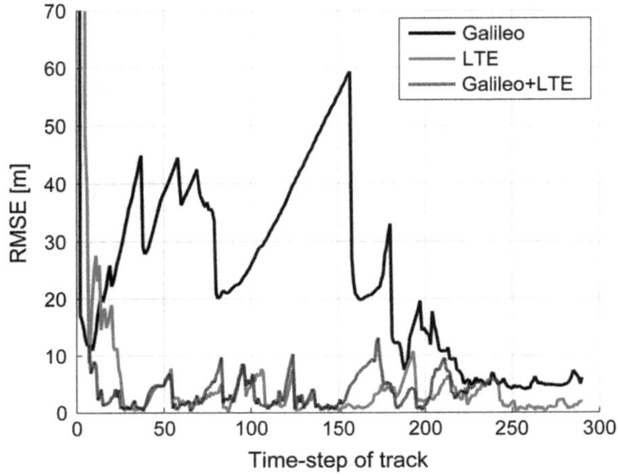

Fig. 8.11 Track through urban canyon: RMSE [m] versus time-step for Galileo only, LTE only, and hybrid Galileo and LTE positioning receivers

abscissa. The CDF was averaged over several tracks for the environment in Fig. 8.9 and several noise realizations (Mensing et al. 2010a).

For Galileo only, the EKF and UKF tracking algorithms in Fig. 8.12 yield the same performance as the assumptions of the EKF are still met quite well. Both filters result in the best performance with $P(x < 10 \text{ m}) \approx 0.8$. In contrast, the static GN and the dynamic PKF only yield $P(x < 59 \text{ m}) \approx 0.8$ and $P(x < 65 \text{ m}) \approx 0.8$. The decreased positioning accuracy is mainly due to the fact that often less than four satellites are available. Hence, the static GN cannot estimate a new position but keeps the previous one. The dynamic PKF obtains no measurement update from the static GN, but relies solely on the mobility model. In contrast, the EKF and UKF can exploit directly the TOA measurements. Thus, they do not rely solely on the mobility model if less than four satellites are visible. Note, for smaller errors, the PKF slightly outperforms the static GN whereas for large errors, the static GN outperforms the PKF slightly.

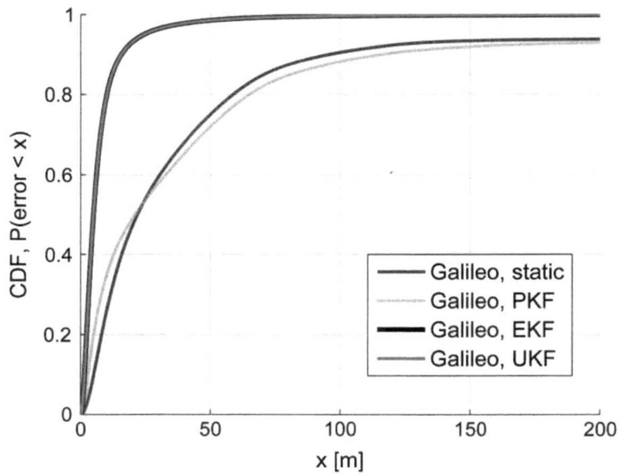

Fig. 8.12 Comparison of snap-shot and tracking based positioning algorithms for Galileo LOS satellites in urban environment: CDF of position error smaller than abscissa x [m]

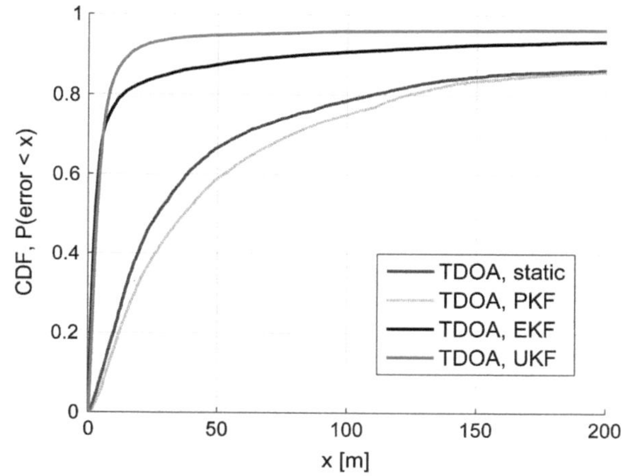

Fig. 8.13 Comparison of snap-shot and tracking based positioning algorithms for LTE TDOAs in urban environment: CDF of position error smaller than abscissa x [m]

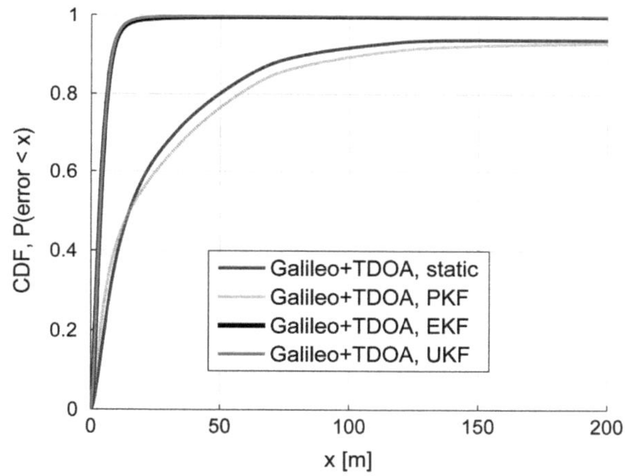

Fig. 8.14 Comparison of hybrid snap-shot and tracking based positioning algorithms for Galileo LOS satellites and LTE TDOAs in urban environment: CDF of position error smaller than abscissa x [m]

Similar to positioning with Galileo only in Fig. 8.12, the PKF and static GN perform worst, $P(x < 123\text{ m}) \approx 0.8$ and $P(x < 108\text{ m}) \approx 0.8$, for LTE-TDOA-based positioning in Fig. 8.13. The major difference between Figs. 8.12 and 8.13 is that the UKF outperforms the EKF, i.e., $P(x < 10\text{ m}) \approx 0.8$ compared to $P(x < 14\text{ m}) \approx 0.8$. Clearly, the UKF can compensate the larger measurement errors of the TDOAs better than the EKF.

The HDF of Galileo TOAs and LTE TDOAs in Fig. 8.14 improves the performance of all positioning algorithms. For instance, the static GN results in $P(x < 50\text{ m}) \approx 0.8$, the PKF in $P(x < 57\text{ m}) \approx 0.8$, and the EKF or UKF in $P(x < 6\text{ m}) \approx 0.8$.

In the case of EKF or UKF, the hybridization of Galileo positioning with localization information from LTE results in a position error of less than 20 m in

99 % of the cases compared to 60 m for Galileo only positioning. Thus, hybridization achieves the goal of more robust and reliable positioning at least for the simulated scenario.

Cooperative Hybridization Methods

In cooperative hybridization methods, mobile stations cooperate by measuring distances between neighboring mobile stations and communicating location information to them. For instance, in Fig. 8.15, we have 4 satellites, 2 fixed stations and two mobile stations. Mobile Station 1 (MS1) has clear sky few and can accurately determine its position with GNSS. However, MS2 is inside a building and cannot determine its position with GNSS due to blocking of the satellite signals. From the two fixed stations, MS2 can compute a TDOA and thus limit the possible locations to a hyperbola in two dimensions, which still means a large location uncertainty. By cooperating with MS1, MS2 can obtain further location information, e.g., through peer-to-peer communication, and determine its location. For instance, MS2 may obtain proximity information, RSS, or RTD from MS1.

The concept of cooperative localization was first proposed in Japan for acquiring real-time positioning information of mobile robots (cf. references in Mensing et al.

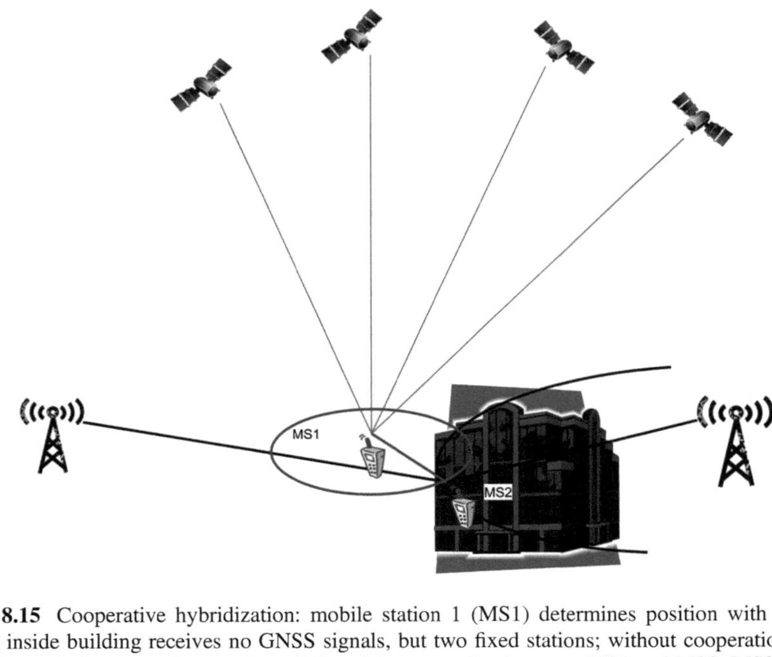

Fig. 8.15 Cooperative hybridization: mobile station 1 (MS1) determines position with GNSS; MS2 inside building receives no GNSS signals, but two fixed stations; without cooperation MS2 location only known to be on a hyperbola; with cooperation between MS1 and MS2, MS2 can be determine its location

2010b). This concept, mostly applied nowadays to wireless sensor networks (WSNs), has been recently introduced to heterogeneous communications networks. Some of the aspects present in the WSN context are also present in general communications networks. However, techniques proposed for WSN cannot be straightforwardly extended to mobile communications networks. This is because these networks usually operate in a very complex wireless environment due to many factors such as shadowing, mobility, communications infrastructure, or multiple air-interfaces. Hence, the heterogeneity of today's wireless communication networks can be seen as an additional problem to be addressed.

In principle, we can distinguish between two different cooperation procedures. In the centralized approach of cooperative localization, e.g., (Mayorga 2007), it is assumed that all information, i.e., the measurements collected by the mobile stations, is provided to one central entity. That could be a location server in a cellular communications system. There, the measurements are jointly processed and the position for each mobile station in the network is determined. Afterwards, this information can be exploited in the network or sent back to the mobile stations. As all measurements are processed jointly in this approach, it is the optimum procedure from a position estimation accuracy point of view. However, the drawback is that all measurements have to be collected at a central entity in advance.

So as to cope with scalability in dense large-scale networks or for mobile-station-centric applications using restricted infrastructure, the distributed cooperative localization approach can also be favored as an alternative to centralized methods, e.g., (Wymeersch et al. 2009). Here, the mobile stations have only the information available that they obtain from their neighbors via peer-to-peer links and the measurements with the fixed stations. Hence, the position estimation complexity is distributed among the mobile stations.

Generally, the communications overhead and extra-signaling is higher for cooperative approaches than for conventional, non-cooperative positioning. Hence, signal-processing complexity and training/signaling overhead are two key problems for existing cooperative localization schemes. This problem can be significant for a wireless network accommodating a large number of mobile stations. Therefore, an efficient cooperative hybridization scheme should achieve the best trade-off between communications overhead and position estimation performance.

Error Mitigation Techniques Through Hybridization

Location dependent measurements are subject to various error sources such as multipath and NLOS propagation, interference and hardware imperfections. With hybridization, we can exploit any redundancy in the location dependent measurements to mitigate these errors. For instance, HDF algorithms can weigh the location dependent measurements according to some side information. Typically this includes signal-to-noise-ratio, signal-to-interference-ratio, and CIR estimates or detected NLOS signals.

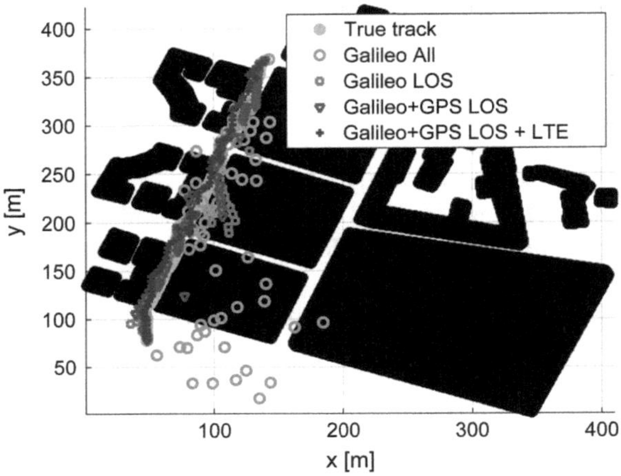

Fig. 8.16 Hybrid positioning with and without NLOS detection for Galileo, GPS, and LTE in Munich urban scenario: position estimates for EKF

For instance in Fig. 8.16, we investigated position estimation of a mobile station equipped with a GNSS and LTE receiver in a typical urban scenario in Munich with an average building height of 26 m for EKF tracking algorithms (Gentner and Sand 2011). In this example, we use the confidence metric

$$\theta_{\mathrm{GNSS}} = \log_{10}\left(\frac{\alpha_1^2}{\alpha_m^2}\right) - \log_{10}\left(\sigma_{\mathrm{N}}^2\right) + 2n\log_{10}\left(\frac{1}{\Delta\tau}\right)\log_{10}\left(\alpha_1^2\right), \tag{8.58}$$

to determine if a received satellite is LOS or NLOS. In (8.58), α_1 denotes the amplitude of the first incoming path, α_m the amplitude of the path with maximum detected amplitude, σ_{N}^2 the noise floor power, n the pathloss exponent, and $\Delta\tau = (\tau_0 + (\tau_m - \tau_1)n_0)$ with τ_0 and n_0 being an initial time offset and normalization factor.

We generated several mobile station tracks for this scenario by applying the gas diffusion mobility model. These tracks include well-behaved GNSS situations with LOS access to several satellites as well as critical scenarios in narrow roads. We assume a fixed GNSS constellation and fixed station constellations with 3 fixed stations where the scenario is at the cell edge. For each position in each track, the CIRs between all sources, i.e., satellites and fixed stations, and the mobile station were determined by raytracing simulations.

Both Figs. 8.16 and 8.17 show that the position accuracy is worst when using all available Galileo satellites including the ones with NLOS biases. Using only the Galileo satellites that are received LOS reduces the maximum error to 25 m. Using both Galileo and GPS LOS signals a further improvement to 18 m is possible as there are fewer cases where less than four satellites are available to estimate the current position and update the EKF tracking filter. Moreover, if the EKF tracking filter uses not only LOS GNSS signals but in addition the LTE signals, 10 m RMSE accuracy can be achieved.

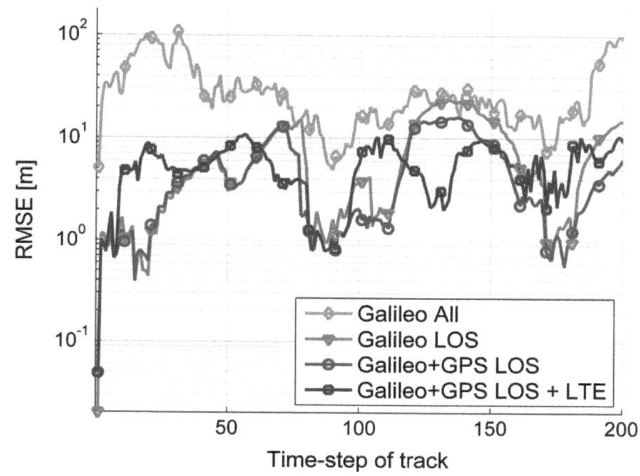

Fig. 8.17 Hybrid positioning with and without NLOS detection for Galileo, GPS, and LTE in Munich urban scenario: RMSE[m] versus time-step of track for EKF

Conclusions

In the previous sections, we discussed different localization information. The simplest localization information is proximity information, i.e., the mobile station communicates with some nearby fixed stations with known positions. While this information has the least requirements on the communication systems, its accuracy can be very limited depending on the communication range of the used communication system. RSS measurements can be more accurate as the measurements are distance dependent. However, the exact distance dependency strongly varies with the environment. Hence RSS measurements are mainly suitable for short- to mid-range communication systems such as WLANs. For mid- to long-range communication systems, RTD, TOA, or TDOA measurements provide better accuracy. However, these measurements also put more stringent requirements on the communication system, e.g., synchronous signal transmission. As modern communication systems such as UMTS, LTE, LTE-Advanced, WiMAX, and WLAN employ MIMO transmission for highest possible data rates, angle of arrival measurement will become feasible for mobile phones.

Whereas the above localization information provides explicit functional relations of the position dependent measurements, fingerprinting is more general. It includes any measurable, location dependent signal characteristic. Fingerprinting stores the measurements in a database and compares current location dependent measurements with the stored values in the database to determine the most likely position. Commercial service providers, e.g., Apple or Google, use already fingerprinting in combination with proximity information and RSS measurements in smartphones. Additionally, smartphones are A-GNSS enabled. Depending on the cellular network operator, this localization information supports a fast time to first fix, weak signal reception, and reduced power consumption in the smartphone.

To realize the benefits of hybridization of Galileo, GPS, or other GNSS signals with localization information from wireless communication systems, several methods exist:

- Static hybridization methods can be used for fast initial or discontinuous localization without requiring tracking and movement information.
- Dynamic hybridization methods exploit the tracking and movement in-formation to improve availability and accuracy of localization services.
- Cooperative hybridization methods extend the concept of hybridization from a single mobile station to several heterogeneous mobile stations that cooperate to obtain robust and accurate position estimates.
- Error mitigation techniques through hybridization help to overcome problems such as interference and NLOS propagation.

The discussions and presented results clearly confirm that the hybridization of Galileo and other GNSSs with localization information from wireless communication systems overcome the impairments of each system and enable mobile phone users to experience seamless LBS.

References

Bailey T, Durrant-Whyte H (2006) Simultaneous localization and mapping (SLAM): part II. IEEE Robot Autom Mag 13(3):108–117

Bluetooth 4.0 (2010) Specification of the Bluetooth System. version 4.0, June

Bruno L, Robertson P (2011) WiSLAM: Improving FootSLAM with WiFi. In: Proceedings of international conference on indoor positioning and indoor navigation (IPIN) Guimaraes, Portugal, Sept 2011

Bucher R, Misra D (2002). A synthesizable VHDL model of the exact solution for three-dimensional hyperbolic positioning system. VLSI Des 15(2):507–520

Chen CL, Feng KT (2005) Hybrid location estimation and tracking system for mobile devices. In: Proceedings of IEEE vehicular technology conference (VTC) Spring, Stockholm, Sweden, 30 May–1 June 2005

Figueiras J, Frattasi S (2010) Mobile positioning and tracking: from conventional to cooperative techniques. Wiley, Chichester

Foy W (1976) Position-location solutions by Taylor-series estimation. IEEE Trans Aerosp Electron Syst 12:187–193

GPP TS 25.215 (2011) Universal mobile telecommunications system (UMTS); physical layer; measurements (FDD). Technical specification, version 10.0.0, April

GPP TS 25.305 (2011) Universal mobile telecommunications system (UMTS); stage 2 functional specification of user equipment (UE) positioning in UTRAN. Technical specification, version 10.0.0, May

GPP TS 36.211 (2011) LTE; Evolved universal terrestrial radio access (E-UTRA); physical channels and modulation. Technical specification, version 10.1.0, April

GPP TS 36.214 (2011) LTE; Evolved universal terrestrial radio access (E-UTRA); physical layer; measurements. Technical specification, version 10.1.0, April

GPP TS 36.305 (2011) LTE; Evolved universal terrestrial radio access network (E-UTRAN); stage 2 functional specification of user equipment (UE) positioning in E-UTRAN. Technical specification, version 10.1.0, April

GPP TS 43.059 (2011) Functional stage 2 description of location services (LCS) in GERAN. Technical specification, version 10.0.0, April
GPP TS 45.008 (2011) Digital cellular telecommunications system (Phase 2+); radio subsystem link control. Technical specification, version 10.0.0, April
Gentner C, Rewadi JM, Munoz Diaz E, Khider M (2012) Hybrid positioning with 3GPP-LTE and GPS employing particle filters. In: Proceedings of ION GNSS 2012. ION GNSS, Nashville, Tennessee, USA, Sept 2012
Gentner C, Sand S (2011) NLOS detection and mitigation based on confidence metric and EKF. In: Proceedings of ION international technical meet (IMT), San Diego, CA, USA, Jan 2011
GRAMMAR D1.4 (2009) Gap analysis of GNSS receivers and technology. Galileo Ready Advanced Mass Market Receiver (GRAMMAR) project Deliverable 1.4, June
GSA (2010) GNSS market report. http://www.gsa.europa.eu/go/download_the_gsa_gnss_market-report, Accessed Oct 2010
Gustafsson F, Gunnarsson F, Bergmann N, Forssell U, Jansson J, KarlssonR, Nordlund PJ (2002) Particle filters for positioning, navigation, and tracking. IEEE Trans Sig Process 50(2):425–437
Heinrichs G, Dovis F, Gianola M, Mulassano P (2004) Navigation and communication hybrid positioning with a common receiver architecture. In: Proceedings of European navigation conference GNSS, Rotterdam (NL), 16–19 May 2004
IEEE 802.11n (2009) Wireless LAN medium access control (MAC) and physical layer (PHY) specifications—Amendment 5: Enhancements for higher throughput. IEEE standard, October
IEEE 802.15.4a (2007) Wireless medium access control (MAC) and physical layer (PHY) specifications for low-rate wireless personal area networks (WPANs)—Amendment 1: Add alternate PHY. IEEE standard, August
IEEE 802.16 (2009) Local and metropolitan area networks part 16: air Interface for broadband wireless access systems. IEEE standard, May
Julier SJ, Uhlmann JK (1997) A new extension of the Kalman filter to nonlinear systems. In: Proceedings of AeroSense: international symposium on aerospace/defense sensing, simulation and controls, multi sensor fusion, tracking and resource management II, SPIE
Kay SM (1993) Fundamentals of statistical signal processing, estimation theory, vol 1. Prentice Hall Signal Processing Series, Engle-wood Cliffs
Levenberg K (1944) A method for the solution of certain problems in least squares. Q J Appl Math 2:164–168
Ma C (2003) Integration of GPS and cellular networks to improve wireless location performance. In: Proceedings of ION GNSS, September
Marquardt D (1963) An algorithm for least squares estimation of nonlinear parameters. SIAM J Appl Math 11:431–441
Mayorga CLF, Della Rosa F, Wardana SA, Simone G, Raynal MCN, Figueiras J, Frattasi S (2007) Cooperative positioning techniques for mobile localization in 4G cellular networks. In: Proceedings of IEEE international conference on pervasive services, July
Mensing C, Plass S (2006) Positioning algorithms for cellular networks using TDOA. In: Proceedings of IEEE international conference on acoustics, speech, and signal processing (ICASSP), May
Mensing C, Sand S, Dammann A (2010a) Hybrid data fusion and tracking for positioning with GNSS and 3GPP-LTE. Int J Navig Obs 2010:12. doi:10.1155/2010/812945
Mensing C, Sand S, Nielsen JJ, Denis B, Maman M, Rodriguez J, Hadzic S, Bastos J, He Z, Ma Y, Zazo S, Vladimir S, Arambasic I, Laaraiedh M, Bernard U (2010b) Performance assessment of cooperative positioning techniques. In: Proceedings of future network and mobile summit, Florence, Italy, 16–18 June 2010
Mensing C, Sand S, Dammann A, Utschick W (2009) Interference-aware location estimation in cellular OFDM communications systems. In: Proceedings of IEEE international conference on communications (ICC '09), 14–18 June 2009
Misra P, Enge P (2004) Global Positioning System: Signals, Measurements, and Performance. Ganga-Jamuna Press, Lincoln

Nanotron Technologies (2011) Symmetrical double-sided two way ranging. http://www.nanotron.com/EN/CO_techn-sdstwr.php. Accessed 14 May 2011

Qi Y, Kobayashi H, Suda H (2006) Analysis of wireless geolocation in a non-line-of-sight environment. IEEE Trans Wirel Commun 5(3):672–681

Rappaport TS (2002) Wireless communications—principles and practice, 2nd edn. Prentice Hall, Upper Saddle River

Ristic B, Arulampalam S, Gordon N (2004) Beyond the Kalman filter, particle filters for tracking applications. Artech House, London

Sarkar TK, Wicks MC, Salazar-Palma M, Bonneau R (2003) Smart antennas. Wiley, New York

Urruela A, Sala J, Riba J (2006) Average performance analysis of circular and hyperbolic geolocation. IEEE Trans Veh Technol 55(1):52–66

Wan EA, van der Merwe R (2000) The unscented Kalman filter for nonlinear estimation. In: Proceedings of adaptive systems for signal processing, communications, and control symposium (AS-SPCC), October

WHERE D3.1 (2009) Physical layer enhancements using localisation data. Wireless hybrid enhanced mobile radio estimators (WHERE) project deliverable 3.1, January

Wymeersch H, Lien J, Win MZ (2009) Cooperative localization in wireless networks. In: Proceedings of IEEE, February

Chapter 9
Putting It All Together: TUTGNSS Receiver Platform

Jussi Raasakka and Heikki Hurskainen

Abstract The following chapter describes the TUTGNSS receiver implementation. TUTGNSS receiver is developed in Tampere University of Technology, Department of Electronics and Communication Engineering. The main target of the TUTGNSS receiver is to give to the GNSS community a free open source GNSS receiver for non-commercial purposes where new algorithms and architectures can be studied. As the TUTGNSS receiver is a host-based system with both hardware and software being targeted as open source, new architectures and algorithms for GNSS receivers can be implemented and tested. Currently, the TUTGNSS receiver supports GPS L1/L5 and Galileo E1/E5a frequencies. Future work will be to add GLONASS support.

TUTGNSS: Introduction

History

The history of the TUTGNSS started from the Advanced Techniques for Mobile Positioning (MOT) project, a project funded by the Finnish Funding Agency for Technology and Innovation (Tekes). MOT project was the beginning of long line of projects leading to the development of the TUTGNSS receiver. In the MOT project acquisition algorithms for GNSS receivers were studied. After completion of the MOT project, a Tekes-funded continuation project called Advanced Techniques for Personal Navigation (ATENA) was initiated. In the ATENA project, lessons from the MOT project were gathered, and fully functional acquisition and tracking units for GPS receivers were implemented.

During the EU FP6-funded Galileo Receivers for Mass Market (GREAT) project, a first receiver prototype using L1/E1 band was created. Then, after successful

J. Raasakka (✉) · H. Hurskainen
Department of Electronics and Communications Engineering,
Tampere University of Technology, Tampere, Finland
e-mail: jussi.raasakka@tut.fi

J. Nurmi et al. (eds.), *GALILEO Positioning Technology*,
Signals and Communication Technology 182, DOI 10.1007/978-94-007-1830-2_9

completion of the GREAT project we decided to continue the development of TUTGNSS receiver by adding E5a/L5 frequency capability. This development was made in the EU FP7-funded Galileo Ready Advanced Mass MArket Receiver (GRAMMAR) project. Adding E5a/L5 support meant restructuring and rethinking of the hardware of the TUTGNSS accelerator. The hardware and software of TUTGNSS receiver capable of handling both L1/E1 and L5/E5a signals is explained in the following chapters.

Hardware Accelerated Software Receiver

In the history of TUTGNSS, many different architectural choices for the TUTGNSS receiver have been investigated. GNSS architectures can be divided into three categories.

1. Pure hardware receiver, where all processing is done using dedicated hardware.
2. Software receiver, where all processing is done using general purpose processor(s) or DSP(s).
3. Host based solution, where most computationally expensive operations are done in hardware. Other operations are done using general purpose processor(s) or DSP(s).

After careful considerations, we ended up with host based solution. As TUTGNSS receiver is mainly build for research activities in mind, host based solution offered both good flexibility through software updates and enough processing power for even the most demanding algorithms through hardware accelerators. Conceptual view of TUTGNSS receiver is given in Fig. 9.1.

The hardware part of TUTGNSS receiver consists of acquisition and tracking units that can be synthesized into Field-Programmable Gate Array (FPGA). The number of the acquisition and tracking units is restricted only by the size of the FPGA device, so TUTGNSS accelerator hardware can be configured to use only a small portion of the FPGA resources, or on the other hand it can be configured to take up all the available resources of the FPGA increasing the parallelism, thus increasing the TUTGNSS accelerator performance.

Software part of TUTGNSS receiver consists of controlling the hardware functionality, decoding satellite data, and calculating user position, velocity, and time (PVT) solution.

TUTGNSS: Platform

TUTGNSS receiver is implemented in Altera Stratix II DSP professional development board (Altera 2005). In principle TUTGNSS receiver can be synthesized to any available Altera FPGA board if the FPGA has enough resources so that the

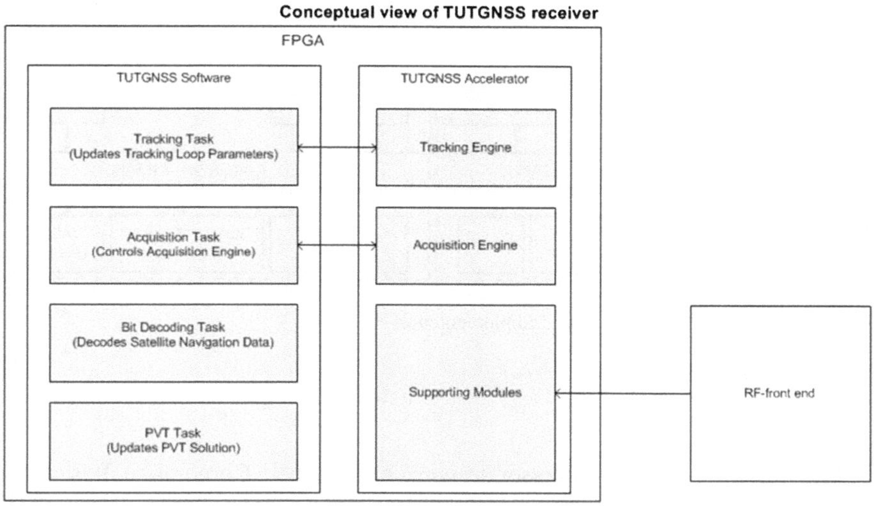

Fig. 9.1 Conceptual view of TUTGNSS receiver

TUTGNSS receiver will fit into corresponding FPGA chip. Additional requirement for the FPGA platform is that it has suitable expansion interfaces to connect the RF-front end to the FPGA platform.

System Configuration

Current implementation of the TUTGNSS receiver uses Altera SOPC builder to build up the system. Figure 9.2 shows the components used to build up the TUTGNSS receiver.

TUTGNSS system uses two Nios II processors. One is used as the host processor for TUTGNSS hardware and the other is for debugging and logging purposes. Design choice was made to separate the debugging part and logging part from the actual TUTGNSS receiver. This enables the use of TUTGNSS receiver independently of the debugging domain and it makes the TUTGNSS hardware working as a standalone receiver. In order to support good visibility and logging options we have included second processor to monitor the status of the TUTGNSS receiver. Other components, such as the performance counter are included in case some part of the software needs to be timed for performance reasons. Timers are needed for the Real-Time Operating System (RTOS).

We decided to put as much of the host processor code into tightly coupled memories to ensure real-time operation. This puts heavy restrictions on the code size that is available for the GNSS application, but the current implementation uses roughly 50 % of the available memory from Stratix II EP2S180 device.

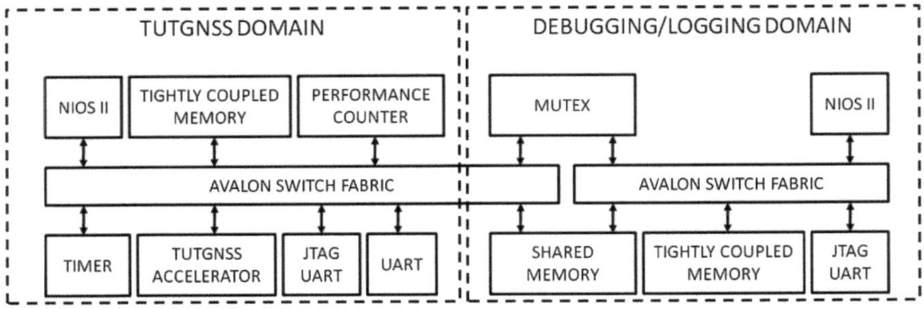

Fig. 9.2 TUTGNSS receiver SOPC implementation

Nios II

Nios II processor is a soft processor designed by the Altera Corporation. It is chosen for controlling the baseband unit since it provides necessary performance for the TUTGNSS application, and also has good support from Altera.

Hardware Software Interface

TUTGNSS accelerator is seen from the NIOS II processor as a memory mapped slave interface. This enables easy interfacing to TUTGNSS hardware with simple read/write commands through NIOS II API. Current TUTGNSS receiver uses 16-bit address space to access its control registers. However it should be noted that the actual address space needed for the TUTGNSS control registers is much smaller. Reason for choosing such large address space was to make sure that future enhancements can be easily fitted into the currently available address space. Larger address space could be needed, e.g. for storing RF data samples either for software processing, or logging of data samples between each state of the signal processing chain within the TUTGNSS accelerator hardware.

Since the TUTGNSS receiver is targeted for open source for academic research and study purposes, TUTGNSS hardware is kept as easily upgradeable as possible. This leads to some overhead in the size of the hardware, but we believe that the justification for this overhead is valid.

TUTGNSS: Accelerator

TUTGNSS accelerator handles computationally expensive operations of the TUTGNSS receiver such as acquisition and tracking of GNSS signals. One of the key principles of the TUTGNSS accelerator design was to keep the hardware as modular as possible. This reduces the effort for the development of future extensions to the system. As an example, all of TUTGNSS accelerator control registers and interface to host processor resides in a single module. Adding new module to the system or even changing communication interface to the host processor can be achieved with modifications to a single module. TUTGNSS accelerator supports Altera Memory-Mapped interface (Altera 2011) and SPI interface to communicate to the host processor.

TUTGNSS accelerator has been implemented to directly support Altera SOPC builder design flow. This means that TUTGNSS accelerator can be directly instantiated from the SOPC builder tool. Using Altera SOPC builder design flow TUTGNSS receiver can be build from the scratch in just few minutes.

Overview of TUTGNSS Baseband Hardware Blocks

As the TUTGNSS accelerator is based on the modular approach, the functionality of the hardware is divided into modules, each having their distinctive job in the signal processing chain of TUTGNSS receiver.

Figure 9.3 shows the top level architecture of the TUTGNSS accelerator. Master Control Unit (MCU) handles communication between TUTGNSS accelerator and host processor. Baseband Converter Unit (BCU) is used to synchronize and convert different RF Radio Front Ends into format suitable for the TUTGNSS accelerator. Sample Counter Unit (SCU) is used to measure time within the TUTGNSS accelerator. Acquisition Unit (AU) handles the acquisition processing and Tracking Unit (TU) handles tracking. Galileo Memory Unit (GMU) is used to store the Galileo memory codes. Phase Measurement Unit (PMU) is used to measure phases of the incoming satellite signals. Tracking Result Unit (TRU) is used to store the results coming from the tracking part of signal processing. In the following subchapters, functionality of each module is described in more detail.

Master Control Unit

The Master Control Unit is the heart of the TUTGNSS receiver. It contains all control registers that control the TUTGNSS accelerator. Current implementation of the TUTGNSS MCU includes both Avalon Memory-Mapped Interface and SPI Interface to access the TUTGNSS accelerator hardware. Both of these interfaces can be used to control the TUTGNSS hardware, though Avalon Memory-Mapped Interface is the preferred one.

MCU includes SPI master so that configuration for supporting hardware such as RF-front ends can be made without any external modules. Currently the SPI master supports simultaneously up to six SPI slaves.

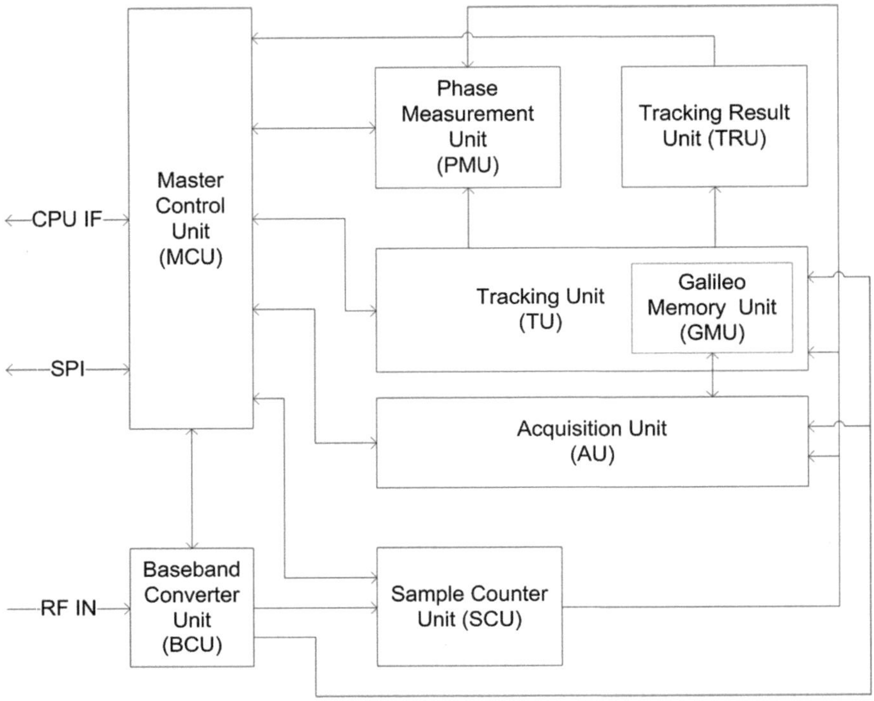

Fig. 9.3 TUTGNSS baseband hardware architecture (*top level*)

Since all the control code is contained in a single module, it is easy to add support for additional control interfaces. Only thing that needs to be changed is the MCU part of the hardware design.

Baseband Converter Unit

The baseband converter unit changes any type of RF-front end to common three bit I&Q signed presentation. Additionally, BCU removes the Intermediate Frequency (IF), so that acquisition and tracking units need only to account the remaining Doppler frequency of the incoming satellite signal. BCU architecture is given in Fig. 9.4. BCU contains filtering and optional decimation for input, separately for both acquisition and tracking. Separate signal paths are required to enable low complexity acquisition using lower sample rate, and high precision tracking using higher sample rate.

BCU has built-in support for several commercially available RF-front ends. This enables switching RF-front end without changes to the TUTGNSS accelerator.

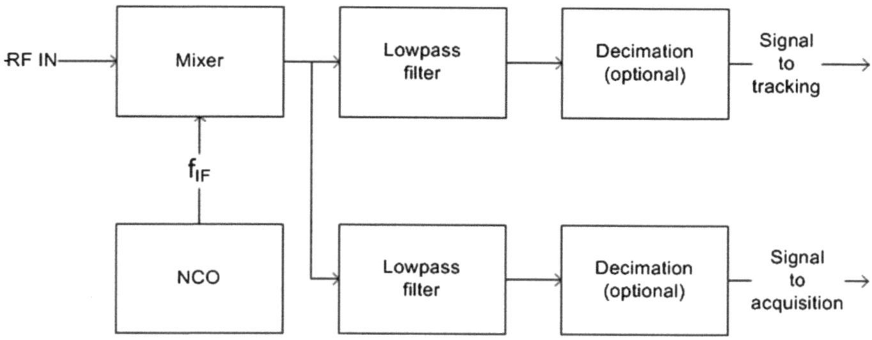

Fig. 9.4 BCU architecture

Additionally BCU includes debugging mode where RF-front end data can be emulated through SPI interface. This enables good debugging capabilities to the TUTGNSS receiver as the input data parameters (Satellites/Doppler/Delays) are known. This enables verification of correct operation of the TUTGNSS receiver. Dataset for RF-Front End emulation can be easily made using software programming environments, such as Matlab®.

Sample Counter Unit

The sample counter unit (SCU) measures the time of the TUTGNSS receiver by counting the samples coming from the RF-front end. SCU is used as local time reference within the TUTGNSS receiver. SCU timing information is also used to synchronize all TUTGNSS accelerator modules together. SCU takes its input from the BCU and updates its counter whenever new sample arrives. When software knows the incoming sample rate this information can be used as very high precision clock, which accuracy is depended on the RF-front end oscillator. If software uses debugging mode of the BCU then time SCU increments only when user drives new samples to BCU input.

SCU is heavily dependent of the sampling frequency of the RF-front end. The sampling frequency defines the smallest time unit in TUTGNSS receiver, which is one SCU increment.

Acquisition Unit

The acquisition unit searches satellites from the incoming data stream. Acquisition unit is used by setting desired satellite search parameters through control register interface and then starting the acquisition by writing any value to acquisition start register. Acquisition result (Amplitude/Doppler/Delay) can be read from the appropriate control registers after acquisition has been completed.

Currently, the AU is implemented as parallel matched filter structure, which we considered to be the best architecture for hardware implementation. There are several discussions between the matched filter approach versus the FFT approach in the literature (see also Chap. 5). The matched filter is more straightforward to implement in hardware, and due to small word length of the incoming RF-front end data (typically 1–3 bits) it can be implemented efficiently.

The architecture of acquisition engine is illustrated in Fig. 9.5. The engine implements parallel code search by using matched filters. Acquisition is implemented only for E1/L1 frequency signal. L5/E5a signal acquisition can be formed from L1/E1 tracking results, since L5/E5a signals share the common dynamic behavior with their L1/E1 counterparts.

The incoming I/Q signal can be decimated in BCU to decrease the complexity and to allow longer integrations with smaller number of correlators. The remaining carrier and Doppler components are removed from incoming stream before it is saved to data memory. Similarly, replica codes are generated (or in Galileo E1 case, read from memory) and saved to PRN memory. Then the contents of memories are correlated in parallel fashion and result is saved to result memory. Coherent integration length is defined to be 4 ms, which is the length of Galileo E1 OS signal epoch length. The number of non-coherent integrations can be set though control register interface.

Tracking Unit

The tracking unit handles the tracking of each satellite found from the incoming data stream. The current implementation of TUTGNSS HW/SW satellite tracking can handle GPS L1/L5 + Galileo E1/E5a signals.

The channels can be combined to work in a master/slave mode, meaning that one Galileo satellite is being tracked by two tracking channels. One channel is tracking the data channel, and the other one is tracking the pilot. The tracking channels are fully software customized meaning that each channel can be assigned as a master or slave and to track any GPS L1/L5 or Galileo E1/E5a signal.

Each tracking channel has 8 correlators to enable advanced algorithms, such as advanced multipath mitigation algorithms to be run on the platform. Also, TUTGNSS accelerator includes other advanced techniques, such as rate and phase feedback for the NCO's enabling state-of-the-art loop filters.

TUTGNSS accelerator produces correlation results every epoch. TUTGNSS software is used increase coherent/non-coherent integration period by simply adding these results coherently/non-coherently.

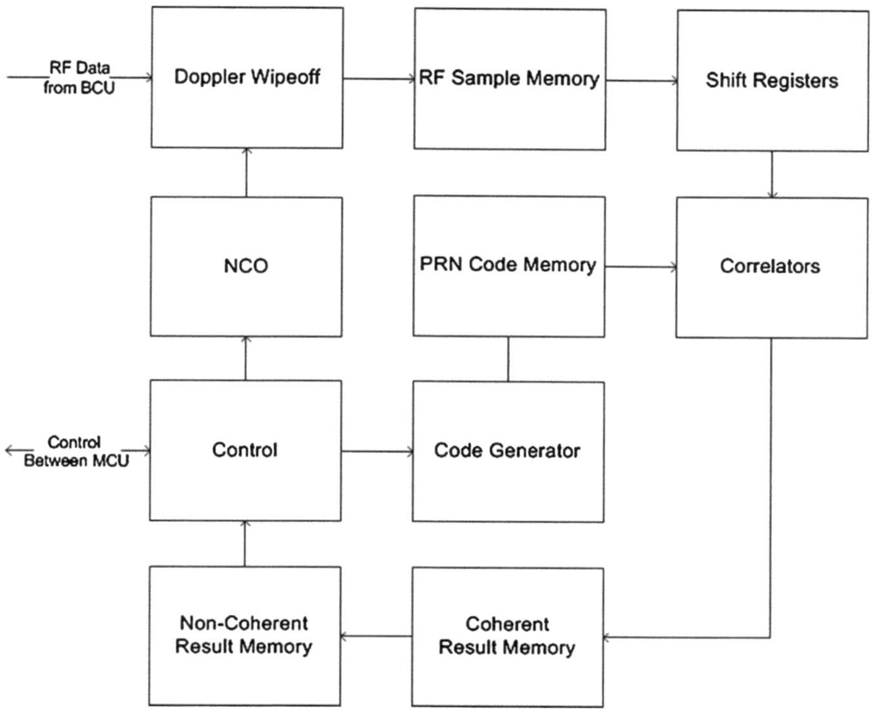

Fig. 9.5 Acquisition unit

Tracking Channels

Software register interface is designed to support maximum of 16 tracking channels. More tracking channels can be added to the TUTGNSS receiver, but it requires modification to the register interface.

When TUTGNSS accelerator is provided with dual frequency input with two ADCs, the scheme in Fig. 9.6 is used. Each channel has its own pre-channel multiplexer, which selects which signal is forwarded to the channel. All channels operate independently, allowing maximum flexibility over number of channel allocated to a certain frequency (Fig. 9.6).

Flexible Tracking Channel

The tracking channels are responsible of producing tracking results from the tracked satellites. As the tracking process is quite straightforward from the hardware point of view, only few control registers are required. Mainly code/carrier Doppler

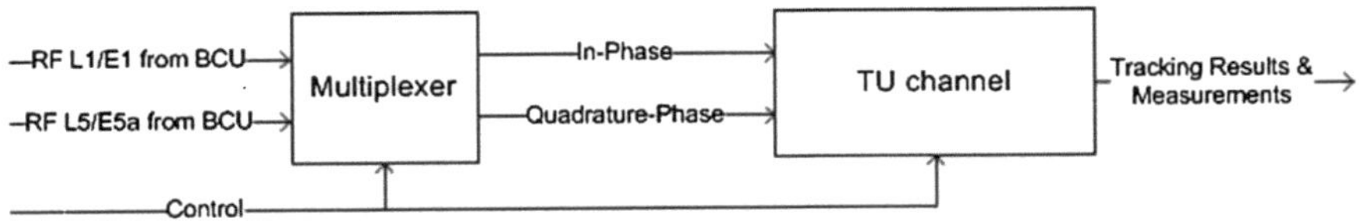

Fig. 9.6 Pre-channel multiplexers (same approach is used to all channels)

updates are required to keep track of each satellite. To ensure proper calculation of correction parameters in the TUTGNSS software, updates to satellite parameters are only checked at each epoch edge.

The architecture of the TUTGNSS flexible tracking channel is given in Fig. 9.7.

Code Generation

The code generation for Galileo E5a (Fig. 9.8), GPS L1 (C/A) (Fig. 9.9), and GPS L5 (Fig. 9.10) are implemented in order to support all signals of interest. This is done by implementing generic Linear Feedback Shift Register (LFSR) structure with long enough registers and reconfigurable feedback taps and registers output point.

Galileo E1 codes are implemented as memory codes. BOC modulation is not saved to the memory, but a simple BOC generator is used instead.

Galileo E5a generator can also produce Galileo E5b codes since the structure of coder in both is same. However, TUTGNSS receiver does not officially support Galileo E5b signals.

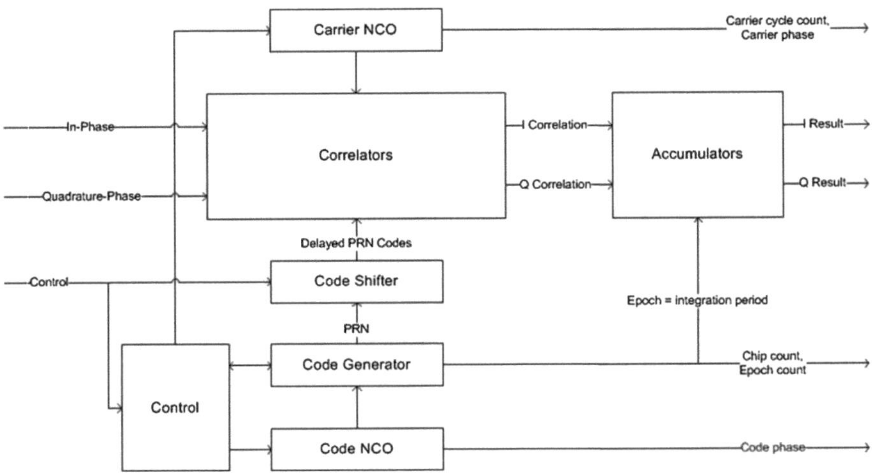

Fig. 9.7 TUTGNSS channel

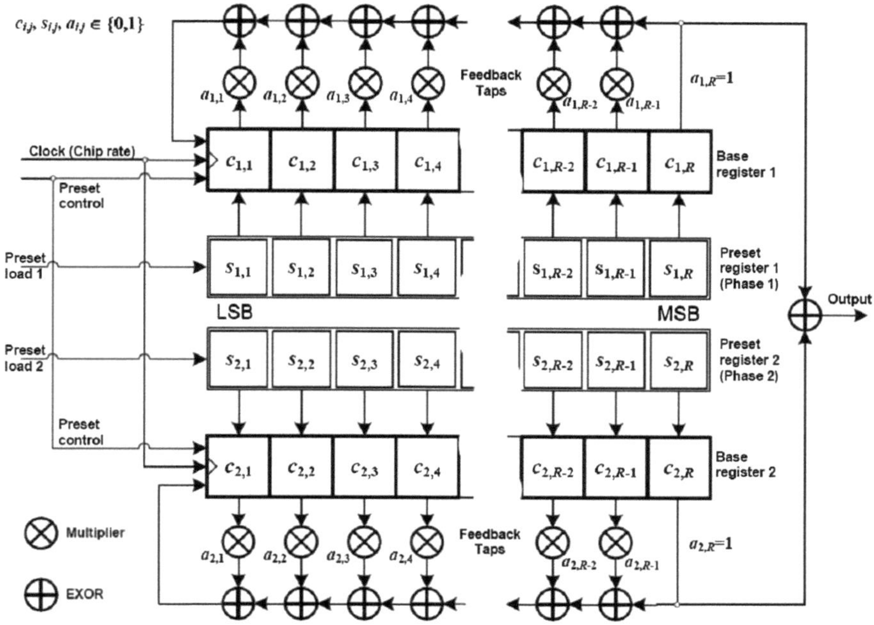

Fig. 9.8 Galileo code generator ©ESA/GSA (Galileo ICD 2010)

Correlators

The correlators of TUTGNSS accelerator tracking channel with used word lengths are illustrated in Fig. 9.11. The structure supports only 1 bit PRN code with 8 delays, and thus CBOC is not currently supported.

GMU

In order to support Galileo E1 OS memory codes, a new hardware entity, Galileo Memory Unit (GMU) is included in the TUTGNSS accelerator. TUTGNSS accelerator contains only one Galileo code memory where all the 16 tracking channels and the acquisition unit fetches the codes on blocks of 32 bits. An arbiter is used to ensure access and predefined priorities for multiple sources. Tracking channels have higher priority over acquisition, based on the necessity to use tracking continuously whereas acquisition is performed only periodically.

GMU communicates with tracking and acquisition via seventeen 16bit address buses. PRN code is fetched as 32 bit data blocks. A block diagram of GMU is given in Fig. 9.12.

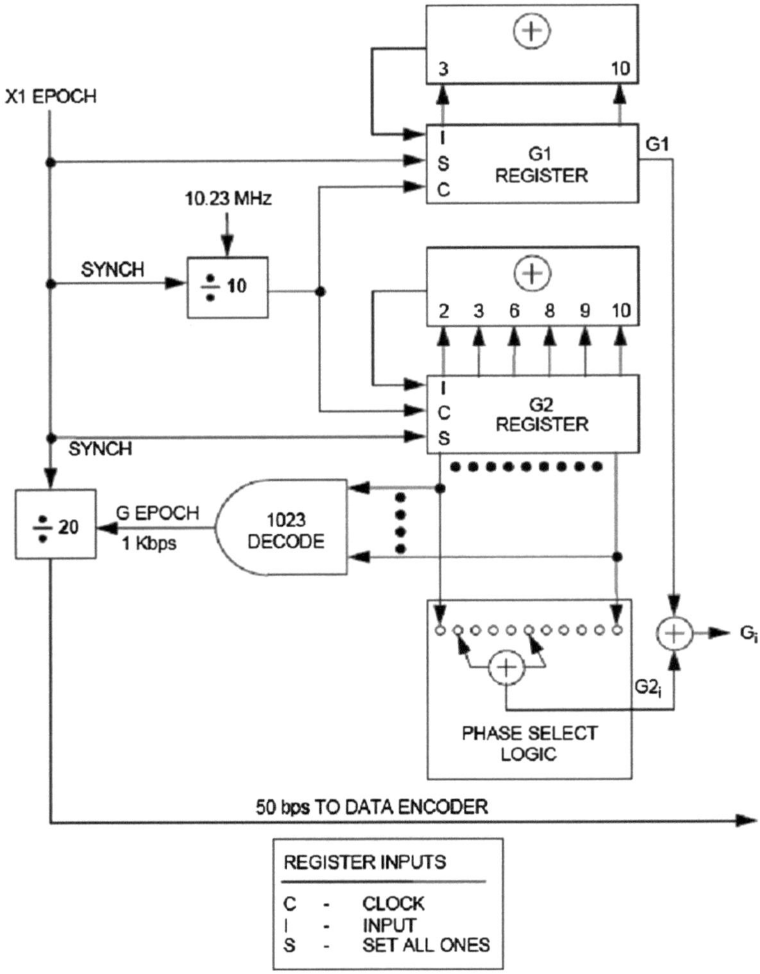

Fig. 9.9 GPS L1 (C/A) code generator (GPS 2010)

GMU is implemented as a single port read only memory containing 32 PRN codes (first 32 from ICD) for both data (E1-B) and pilot (E1-C) component of the Galileo E1 signal. Four '0' bits are padded after each PRN code to align the starting point for a code better (i.e. for 4,092 chips 4,096 bits of memory are used). Codes are stored without any additional modulation.

Tracking Result Unit

The tracking result unit collects all results from the TU. It is functionally a FIFO which collects all the integrated results from the TU so that software can later read

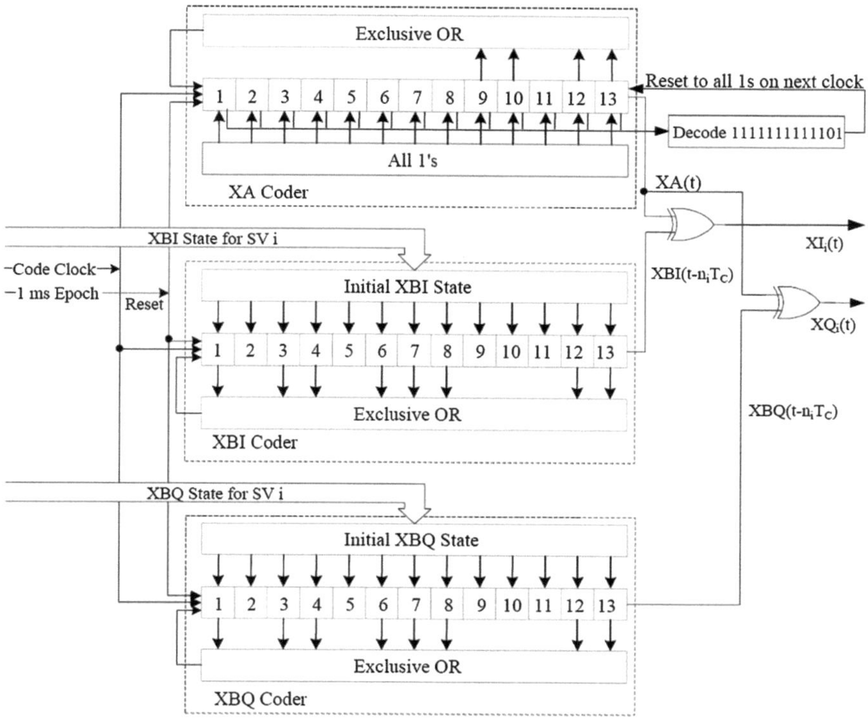

Fig. 9.10 GPS L5 code generator (GPS 2011)

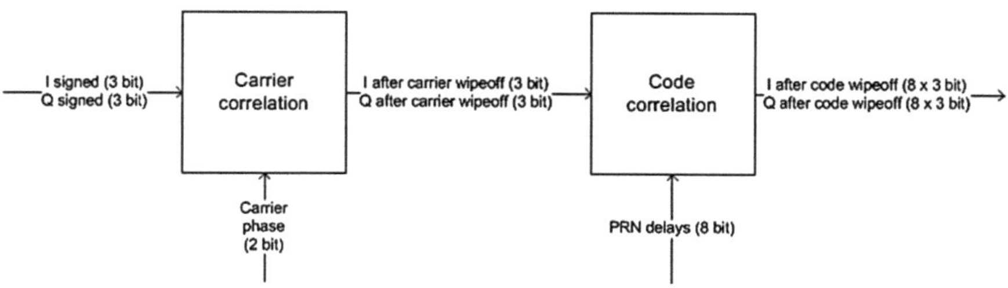

Fig. 9.11 TUTGNSS correlators

the results when it has time to do it. The software part of the TUTGNSS receiver uses these results mainly to update TU NCO registers and demodulate data information concerning each satellite signal. The results are generated at each epoch edge so regarding to different GNSS systems/signals, the integration time may be different. The interface is specified to support up to 8 correlators, but the hardware may have limitations on how many correlators there are in the TUTGNSS accelerator.

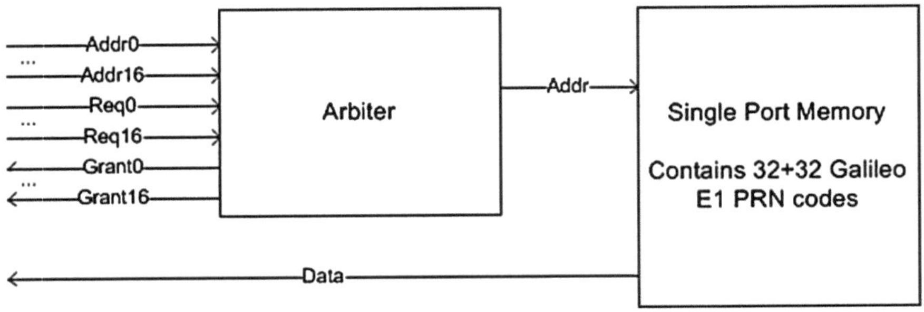

Fig. 9.12 Galileo memory unit

Phase Measurement Unit

The phase measurement unit measures the phases of the tracking channel NCOs at software specified intervals. Measurements are made for both the code and carrier phases. This phase measuring is related to the local time information received from the SCU. The phases are measured from every tracking channel simultaneously, so the software can derive pseudorange information for each satellite with the aid of demodulated data from the satellites. PMU stores additional information about code and carrier phases so advanced methods on pseudorange calculation are also possible.

TUTGNSS: Software

Operating System

The TUTGNSS software is designed to run on top of a RTOS. Since Altera already provides good support to the MicroC/OS II RTOS (Micrium 2012) it was chosen for implementation.

There were several reasons why the decision was made to use RTOS in the host processor. We concluded that it eases up the development of the control software, since many of the GNSS receiver functionalities are inherently parallel (such as acquisition and tracking). Using tasks running in parallel made the control software development much simpler. Making our own scheduler would cost us significant overhead, not to mention the testing cost for such software.

Another benefit from the RTOS software is that we can easily extend the host processor software to include additional features, such as TCP/IP communication through Ethernet or data logging capabilities to memory cards. Adding these features would need significant coding effort from our coding team. So keeping the enhancements/upgrades as simple as possible we decided to go for MicroC/OS II RTOS.

Interrupts

Like many typical embedded applications, TUTGNSS software is mainly interrupt-driven software. TUTGNSS hardware uses three different interrupts to alert control software that the processing of RF-front end data has been completed. Main purpose of the Interrupt Service Routine (ISR) is to forward data from the TUTGNSS accelerator to appropriate task depending of the interrupt type. Data transfer between ISR and tasks is handled via message queues. The activity diagram of ISR is presented in Fig. 9.13.

The acquisition interrupt is generated when the acquisition engine has finished processing the acquisition of the Doppler, and delay bins of the chosen satellite.

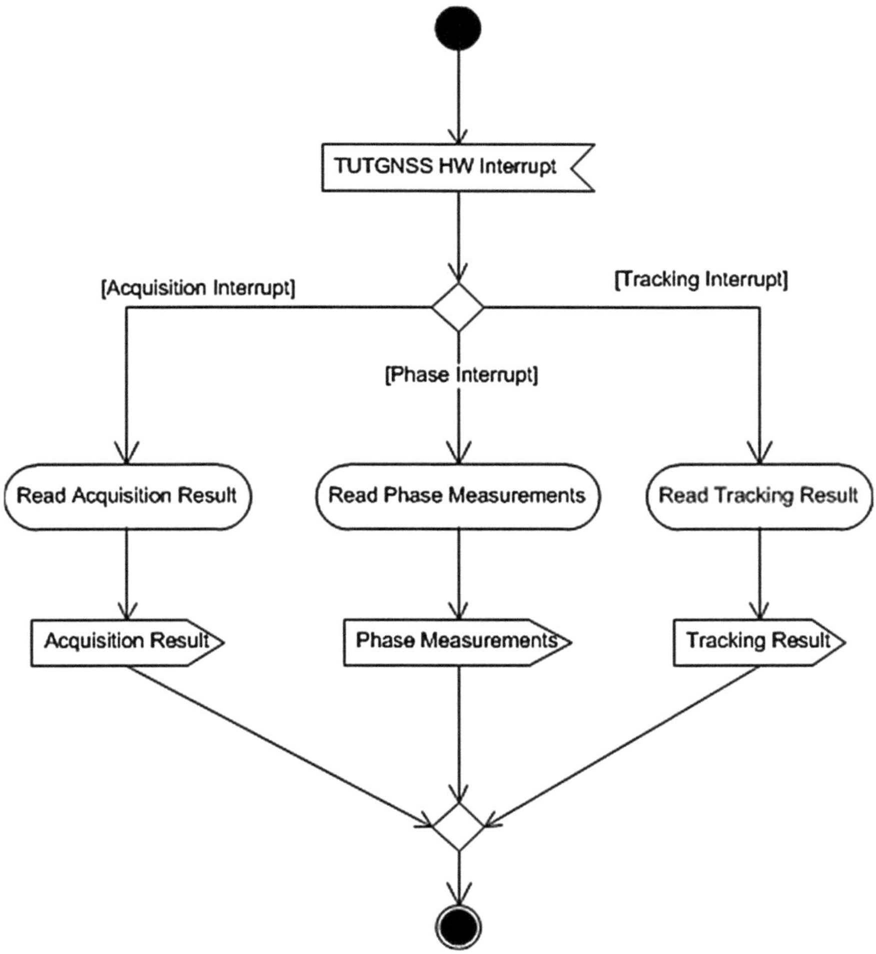

Fig. 9.13 TUTGNSS interrupt service routine

The tracking interrupt is generated whenever any active channel has finished processing one epoch of incoming radio front end data.

The phase interrupt is generated at variable rate decided by the PVT task through control register interface. The phase interrupt is raised when a new phase measurement from each channel is ready for further processing.

Tasks

TUTGNSS software is divided into different tasks mainly handling the information provided by TUTGNSS accelerator trough ISR. All individual tasks are specified to run during the whole lifetime of TUTGNSS receiver application.

Acquisition

The acquisition part of the TUTGNSS hardware is controlled by the acquisition task. Activity diagram showing the main functions of the acquisition task is shown in Fig. 9.14.

The first acquisition task is to determine the need to perform a new acquisition (if any). Initially, the acquisition control software determines the need to perform

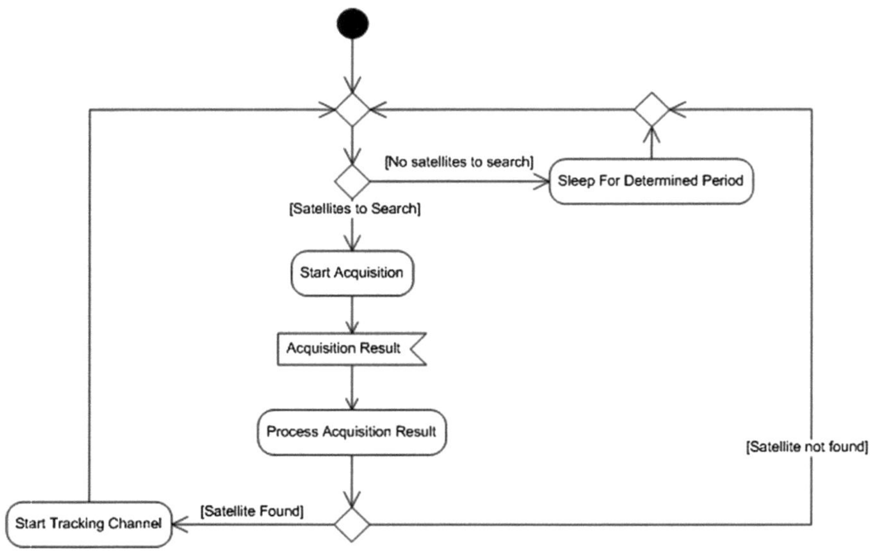

Fig. 9.14 Acquisition task

new acquisition (if any). If all the tracking channels are in use, i.e., tracking satellites, or if the acquisition control software determines that all currently visible satellites are being tracked, the acquisition control software goes to idle mode. If the acquisition control software goes idle, it checks after a predetermined time interval whether the situation has changed and acts accordingly.

If the acquisition control software decides that it should start the acquisition procedure, it updates the TUTGNSS accelerator control registers values, determined by the acquisition control software logic. Then, the acquisition task goes idle until it is awoken by the signal sent by the interrupt service routine, notifying that TUTGNSS accelerator acquisition engine has finished processing the data.

If the acquisition control software logic decides that the searched satellite is in view (based on the acquisition results), the TUTGNSS tracking channel is initialized with the acquisition results. This process keeps repeating during the whole lifetime of the TUTGNSS receiver application.

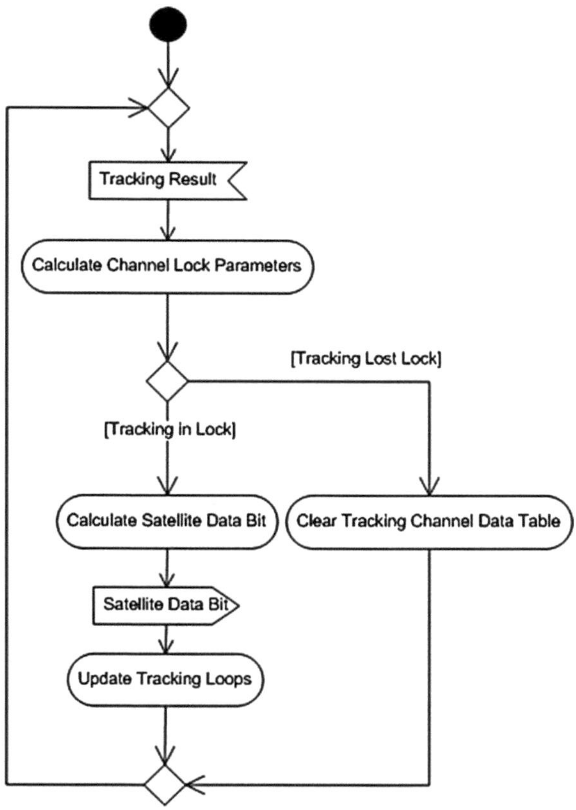

Fig. 9.15 Tracking task

Tracking

The tracking part of the TUTGNSS software is responsible of the tracking of the acquired satellite signals from the sky. Currently the tracking engine for the TUTGNSS receiver supports GPS L1/L5 and Galileo E1/E5a signals. The tracking module follows the tracking procedure shown in Fig. 9.15.

After getting the tracking result, the software decides whether we are still tracking a valid GNSS satellite, so that we can provide pseudorange measurements. If the tracking procedure still believes that it is tracking a valid satellite, then it decodes the satellite data parameters according to each GNSS ICD. After validating that the tracking channel is tracking a valid satellite, its parameters are updated for the next integration round.

Bit Decoding

The bit decoding gets the determined data bits from tracking. Data bits are then further processed depending on the satellite carrier type. In the case of the GPS L1 signal, only the search for the preamble and the check of the parity bits are necessary. However when processing Galileo E1, more advanced methods are necessary. These methods include block interleaving, CRC calculation, and Viterbi

Fig. 9.16 Bit decoding task

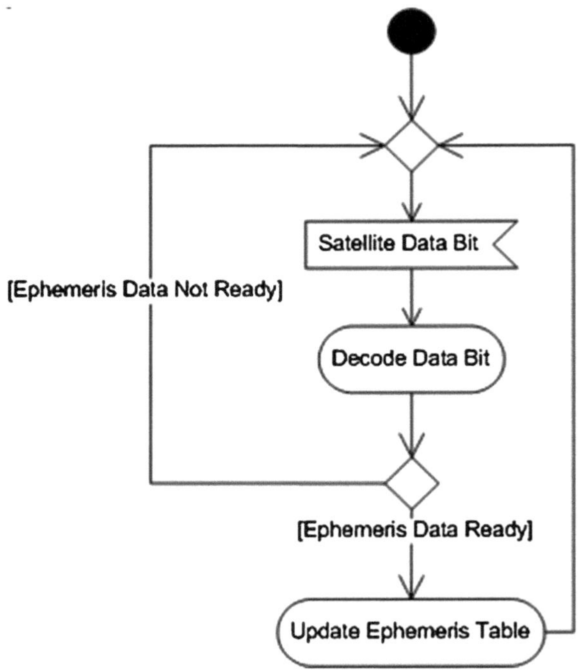

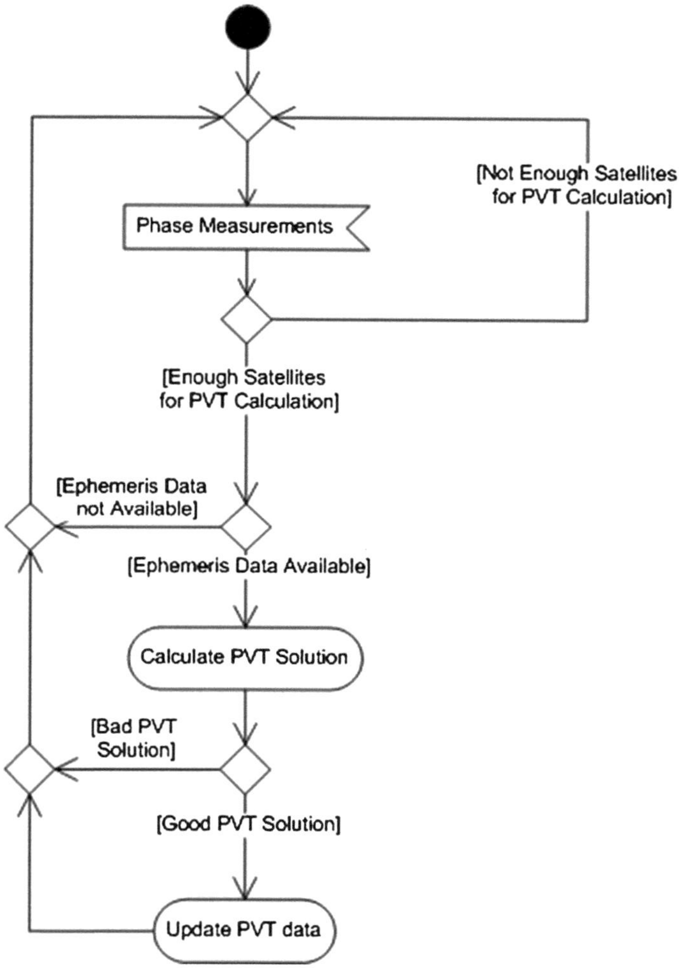

Fig. 9.17 PVT task

decoding. Even though this greatly enhances the integrity of the navigation data reception, it causes much strain on the decoding of the navigation data bits for Galileo receivers. In TUTGNSS receiver it was chosen to implement the Viterbi decoding in hardware, in order to enable real-time processing, without causing too much load to the NIOS II core. Figure 9.16 shows the simplified flow of the bit decoding task.

PVT

The Position, Velocity, and Time (PVT) task is used to update the user PVT solution. This task periodically checks if enough satellites are in the tracking mode, so that the calculation of PVT solution is possible. Depending on the situation, 2–4 satellites need to be in tracking mode in order to proceed with updating the user PVT solution. After confirming that the PVT engine has enough satellites to calculate PVT solution, The PVT engine checks whether all necessary parameters for the satellite position calculation have been obtained (ephemerides parameters). If both of these checks pass, PVT engine calculates the new PVT solution. The calculated PVT solution is verified using the RAIM/FDE checks, which are built into the TUTGNSS receiver. If the PVT solution passes RAIM/FDE checks, the TUTGNSS receiver PVT information is updated. Figure 9.17 shows the activity diagram for PVT calculation.

References

Altera (2005) Stratix II EP2S180DSP development board. http://www.altera.com/literature/manual/mnl_stx2_pro_dsp_dev_kit_ep2s180.pdf

Altera (2011) Avalon interface specifications. http://www.altera.com/literature/manual/mnl_avalon_spec.pdf

Galileo ICD (2010) Galileo open service. Signal in space interface control document. European space agency/European GNSS supervisory authority

GPS (2010) Systems engineering & integration interface specification IS-GPS-200. Global positioning system directorate (2011)

GPS (2011) Systems Engineering & Integration Interface Specification IS-GPS-705. Global Positioning System Directorate (2012)

Micrium (2012) μC/OS-II real time kernel, http://micrium.com/page/products/rtos/os-ii

Chapter 10
Fully Software Implementations for Galileo Receiver

F. Dovis, M. Fantino, M. Pini and A. Molino

Abstract The last decade has seen the success of satellite based navigation applications in the ordinary people's life. New services with demanding performance are boosting also the development of improved technologies for navigation receivers. Although the navigation technology also rapidly evolved towards more complex signal processing techniques, when compared to communication receivers, GNSS receivers are dealing with signals of smaller bandwidth and much lower data-rates, thus making them appealing for Software-based implementations, considering both Hardware/Software platforms and fully software implementations. This chapter will provide a discussion of the technological challenges for the implementation of software positioning receivers, also discussing as examples of general validity the implementation of acquisition and tracking stages in a fully software receiver.

Software-Defined or Fully Software GNSS Receivers: Challenges and Opportunities

Software radio is a promising technology that generates interest in the receiver industry for some applications and it provides a useful simulation and testing environment. Software Radio (SR) or Software-Defined Radio (SDR) implementations allow testing of algorithms in realistic conditions by means of code updating without replacing hardware components.

F. Dovis (✉)
Department of Electronics and Telecommunications, Politecnico di Torino, Torino, Italy
e-mail: fabio.dovis@polito.it

M. Fantino · M. Pini
Navigation Technologies Research Area, Istituto Superiore Mario Boella, Torino, Italy

A. Molino
CSP, Torino, Italy

J. Nurmi et al. (eds.), *GALILEO Positioning Technology*,
Signals and Communication Technology 182, DOI 10.1007/978-94-007-1830-2_10

The concept of SR or SDR was introduced for efficient implementation of communication transceivers (Mitola 1995) and the implementation of Satellite Navigation Receiver according to such a philosophy represent a specific implementation case. The distinction between SR and SDR depends on where the digitization of the signal takes place. The current technology does not allow, in general, to sample the Radiofrequency signal directly, and a first stage of demodulation of the signal towards an Intermediate Frequency (IF) is always needed. Thus, in most of the cases we face SDR implementations. Nevertheless, the bandwidth of the signal to be digitized still put constraints on the feasible implementation of the digital part of the receiver. In most of the cases SDR implementations targets a mixed solution where programmable hardware (Field Programmable Gate Arrays, Digital signal processor, etc.) interacts with software modules running over general purpose processors or microcontrollers.

The signal processing involved in GNSS receivers usually require a computational burden lower than other radio terminals, such as 3G mobile phones. In fact, even if GNSS employs Direct Sequence/Spread Spectrum modulations, the open services signals have the chip-rate of few Mega-Hertz, while the data rate is less than one kilo-hertz. Then, some of the technological issues regarding the development of SDR terminals have a lower impact in the case of GNSS at least for some of the services.

Algorithms able to increase sensitivity and performance of the GNSS receiver as well as multipath mitigation techniques are just some of the most common examples of advanced signal processing algorithms being studied and designed. Such new signal processing techniques need to process signals usually sampled at a high rate, thus dramatically increasing the computational load. However, the development of SDR GNSS receivers is even more appealing, mainly for the high level of flexibility. During the design phase it will be necessary to match the strategies for algorithms implementation with the hardware capabilities in order to carefully balance performance versus complexity, looking for the best trade off.

Review of Existing Software Receivers

The strong interest in positioning and navigation applications together with the development of new satellite navigation systems (Dovis et al. 2011) is leading to an intensive research work for developing algorithms able to take advantages of the format of the new signals. The future generation of GNSS receivers will be asked to provide at the same time improved positioning performance at lower costs. Due to the growing system complexity and the rapid enhancements in signal processing algorithms, the possibility of having reconfigurable SDR terminals can decrease development costs for both users and manufacturers.

From a technological point of view, many of the considerations performed on traditional wireless terminals apply also to satellite navigation receivers. In fact, modern mass-market GPS receivers are System-on-Chips, composed of a RF Front

End and a digital processor integrated in a single chip with CMOS technology. The digital processor is composed of an ASIC and a microprocessor. The ASIC performs all the signal processing at IF, and in particular it acts as correlation engine for both acquisition and tracking units. The processor tasks include tracking discriminators and filter loops, data decoding, pseudo-range measurements, position and time computation and Human Machine Interface functionalities.

In the following discussion about the HW/SW mixed implementation versus fully software will be introduced, then focusing on fully SW implementation and presenting a specific case of fully SW receiver designed and implemented by the authors. Remarks on advantages, computational bottlenecks, design choices, relative to such specific case-study can be extended to any other fully software implementation and can be useful for any readers willing to face the implementation of a fully SW GNSS receiver.

HW/SW Mixed Implementations

Recent improvements on the technology of Field Programmable Gate Arrays (FPGAs) lead to the application of this technology to different application fields such as GNSS. These improvements have been achieved mainly in terms of performance and power consumption, making these devices an attractive trade-off between the "pure hardware" approach and the "fully software" one. Also, being FPGA devices programmable components, they are usually considered an effective way to implement SDR terminals. Therefore, many recent SDR-targeted receivers are implemented in mixed HW/SW modules, having HW modules implemented on reconfigurable logic.

The most challenging aspect of these receivers is usually the design of how the different parts of the receiver are implemented as software modules or as hardware blocks.

The overall performance of the receiver and its degree of flexibility strongly depends on the HW/SW partitioning, that usually is decided taking into account the three following design rules:

- HW blocks are ad hoc circuits, therefore they show better computation speed and power consumption, and for this reason highly computation intensive tasks must be done in hardware;
- HW blocks are not as flexible as the SW ones, it is therefore better to keep in software all the algorithms that need to be deeply configured or that can be upgraded;
- HW/SW borderline is usually the system bottleneck. Hence, it is better to operate the logical "cut" (between HW and SW) in a place where the communication bandwidth is relatively low.

In the structure of typical GNSS receivers, all the three mentioned criteria can yield to one straightforward solution. Such an architectural design foresees the HW implementation of the Acquisition Unit with the exception of the correlators of the tracking loops, while all the other functions of the receiver, including pseudo-range and PVT computation are devised as SW routines. In the proposed partitioning, HW tracking unit includes only parallel correlators, so that the tracking loop filters, that requires floating point arithmetic, are easily implemented in software. As an example, this solution has been adopted by the authors in a mixed HW/SW receiver prototype that was implemented in a development board based on ARM9 microprocessor and a Xilinx Virtex-II FPGA (Girau et al. 2007). Another complete receive, designed with a similar approach is described in Hurskainen et al. (2009) where the authors exploited a multicore processor to cope with the processing speed requirements. Other examples of implementations can be found in Heckler and Garrison (2004), Paakki et al. (2010) and Stöber et al. (2010)

Fully Software Digital Processors

Fully SW SDR GNSS receivers feature a digital processor based on a general purpose processor, receiving the digitized data collected by a front-end interfaced via a standard communication port such as a Universal Serial Bus (USB) 2.0. Fully SW approach allows to achieve, for the digital processor, many of the SDR goals and advantages: high reconfigurability and flexibility, extensive use of high-level programming languages, hardware platform independence. Even though Fully SW based receivers are already available, several issues make the commercial diffusion of this technology quite improbable in the short/medium term.

Size, power dissipation and cost are not comparable with what offered by commercial mass-market products. A single chip, ASIC based receiver, which include also the front-end, requires an area of about 7×7 mm^2, while an x86 processor, in order to work, requires a motherboard with several peripherals and it cannot work without cumbersome cooling device. Then the total area is of the order of tens of square centimeters. Mass-market products, have, nowadays, power dissipation lower than 100 mW, whereas for example, x86 systems, designed for notebook, require about 50 W. Furthermore, mass-market GPS receivers chip have a lower cost per unit.

In fact, the cost of Fully SW receivers has to include also the processor board (plus the external front-end chipset), even if the cost of the processing capabilities could be shared with other applications. Some of these issues would be partially solved by moving the Fully SW based receivers on smaller processors like the ones equipping modern PDA handsets. These processors feature lower size and power consumption, which leads to longer battery duration and the absence of cooling device, with a consequent reduction of the total size. However, the achieved elaboration power is at least one order of magnitude less than x86 processors. In addition, even the elaboration power being equal, the program implementing the GNSS receiver would occupy almost all the computational power of the processor, preventing the correct behavior of the other PDA functionalities.

Fully SW based receivers able to decode in real-time the signal for a dozen of GPS C/A SVs require about 30–60 % of the computational power of a modern x86 based processor.

Fully SW based SDR GNSS receiver have been so far limited to the use of open service signals, featuring lower chip rate and sampling frequency. Then the lack of elaboration power will be more evident in presence of signals such as Galileo E5a, E5b, or GPS L5, which involve a significant increment of the sampling frequency. A possible solution to partially benefit from the enhanced performance of these signals consists in decoding them only for short time periods in correspondence of the PVT computation, once the navigation message is completely known. This strategy avoids the great computational burden involved in the real-time parallel channel tracking, as performed in ASIC based receivers. As a matter of fact, the possibility to substitute current ASIC-based with Fully SW reconfigurable receivers, strictly depends on the future availability of more powerful, low-power and low-area processors. However, as remarked in Fantino et al. (2009) they represent a valuable tool for research and development purposes. They are also very popular as training tools for teaching, allowing an inspection of the signals all along the processing chain. For such purposes the real-time processing of the signals is not always required, and SR implementations in interpreted languages such as Matlab are suitable to the scope (Borre et al. 2006). Nevertheless, the implementation of a real-time GNSS receiver has to match requirements that are often driving the architectural design and force the author to implement smart customized solutions.

The first work on SDR applied to satellite navigation was proposed in (Akos 1997), and such work represented the starting point for what is commonly believed to be the first implemented real-time processing-capable Fully SW SDR receiver for the GPS L1 C/A signal. However, the first commercial SDR GPS receiver was delivered later in 2004 (Nordnav 2004). In such a receiver the RF signal was processed by a customized front-end that, during the A/D conversion, was also watermarking the data for licensing purposes, making the software part able to work only with the corresponding front-end The receiver was designed to process the civil GPS code on L1, and the front-end embedded a filter at IF with a bandwidth of about 4 MHz, thus sampling the signal at about 16 Msample/s.

The Fully SW approach became quickly popular in universities and research centers, and in the following years several other receivers were implemented, as for example (Ledvina et al. 2006; Charkhandeh et al. 2006; Falk et al. 2010; Fantino et al. 2009).

The first implementations of software receiver used front-ends embedding commercially available Analog-to-Digital Converters (ADCs) that were connected directly to the PC, generally via the PCI bus. For practical reasons, especially when the SR receiver aimed at becoming a commercial product the front-end started to be coupled to a USB 2.0 interface.

In the next Section we will focus on the description of implementation of the Fully software receiver N-Gene, described in Fantino et al. (2009), as a paradigm of the different design steps and possible implementative solutions.

N-Gene Fully SW Receiver

N-Gene processes the sample stream, at the ADC output, quantized over 8 bits, with the ability of processing up to 12 channels in real time. N-Gene receives the GPS Coarse Acquisition (C/A) code on L1 and tracked live GIOVE-A and GIOVE-B signals transmitted on the E1 band, and it is able to receive the signal from the Galileo operational satellites. The receiver guarantees an improved positioning accuracy through the (optional) use of differential correction broadcasted by the Wide Area Augmentation System (WAAS) and the European Geostationary Navigation Overlay System (EGNOS) systems. A USB connection toward the RF front end makes the receiver particularly versatile. The receiver works either with low cost front ends, representing the analog signal with few bits per samples at a low rate, or with customized front-ends, characterized by higher sampling frequency and a digitization over 8 bits. The 8-bits quantization becomes fundamental in case a fine resolution representation of the signal is needed, as in the implementation of innovative interference mitigation.

The main operations of N-Gene are controlled by an activity manager, which after an initialization phase allocates all the necessary resources for the receiver functioning. The main task of the activity manager is to retrieve the data from the USB and then call one by one each channel to process such data.

The complexity of the functions allocation can be described by means of a channel state diagram. After the receiver initialization, each channel is set in idle mode. When a PRN code is assigned to the channel the acquisition phase begins. The transition from the Acquisition to the Tracking is different and independent for each channel. In fact first the presence of the PRN in the received signal is confirmed using a Bernoulli like test and if the PRN is declared present then the code delay and Doppler shift are refined in order to allow a correct signal lock. From this point, the channel is in tracking mode until the bit as well as the subframe synchronizations are not achieved leading to the Navigation state and to the pseudorange evaluation. If, during the tracking phase, the signal is lost, a fast re-acquisition is attempted and depending on the results the channel can return in tracking mode or in Idle mode waiting for another PRN to be assigned.

The above mentioned operations are shown in a graphical form in the Finite State Machine reported in Fig. 10.1.

Specific Issues of Software Architectures

In a software receiver implementation several other issues must be considered in order to achieve the real time requirements. The algorithms must be designed around the capability of the programmable processor to operate on the data being received in an efficient way. In other words the algorithms must be adapted somehow and somewhat around the instruction sets made available by the specific

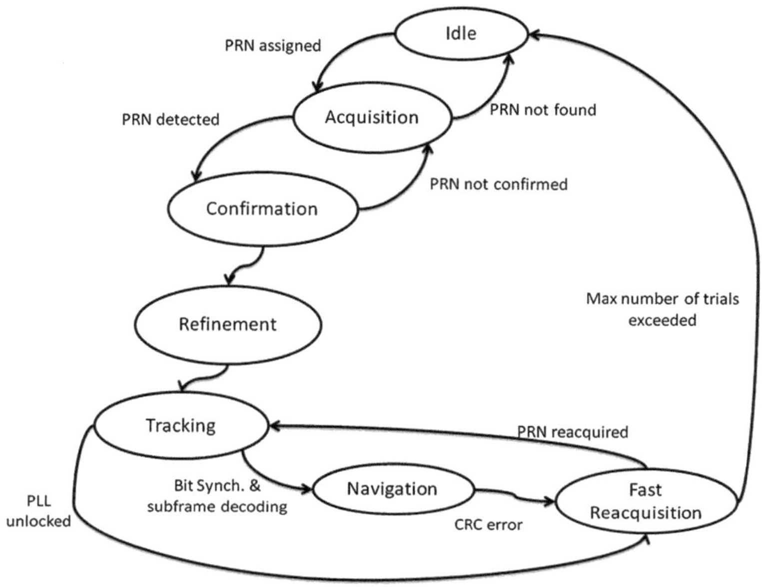

Fig. 10.1 Channel state diagram

processor being used. However, independently from the class of target processors a general rule is to group the data in blocks and to operate the same operation in parallel to all of those elements contained in a block. In this way the receiver algorithms must operate always on a block of data and not on the single sample coming at that moment on the front end. A buffer is then necessary to store at least the minimum amount of data to be processed in block increasing the memory requirements of the hardware being used. Nevertheless, memory size might become a issue in terms of gates being consumed inside an FPGA. In fully software receivers the availability of Gb of RAM relax quite a lot this constraint.

Software Implementation of Acquisition Stage

In a CDMA based GNSS system, each satellite continuously transmits a periodic code signal, which is modulated by information symbols. The code signal is a spreading sequence made of L_C chips and the sequence length (or repetition period) is denoted by T. Each satellite is characterized by a unique code. The cross correlation properties of such codes allow the receiver to efficiently separate satellite signals which, when received, are superimposed in the time domain.

The first operation performed by any GNSS receiver, when switched on, is called *signal acquisition* (see Chap. 5). The receiver signal acquisition stage provides to the following *tracking stage* a list of the acquired satellites, and for each of them, a coarse estimation of the code delay as well as a rough estimation of the Doppler frequency shift. The code delay τ is the offset of the received code with respect to a locally generated instance of it; the Doppler shift f_d is due to the relative dynamics between satellite and receiver.

After being acquired a satellite needs to be tracked: its associated received code delay and its Doppler frequency shift need to be accurately estimated and their variations over time need to be continuously updated.

In the acquisition procedure, the declaration of presence or absence of a satellite and, consequently, the determination of both code phase and Doppler shift are obtained by computing a two dimensional matrix called search space. Each entry of such a matrix contains the value of the two-dimensional correlation (cross ambiguity function) between the received signal and a locally generated copy, the latter characterized by a specific value of code delay, $\hat{\tau}$, and Doppler shift, $\hat{f}_d$.

It is well known the one of the most efficient way to obtain the search space is by implementing a fast parallel acquisition technique, where the correlation function is evaluated by means of FFT operations (Alaqeeli et al. 2003; Weill 2007). Unfortunately, the sampling rate of modern GPS receivers might be on the order of 10–20 MHz, depending on the receiver applications, thus leading to large FFT processing blocks, which increase the cost of the receiver. For example, assuming a realistic sampling frequency of 16.3676 MHz and an integration period of 1 ms (corresponding to one code repetition period in GPS systems), the FFT would have to be evaluated over 16,368 points.

In Starzyk and Zhu (2001), the authors introduce a more efficient technique for the acquisition process, called Averaging Correlations (AC), which requires shorter FFT blocks. The basic idea presented in Starzyk and Zhu (2001) is to pre-process the signal by averaging the samples over each single chip, thus reducing the FFT to a number of points equal to the number of chips in the given integration period. In the GPS case averages are then stored in vectors of size 1,023. The concept described above is depicted in Fig. 10.2, where a section of the sampled signal is shown on the left. Each sample is denoted by a bullet. For different phases ($\phi = 0, 1$ in Fig. 10.2) the samples are averaged over time intervals corresponding

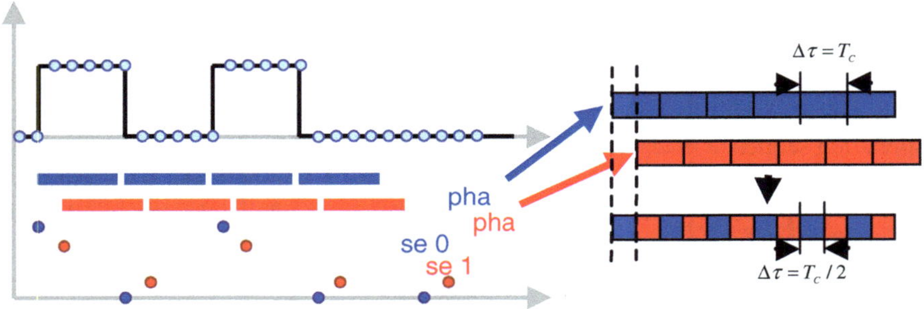

Fig. 10.2 Average or compression operation

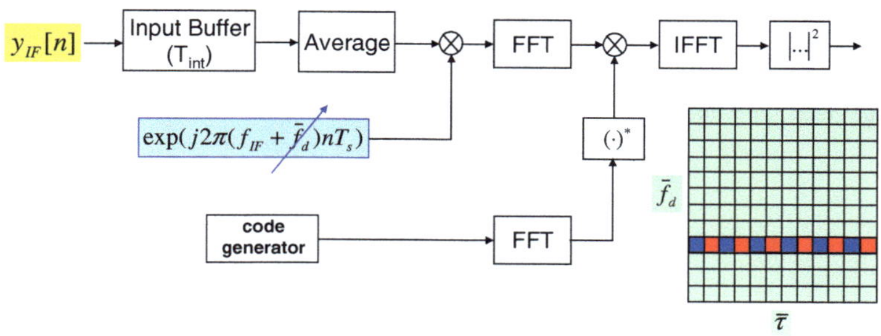

Fig. 10.3 Fast acquisition method with signal compression

to the chip duration represented by the blue and red bars, respectively. The vectors of averages are shown on the right.

The idea of adopting the AC operation can reduce in an efficient way the number of points to be explored in parallel with a relevant seep-up of the conventional fast acquisition technique. In practice, with this technique the fast acquisition is applied over the vectors of averages instead of on the original signal. By consequence the FFT can be applied over much shorter signal blocks thus improving the computation speed (Molino et al. 2008). However, since the incoming signal is masked by noise, there is not enough information to determine the first sample of the chip where starting the average (i.e. determining the right phase ϕ). This problem forces the algorithm to repeat the FFT several times, one for each required phase. When the right phase is chosen the correlation reaches a maximum.

The acquisition scheme, shown in Fig. 10.3, is very similar to the classical fast acquisition, where the cross ambiguity function is obtained performing a circular correlation in time domain with two direct and one inverse Fourier transform, as. The scheme is different from the classical fast acquisition because of the presence of the "Average" block implementing the AC technique.

This technique is particularly appealing for fast signal acquisitions in real time GNSS software receivers, since it provides a fast, flexible, and reconfigurable signal acquisition block. The averaging technique, reducing the number of samples induces a loss in the sensitivity performance of the acquisition stage. However, the large saving in the processing time allows to implement statistical tests on the acquisition outputs, more complex than in usual schemes, yielding to an improved overall detection probability.

It is particularly interesting to compare the detection rate versus the Carrier to Noise Power spectral density (C/N_0) given a fixed false detection rate of the classical method and the one implementing the AC. Figure 10.4 shows the case where a signal sampled at 16.3676 MHz is processed using a classical approach (red curve) and the case with complexity reduction (blue curve). Due to the reduction of complexity deriving by the AC (that for this specific case of sampling frequency it is of about 22 times) the saving in terms of operation is used to perform a statistical improvement.

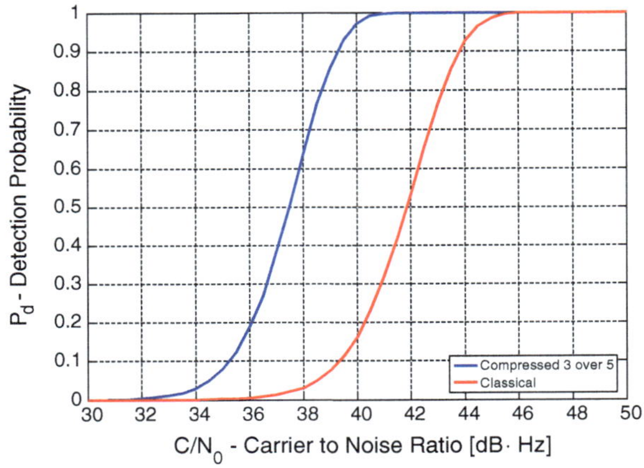

Fig. 10.4 Detection comparison between fast technique and AC fast technique (false detection rate 10-4)

The results is quite interesting because simply adopting an M over N Bernoulli test, the AC fast acquisition methods outperforms of about 4 dB-Hz the classical fast acquisition (see Fig. 10.4) and still the number of operations to be executed is significantly lower than the multiplications necessary for the fast acquisition based on a full FFT.

Software Implementation of Tracking Stage

Modern digital GNSS receivers sample the analog signal at the output of the IF filter and split the signal over different digital channels. As described above, the first step in GNSS processing is the signal acquisition: the satellites in view are detected and a first rough estimation of Doppler shift and code phase is performed. The signal tracking follows the signal acquisition. Over each channel of the receiver, a Delay Lock Loop (DLL) is used to synchronize the received spreading code and a local replica (Borio et al. 2008), while a Phase Lock Loop (PLL) is generally employed to track the phase of the incoming carrier. The signal tracking relies on the properties of the signal correlation and is fundamental to demodulate the navigation message and estimate the range between the user and the satellites. Conventional receiver architectures generally include a Frequency Lock Loop (FLL) to refine the rough estimate performed by the signal acquisition. The FLL eases the PLL lock, reducing the transient time between the signal acquisition and the steady-state carrier/code tracking. Recently new techniques based on digital signal processing have been developed in order to obtain higher precision and reduced computational load, improving the robustness against noise and interference. In this section, we recall some fundamentals of the tracking loops before describing a software implementation of a standard tracking system.

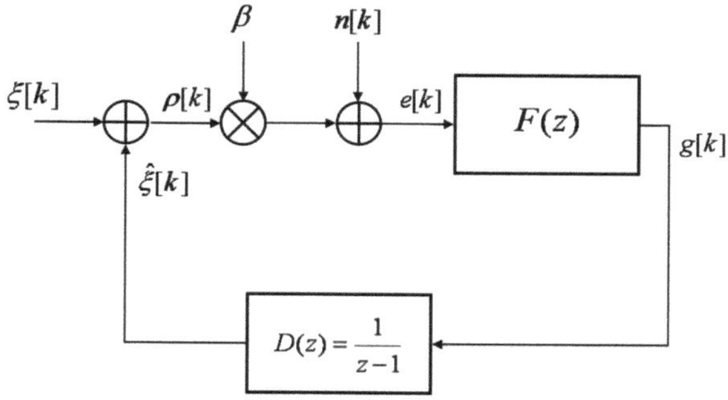

Fig. 10.5 Linear model for digital tracking loops

Digital Model

The theory of the tracking loops for GNSS receivers is well known and reported in many books such as Kaplan and Hegarty (2006) and Misra and Enge (2006). Although most of the references model the tracking loops as continuous time systems, one must note that modern GNSS receivers process stream of samples at precise time instants. Therefore, in the design of tracking loops, the most appropriate approach consists in deriving a model in the discrete time domain. Introducing a terse analysis, an example of digital tracking model is given below.

In Fig. 10.5 $\xi[k]$ is the parameter to be estimated and represents the phase of the input signal, whereas $\hat{\xi}[k]$ is the estimated phase at the kth step. Neglecting the noise (i.e. $n[k]$), within any tracking loop a phase error signal $\rho[k] = \xi[k] - \hat{\xi}[k]$ is generated on the basis of a non linear discrimination function, which is also known as S-curve. GNSS receivers can use different discriminators, but in all cases, for small errors the nonlinear function can be linearized as:

$$e[k] = \beta(\xi[k] - \hat{\xi}[k]) = \beta\rho[k] \tag{10.1}$$

where β is the slope of the discrimination function around the origin. The discriminator output is then sent to a digital filter with impulse response equal to $F(z)$, before being numerically integrated in order to get a new estimation of the phase. The loop filter is a low pass, suppressing higher frequencies. The type of filter used and the number of poles determines the tracking loop robustness to dynamic stress. The signal at the filter output is then numerically integrated:

$$\hat{\xi}[k+1] = \hat{\xi}[k] + g[k] \tag{10.2}$$

Note that in real systems the noise is superimposed to the input signal, but the proposed model considers the noise added to the error signal. Referring to both Fig. 10.5 and to Eq. (10.2), it is possible to derive the z-transform of the estimated phase:

$$\hat{\xi}[z] = D(z)F(z) \cdot \{n[z] + \beta(\xi[z] - \hat{\xi}[z])\} \tag{10.3}$$

Equation (10.3) is the starting point in the derivation of the noise equivalent loop bandwidth, which determines the tracking system performance. The filter coefficients can be chosen in order to implement the desired loop bandwidth and give specific features to the tracking structure. However, there is a trade off between the noise bandwidth and the response of the loop. In fact, a narrow loop bandwidth would lead to a less noisy NCO control, but the loop will never be able to follow dynamics of the signal without aiding.

The equivalent bandwidth can be derived from $H_N[k]$, which represents the transfer function between the additive noise and the error $\rho[k]$ in case $\xi[k] = 0$.

$$H_N[z] = \frac{-\xi[z]}{n[z]} = \frac{\rho[z]}{n[z]} = \frac{-F(z)}{(z-1) + \beta F(z)} \tag{10.4}$$

As it is possible to note from (10.4), the order of such a transfer function depends on $F[z]$. When the filter is a simple gain factor $H_N[z]$ becomes a first order transfer function. On the other hand, when $F[z]$ is a first order function, the tracking loop can be seen as a second order system. In this case there is an infinite set of loop filter coefficients that can be chosen to have the desired loop bandwidth. Generally, designers add another equation to determine the best coefficients, setting the damping factor (which depends on the loop filter coefficient as well) to $1/\sqrt{2}$. This solution corresponds to the shortest transient time of the tracking loop. Finally it is worth mentioning that the design of robust tracking loops in weak signal conditions is challenging. It becomes particularly important for those applications where the receiver is expected to work with C/N_0 ratios lower than 30 dB-Hz. Several approaches have been proposed in the past to enhance the ability of GNSS receivers to track weak, attenuated or corrupted signals. Some of them consist of appropriate antenna designs implementing advanced beam steering, null forming and adjusting RF front end gains. Other techniques focus on the base band signal processing and aim at tuning the tracking loop parameters described above.

Software Implementation

Figure 10.6 shows the block diagram of a tracking system commonly used in digital GNSS receivers. It is composed by a PLL for carrier phase tracking and a DLL for the synchronization of the incoming and the local codes. Figure 10.6 shows the tracking system of one single channel, but the same architecture is repeated over all channels to track different satellites.

Software receivers store the signal samples at the ADC output in a First In First Out (FIFO) input buffer. Such samples are then processed by the tracking systems of all channels. Considering the operations within a single channel, the number of

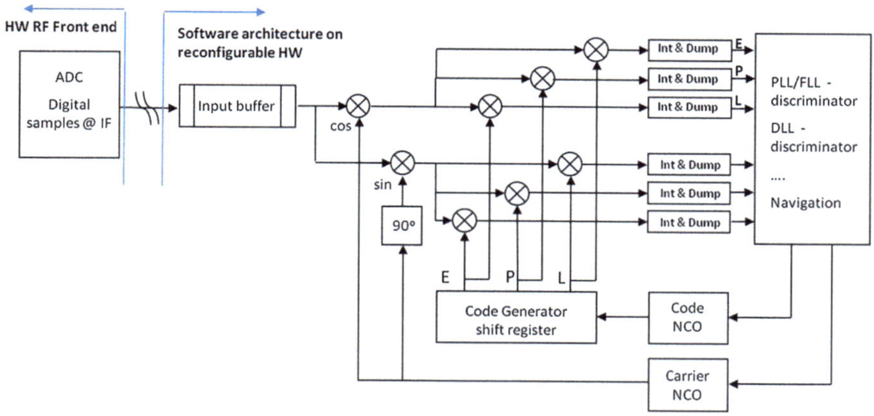

Fig. 10.6 Block diagram of a code and carrier tracking loop for GNSS receivers

samples corresponding to a predefined integration period (e.g. taking the integration period equal to the spreading code period, we have 1 ms for the GPS C/A code and 4 ms for the Galileo E1-B) is correlated with a digital sequence obtained from local replicas of carrier and spreading code. The values of correlation are then used to produce feedback control signals, on the basis of proper discrimination functions, one for the PLL, one for the DLL. As seen in the loop model described in the previous section, such control signals are filtered and used to steer the code and carrier generators that prepare the local replicas for the next loop iteration. The process continues and the system follows the input signal variations over time. Note that the described synchronization process corresponds to finding the best estimate of the local carrier frequency/phase phase and local code delay that maximize the correlation between the incoming and the local replicas.

Non coherent tracking systems, like that shown in Fig. 10.6, use two branches, one In phase (I) and another in Quadrature (Q). Generally speaking, non coherent tracking loops are more robust and do not require the estimate of the carrier phase (i.e. they do not necessarily need a PLL; an effective system can be designed combining a FLL and a DLL). For example, right after the signal acquisition, when the tracking phase starts, the system has not yet recovered the phase of the incoming carrier and part of the power goes on the quadrature branch. Differently from coherent tracking loops (that use only the I branch), in this case using both the branches the discriminators are still able to produce feedback signals. If a PLL is used, after an initial transient time, the incoming carrier is synchronized with the local one and the received signal is completely converted on the I branch.

In software implementations, performing the signal tracking operations in real time, avoiding input buffer overflows, is a challenge. For real time software receivers, the input buffer control, the channel schedule as well as the receiver logic must be carefully designed. The input buffer stores samples at the sampling rate, that is equal to some MHz for narrow band receivers. The generation of the local sequence of samples and the correlation process also require computational resources, because they come at the same rate. Some designers code this part of the receiver in low-level programming language (e.g. Assembly language), rather than general purpose languages, like the C. Others optimize the generation of the local signals through efficient algorithms. For example some designers approximate the local carrier to a sinusoidal with techniques demanding moderated computational burden. On the other hand, the post-correlation operations (i.e. generation of the feedback control signal and filtering) are performed at the end of each integration period and are generally not critical for the real time constraints. Most of the real time GNSS software receivers, running on general purpose processors, implement this part in C language.

Pseudorange Creation and PVT

When both the DLL and PLL are locked, the incoming signal is despread and converted to baseband. The navigation data bits appear at the output of the in-phase Prompt correlator and can be decoded. In addition, with the DLL locked, the local and the incoming codes are aligned. Referring to the local code, the receiver knows exactly when a new code period starts and is able to recognize navigation data bits and boundaries of the navigation message. The receiver stays synchronized to the tracked satellites, continuously counting the number of received chips, full code periods, navigation bits and message frames. These counters are fundamental to measure the misalignment over different channels, tracking different satellites, and are used to compute the pseudoranges.

In this section we focus on the measurements of the pseudoranges, describing one method commonly used to estimate the distance between the satellite and the user's receiver. The starting point for decoding the navigation message is the detection of a preamble that allows the receiver to recognize the beginning of a subframe. One of the approaches that is commonly adopted to perform the pseudoranges estimation, is based on setting a common receiving time t_u^R over all the channels. In this case, the reference channel is the one that receives the subframe transmitted at t_{tx}^{GPS} first. For all the tracked satellites (including the reference one), the receiver counts the elapsed time between the reception of subframe and t_u^R. This means that the receiver measure the delays as:

$$\delta_i = t_u^R - t_{rx,i}^R \quad \forall i = 1, \ldots, 4 \tag{10.5}$$

Figure 10.7 depicts the method of fixing a unique time of reception for four GPS satellites in view

Once the δ_i has been computed, the receiver is able to calculate the pseudorange. This can be accomplished by evaluating the delta-difference (Δ_i) of delays respect to the reference channel. The aforementioned difference Δ_i is stated as:

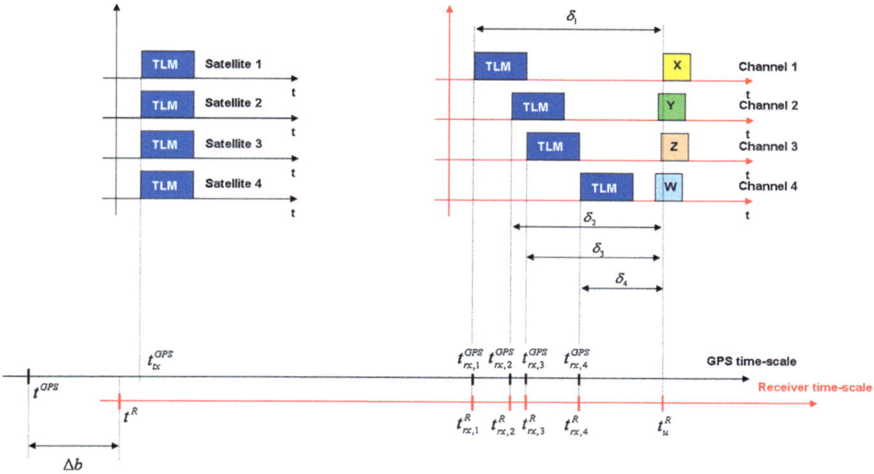

Fig. 10.7 Pseudorange computation based on "common reception time", evaluating the start of the subframe for a GPS system

$$\Delta_i = \delta_i - \delta_1 \quad \forall i = 1, \ldots, 4 \tag{10.6}$$

and consequently, with these time differences, the set of distances between the receiver and the satellites can be written as following

$$\rho_i = \rho_1 + c \cdot \Delta b + c \cdot \Delta_i \quad \forall i = 1, \ldots, 4 \tag{10.7}$$

where:

- ρ_1 is the pseudorange associated to the reference channel;
- Δb represents the clock bias between the receiver and the GPS time scale;

This technique for pseudoranges' estimation is usually employed in commercial GPS receivers. The main reason behind this choice is the relative simplicity and suitability of that approach in real-time implementations, since it does not require to wait until all the channels have received the same data bit (e.g. the beginning of the same subframe) to compute the pseudoranges. This concept gets more clear if we consider that, during the data demodulation and the tracking process, the receiver continuously counts the number of samples processed on that channel, as well as the number of frames, subframes and data bits decoded. As a consequence, through a system of counters, it becomes easy to compute the time difference Δ_i among the channels at a certain t_u^R.

In order to compute the satellite position, the receiver needs the ephemeris and the time of transmission, which is referred to the beginning of the subframes. All the information the receiver needs is embedded in the navigation message. The user solves a system of at least four equations to recover its three spatial coordinates. In fact, in addition to x, y and z, there is the unknown bias between the user's clock and the satellite time scale. When four satellites are correctly tracked, the system of equations can be written as follows:

$$\begin{cases} \sqrt{(x_1 - x_u)^2 + (y_1 - y_u)^2 + (z_1 - z_u)^2} = \rho_1 + c \cdot \Delta b \\ \sqrt{(x_2 - x_u)^2 + (y_2 - y_u)^2 + (z_2 - z_u)^2} = \rho_1 + c \cdot \Delta b + c \cdot \delta_2 \\ \sqrt{(x_3 - x_u)^2 + (y_3 - y_u)^2 + (z_3 - z_u)^2} = \rho_1 + c \cdot \Delta b + c \cdot \delta_3 \\ \sqrt{(x_4 - x_u)^2 + (y_4 - y_u)^2 + (z_4 - z_u)^2} = \rho_1 + c \cdot \Delta b + c \cdot \delta_4 \end{cases} \tag{10.8}$$

where ρ_1 is the pseudorange of the reference channel, which corresponds to the shortest travelling time.

After having removed the satellite offset and atmospheric effects the Position Velocity and Time (PVT) computation can be performed. One of the simplest algorithm for the position estimation is based on the least-squares (LS) method. This is commonly used and widely described in literature. However, many software receivers, as well as commercial receivers, base the PVT on Kalman (or extended-Kalman) filters, that are able to smooth the LS solutions, providing higher performance. The description of the Kalman filter is out of the scope for this chapter, interesting readers can find a detailed theory in many sources available in literature.

Conclusion

In this chapter we discussed the implementation of a GNSS receiver by as software defined or fully software architectures. At present most of the implementations have been designed for research purposes and their application to the whole set of GNSS signals is limited by the processing capabilities of the hardware architectures. Nevertheless, the improvements of the technology and the design of more efficient algorithms makes likely their use, in a mid-term perspective, in mass-market implementations or as core of embedded systems.

References

Akos DM (1997) Software radio approach to global navigation satellite system receiver design. Ph.D. dissertation, Ohio University

Alaqeeli A, Starzyk J, van Graas F (2003) Real_time acquisition and tracking for GPS receivers. In: Proceedings of the 2003 international symposium on circuits and systems, pp 500–503

Borio D, Fantino M, Lo Presti L, Pini M (2008) Robust DLL discriminator functions normalization in GNSS Receivers. In: Presented at 2008 IEEE/ION position location and navigation symposium, 3–8 May, Monterey (CA), USA

Borre K, Akos DM, Bertelsen N, Rinden P, Jensen SH (2006) A software-defined GPS and Galileo receiver a single-frequency approach. Birkausen, Boston. ISBN: 0817643907

Charkhandeh S, Petovello MG, Lachapelle G (2006) Performance testing of a real-time software-based GPS receiver for x86 processors. In: ION GNSS 2006, Fort Worth, Tex, USA

Dovis F, Mulassano P, Dominici F (2011) Overview of global navigation satellite systems. In: Zekavat SA, Buehrer RM (eds) Handbook of position location: theory, practice, and advances. John Wiley & Sons, Inc., Hoboken

Falk N, Hartmann T, Kern H, Riedl B, Pany T, Wolf R, Winkel J (2010) SX-NSR 2.0: a multi-frequency and multi-sensor software receiver with a quad-band RF front end. ION GNSS 2010, Portlan (OR), USA

Fantino M, Molino A, Nicola M (2009) N-gene GNSS receiver: benefits of software radio in navigation. In: Proceedings of ENC

Girau G, Tomatis A, Dovis F, Mulassano P (2007) Efficient software defined radio implementations of GNSS receivers. ISCAS 2007, IEEE international symposium on circuits and systems, 2007, pp 1733–1736, 27–30 May 2007. doi:10.1109/ISCAS.2007.377929

Heckler GW, Garrison JL (2004) Architecture of a reconfigurable software receiver. In: ION GNSS 2004, Long Beach, CA, USA

Hurskainen H, Raasakka J, Ahonen T, Nurmi J (2009) Multicore software-defined radio architecture for GNSS receiver signal processing. EURASIP J Embed Syst 2009:543720

Kaplan ED, Hegarty C (2006) Understanding GPS: principles and applications. Artech House mobile communications series. Artech House, London

Ledvina BM, Psiaki ML, Humphreys TE, Powell SP, Kintner PM (2006) A real-time software receiver for the GPS and Galileo l1 signals. In: ION GNSS 2006, Fort Worth, (TX), USA

Misra P, Enge P (2006) Global positioning system. Signal, measurements and performance, 2nd edn. Ganga-Jamuna Press, Lincoln. ISBN 0-9709544-1-7

Mitola J (1995) The software radio architecture. IEEE Commun Mag 33(5):26–38

Molino A, Girau G, Nicola M, Fantino M, Pini M (2008) Evaluation of a FFT-based acquisition in real time hardware and software GNSS receivers. International symposium on spread spectrum, Bologna, Italy, 25–28 Aug 2008

NordNav Technologies (2004) NordNav R-30 GPS receivers user manuals

Paakki T, Raasakka J, Della Rosa F, Hurskainen H, Nurmi J (2010) TUTGNSS University based hardware/software GNSS receiver for research purposes. Ubiquitous positioning indoor navigation and location based service (UPINLBS), pp 1–6, 14–15 Oct 2010

Starzyk JA, Zhu Z (2001) Averaging correlation for C/A code acquisition and tracking in frequency domain. In: Proceedings of the 44th IEEE 2001 Midwest symposium on circuits and systems, pp 903–908

Stöber C, Anghileri M, Ayaz AS, Dötterböck D, Krämer I, Kropp V, Jong-HoonW, Eissfeller B, Güixens DS, Pany T (2010) ipexSR: a real-time multi-frequency software GNSS receiver. In: Proceedings of ELMAR, pp 407–416, 15–17 Sept 2010

Weill LR (2007) Lightening the data processing load (Signal compression in GNSS receivers). Inside GNSS, vol 2 no 7, pp 37–45, Fall 2007

Part III
Advanced Galileo Positioning Technologies

Chapter 11
Galileo Receiver Research in Europe

Ignacio Fernández Hernández

Abstract Since the Galileo program was conceived by the EU, a high priority has been given to the development of the user segment. As a complement to other more technical chapters of the book, this chapter presents a global view of private and public investment in user localization technologies research and development (R&D), with a focus on Galileo receivers in Europe. The chapter starts by presenting some information on GNSS higher education in Europe. It then presents some metrics on R&D expenditures of major GNSS-related companies and government organizations, as per publicly available figures. It also describes the past and current Galileo receiver research programs financed with EU budget. Later, some general indicators of technology transfer and innovation are presented. The chapter finalizes with some trends and challenges that the sector may face in the following years.

Introduction

This chapter describes the overall research landscape in the EU, in particular for satellite navigation receiver technologies. It is supported by some figures and a benchmarking exercise with other regions in the world, using the most recent public data found at the time the chapter was written. The reader should notice that it is difficult to isolate Galileo receiver research and development (R&D) from localization technologies R&D in general. It is also challenging to divide R&D investment among regions, particularly in the overly dynamic ecosystem of the location business, where startups or already consolidated firms change hands swiftly from

Note: The information and views set out in this article are those of the author and do not necessarily reflect the official opinion of the European Union.

I.F. Hernández (✉)
European Commission DG Enterprise and Industry, Brussels, Belgium
e-mail: ignacio.fernandez@gmail.com

J. Nurmi et al. (eds.), *GALILEO Positioning Technology*,
Signals and Communication Technology 182, DOI 10.1007/978-94-007-1830-2_11

one region to another, as it is expected to occur in a hi-tech, highly competitive industry operating in a worldwide market.

First, the education landscape including the competences available in the EU is outlined. Then, some figures on research spending by both the private and public sector are provided for both the EU and other regions, followed by some measures of innovation, technology transfer and some highlights on standardization aspects of the GNSS user segment, and in particular Galileo.

As mentioned earlier in this book, during mid and late 2000s, a considerable effort was done by firms, research centres and governmental organisations to demonstrate the feasibility and performance of Galileo receivers. As the Galileo satellite launch plan has been delayed over the years, R&D investment have diverted to other aspects not specific to Galileo, as multi-GNSS, cost-size-power reduction, or hybridisation with other sources of positioning and guidance information. Public investment has followed these lines but without forgetting the initial goal to prepare the ground for Galileo when its signals are available.

GNSS Education in Europe

A high-quality education is required for a successful outcome of research and development, as it furnishes organizations with the required workforce knowledge and skills. Figure 11.1 has been generated using indicators from the "World Economic Forum—Global Competitiveness Report 2010–2011" (Schwab 2011), and presents a generic quality comparison of research institutions for some EU and non-EU countries. This indicator does not refer exclusively to the GNSS field but can be used as a proxy for it. It shows that research quality in the richest EU economies is as high as in other regions as U.S. or Canada.

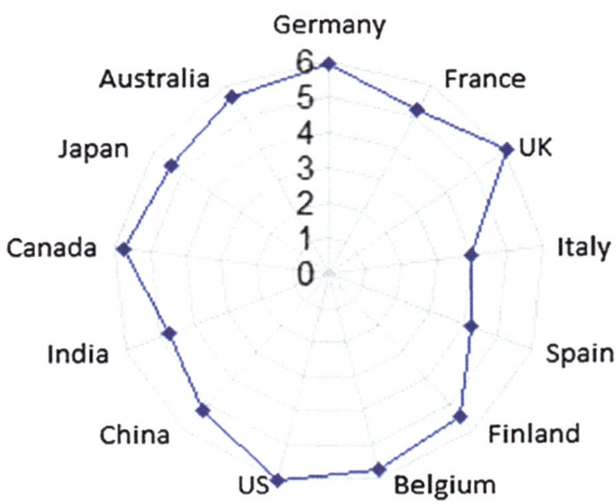

Fig. 11.1 Quality of research-scientific institutions per country, 2010

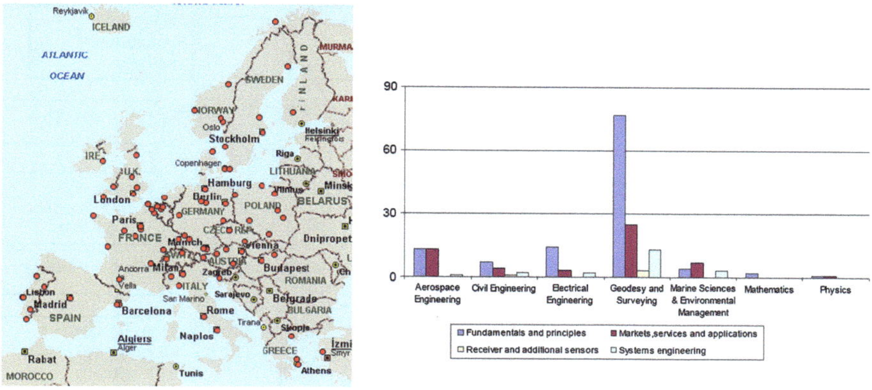

Fig. 11.2 GNSS-related courses offering in EU and related degrees

Satellite navigation principles have been taught for years in universities worldwide, especially in the fields of geodesy and electrical engineering. At the moment, education in satellite navigation is offered by European universities in many countries. Most of it is part of M.Sc. or B.Sc. degrees, but few specific degrees exist as well, as the M.Sc. in GNSS offered by Politecnico di Torino (Italy). GNSS-specific courses for students and professionals are offered periodically in Europe as the yearly ESA GNSS Summer School or the NavtechGPS seminars.

The graphic and chart in Fig. 11.2 present the locations in the EU where GNSS-related courses are offered, and the main disciplines and related degrees (Davies 2008).

Most of the GNSS university courses are related to Geodesy and Surveying disciplines, followed by electrical and aerospace engineering, and are focused on the GNSS fundamentals, followed by 'markets, services and applications', possibly due to the study of practical applications of GNSS related to different scientific and technical disciplines. Receiver technologies are covered as well although to a lesser extent, probably because they are more related to practical implementation aspects dealt with by receiver manufacturers, and also because many aspects related to GNSS receivers, as signal processing and analog or digital electronics, are covered in general telecommunication and electronic engineering courses.

Figure 11.3 presents a non-exhaustive map covering GNSS receiver research competences across Europe (Seco-Granados 2007). All relevant disciplines related to GNSS receivers seem to be covered: antennas, radio frequency hardware, clocks, signal processing, multipath, interference, integrity, high precision, indoor applications, geodesy, hybridization, etc. It can be concluded that GNSS-related disciplines (telecommunications, geodesy, statistics, aerospace engineering…), and especially those related to GNSS receivers (signal processing, control theory, computer programming, analog and digital electronics…), seem to be well covered by education organizations in Europe through generic or specific degrees and training activities.

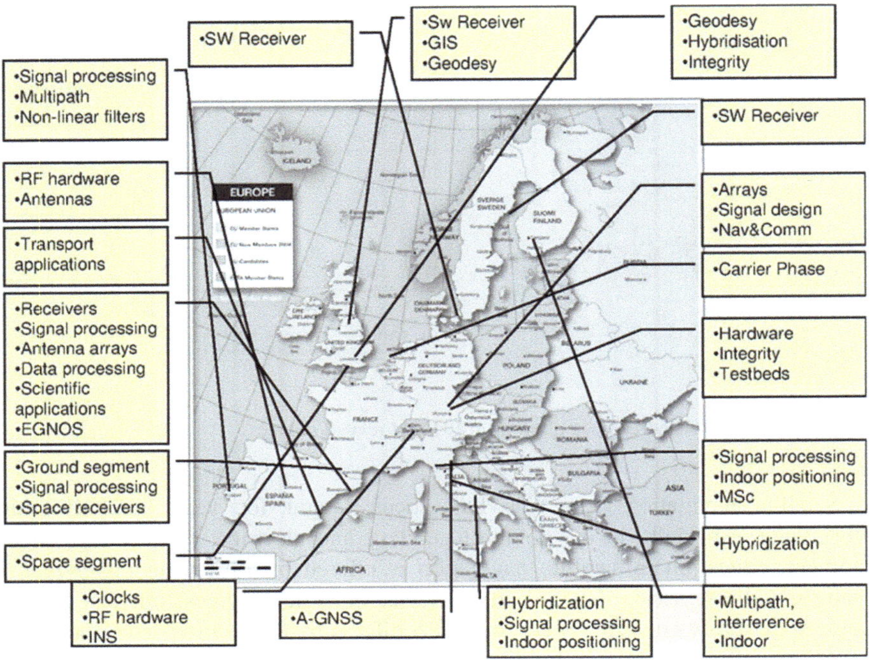

Fig. 11.3 Orientative GNSS R&D competences in Europe

Private Research

In order to provide a global view of the GNSS research activities, investments by receiver manufacturers competing in the different satellite location and navigation markets cannot be neglected in this chapter. This section casts some numbers and provides some figures on the yearly R&D investment in GNSS receiver technologies in the EU and worldwide. One must be cautious with the conclusions due to the following reasons:

- Firms with the highest revenues from GNSS activities, mostly in the semiconductors or IT equipment markets, have other business lines. It is not possible to quantify how much R&D spending is GNSS-related or which are the GNSS R&D strategic lines pursued, with just the public data available.
- For the mobile handheld location market, which represents the majority of the receivers sold, GNSS is a building block of a multi-technology location solution based on GNSS but also Wifi, mobile networks and inertial sensors.
- The value chain of GNSS receiver-based products and services is complex. It involves hardware manufacturers, OEM (original equipment manufacturer) integrators, navigation and application software developers, etc. It is therefore difficult to define the market and its size.

- The industry ecosystem of location technologies is very dynamic, and mergers and acquisitions occur frequently—actually, some may have already occurred since the time this text has been written.

The top five GNSS-related firms in the EU and abroad in the consumer chipset market (2009) are presented in the Table 11.1 and Fig. 11.4. They are based on data from the European Commission Joint Research Center's "*2010 EU Industrial R&D Investment Scoreboard*" (EU 2010). A chart below shows the absolute R&D investment and the R&D investment/sales ratio for these companies. Other firms with sales in Aerospace and defense and other sectors are analyzed in separate tables. The field 'Rank' represents their overall position in the top-1000 EU and non-EU firms ranked by R&D investment in absolute terms in 2009.

Some observations can be inferred from the figures:

- Although with some variability, the (non-weighted) average in both regions is comparable and close to 20 %, although somewhat higher in the U.S. (22.6 %).
- Nokia seems an outlier in overall (i.e. not GNSS-only) R&D spending due to higher sales (2009), but its R&D spending ratio is the lowest of all firms.

As regards R&D expenditures in other sectors (Table 11.2), the firms Garmin, Tomtom, Trimble and Hexagon/Novatel have been used as a reference as, except in the case of Hexagon, their main business line is related to satellite navigation, and

Table 11.1 R&D statistics of principal GNSS chipset manufacturers

Rank	Name	Sector	Country	Net sales 2009 [€m]	R&D investment 2009	R&D/ sales ratio (%)
EU GNSS chipset manufacturers						
2	Nokia	Telecommunications equipment	Finland	40,984	4,997	12.2
20	STM	Semiconductors	Netherlands	5,931	1,547	26.1
50	Infineon	Semiconductors	Germany	3,674	515	14.0
47	NXP	Semiconductors	Netherlands	2,678	535	20.0
156	CSR	Telecommunications equipment	UK	419	116.82	27.9
					Average	20.0
Non-EU GNSS chipset manufacturers						
36	Qualcomm	Telecommunications equipment	USA	7,259	1,700	23.4
50	Broadcom	Semiconductors	USA	3,130	1,069	34.2
52	Texas instruments	Semiconductors	USA	7,267	1,028	14.1
97	Freescale semiconductor	Semiconductors	USA	2,445	580	23.7
290	Atmel	Semiconductors	USA	848	147	17.3
					Average	22.6

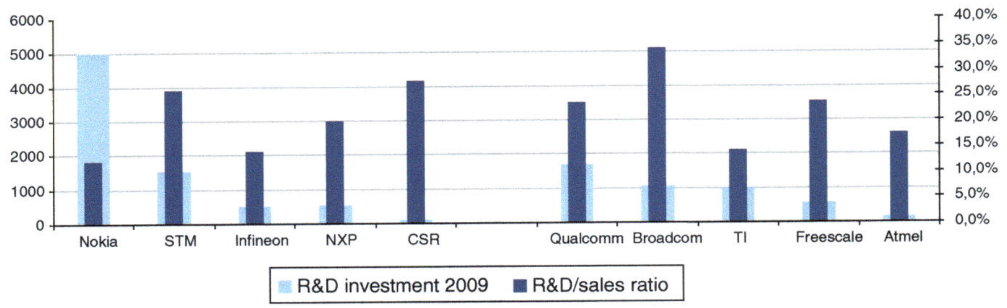

Fig. 11.4 R&D investment of principal GNSS chipset manufacturers

Table 11.2 R&D statistics of GNSS firms in other sectors

Name	Sector	Country	Net sales 2009 [€m]	R&D investment 2009	R&D/ sales ratio (%)
Garmin	Leisure goods	Cayman islands	2,054	166	8.1
TomTom	Electronic equipment	Netherlands	1,480	146	9.9
Trimble navigation	Electronic equipment	USA	785	95	12.1
Hexagon	Industrial machinery	Sweden	1,153	111	9.6
				Average	9.9

in particular GNSS receivers and services in the case of Garmin and Trimble, which seem the only two GNSS-specific companies appearing among the top-1,000 s (EU 2010). Concerning aerospace and defense companies, the three most active firms in GNSS receivers in both the EU and other regions are presented in Table 11.3. Figure 11.5 shows the average R&D/net sales ratio for the three different sectors: consumer GNSS chipsets (semiconductor/telecom equipment), 'others', including PNDs and personal or geodetic receivers, and aerospace and defense.

We see that firms categorized as 'other' seem less R&D intensive than consumer chipset manufacturers, with an average ratio of around 10 %, and aerospace and defense even less R&D intensive with a ratio of around 6 %. It should also be noted that the aerospace and defense industry relies strongly on the procurement of prototypes whose development is close to be considered R&D, even if not so classified.

We can also see that R&D relative expenditures look therefore more sector-dependent than region-dependent, as no significant trend or difference is observed between EU companies and non-EU (mainly US) companies, which is what one could expect from a global economy in which firms compete in worldwide markets (except in the case of aerospace and defense, which is strongly based on national contracts). However, this result may not be extrapolated on a per-country basis: according to Fig. 11.6, based on data from the World Economic Forum (Schwab 2011), some EU countries lag considerably behind U.S. or Japan in R&D spending.

Table 11.3 R&D statistics of GNSS firms in the aerospace and defense

Aerospace and defence						
Rank (EU)	Name	Sector	Country	Net sales 2009 [€m]	R&D investment 2009	R&D/ sales ratio (%)
10	EADS	Aerospace and defence	Netherlands	42,422	2,878	6.8
16	Finmeccanica	Aerospace and defence	Italy	16,501	1,926	11.7
38	Thales	Aerospace and defence	France	12,731	664	5.2
Rank (non-EU)	Name	Sector	Country	Net sales 2009 [€m]	R&D investment 2009	R&D/ sales ratio (%)
109	Lockheed Martin	Aerospace and defence	USA	3,1494	523	1.7
141	Raytheon	Aerospace and defence	USA	17,341	394	2.3
206	Rockwell Collins	Aerospace and defence	USA	3,115	247	7.9
					Average	5.9

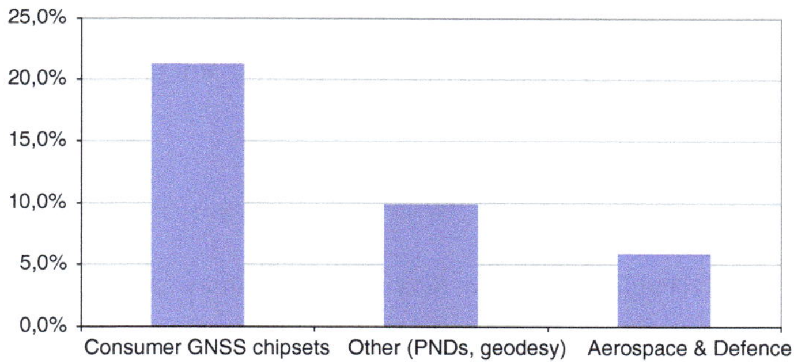

Fig. 11.5 R&D/sales ratio for companies in different GNSS user sectors

With the aim to determine how much the European GNSS receiver industry spends in R&D in total, we can combine the numbers above with the GSA Market Report (GSA 2010), which mentions that the estimated "core GNSS" worldwide market in 2010 amounted to 58 b€, and its estimated EU market share is 20 %. Assuming a R&D/sales ratio of 10 %, which is conservative taking into account some of the data above, the estimated private R&D investment in satnav-related user technologies in the EU would be around €1,1 billion per year. These figures also give an idea of the magnitude of the economy generated by satellite navigation, the positive impact in value creation thanks to GPS, not only in the U.S. but also worldwide, and the current dependence on it from all regions in the world, including Europe.

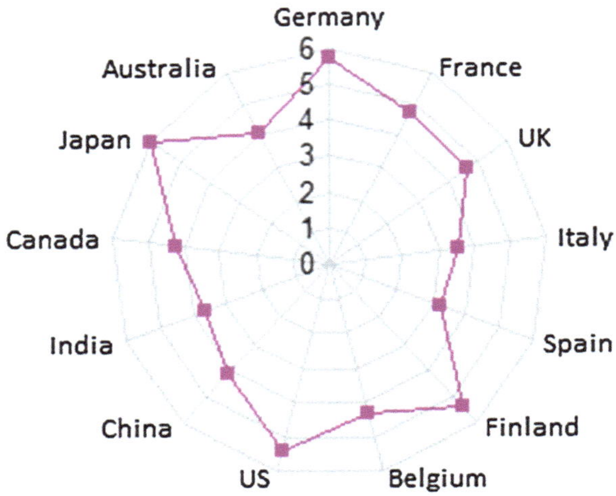

Fig. 11.6 Company spending in R&D per Country, 2010

Public Research

Research and development in GNSS has historically been strongly supported by the public sector for the development of governmental applications. Nowadays, and despite the fact that basic GNSS receiver technologies have been proven for decades, publicly driven R&D in the GNSS user segment still is an important part of policymaking, as exemplified in the following cases:

- Safety and Security:

- Improvement of location based safety services as E911 (Enhanced 9-1-1), E112 or e-call.
- Galileo PRS (Public Regulated Service), for military or other civil security forces.
- Interference detection and mitigation.
- Safety-of-life services, particularly those involving receiver technologies for publicly regulated sectors.

- Transportation:

- Traffic and road use management.
- Improvement of automotive technologies through Advanced Driving Assistance Systems (ADAS).

- Research and industrial policy:

- Technology transfer by cooperation of industry and academia or between research organizations (e.g. through EU cooperative projects or networks of excellence).

- Development of standards.
- Maintenance of know-how and competences in GNSS infrastructure and user technologies.
- Development and improvement of testing infrastructure facilities for academia, research organizations, new ventures, etc.

The case of European publicly funded research is a very specific one compared to other regions in the world, as research opportunities are quite fragmented between European, national and regional initiatives. A comparison of annualized public R&D spending among different regions in the world involved in GNSS was performed as part of the ERIG FP7 project (Davies 2008), estimating an average annual public R&D spending in GNSS of 75 m€ in the EU, including European, national and regional initiatives.

When compared to the U.S. during GPS's development phase before the system was declared operational in 1995 (Pace et al. 1995), we see that the numbers are in the same order of magnitude, especially noting that Fig. 11.7 provides the R&D spending in the GPS user segment only.

Figure 11.8 shows how GNSS R&D is divided in different categories in EU and in other regions.

We could say that about 21 m€ of public R&D spending in Europe (i.e. 28 % of 75 m€) was spent in the user segment annually in average in the late 2000s, period when the first Galileo receiver prototypes were developed.

One differentiator of the EU R&D public spending is the strong focus in 'market, services and applications' and 'business engineering', more than 50 % altogether, when compared to the United States. This may be linked to the focus on preparing

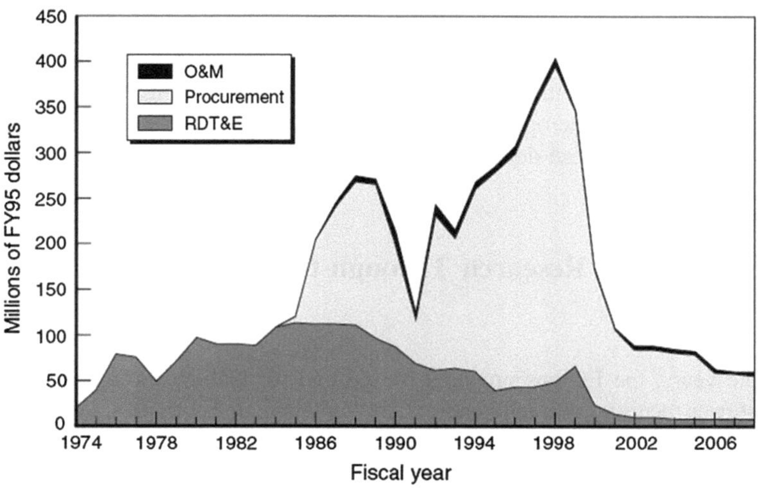

Fig. 11.7 U.S. spending in GPS user segment. Selected Acquisition Reports are sent from DoD to U.S. Congress and reflect major U.S. program expenditures. RDT&E stands for research, development, testing and evaluation

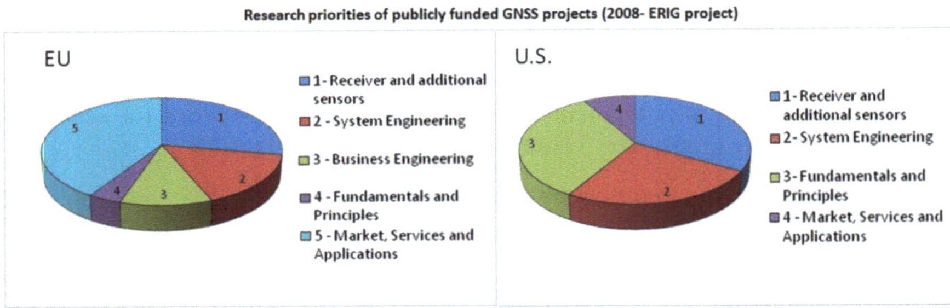

Fig. 11.8 Balance of GNSS research in Europe versus other regions, 2008

the ground for Galileo as a civil system conceived to provide economic growth, new services, markets, etc. Also remarkable is the smaller EU investment on 'fundamentals and principles' (about 4 %).

Concerning European public organizations financing R&D in GNSS, the European Commission and ESA are in the lead. ESA R&D in GNSS is channeled through programmes like the European GNSS Evolutions Programme (EGEP) and is mostly oriented to the "upstream" activities to modernize ground and satellite infrastructure, as for example through inter-satellite links, improvement of Galileo satellite clocks, study of new radiofrequency bands for GNSS, as the C or the S bands, improvement of ground monitoring station receiver capabilities, and the like. Concerning GNSS receivers, ESA also covers activities in their principal domains of expertise, as space and scientific applications.

ESA projects have also developed GNSS receivers required for the Galileo infrastructure, as in the case of the GETR (Galileo Experimental Test Receiver), aimed at receiving the early Galileo signals from GIOVE-A and GIOVE-B experimental satellites, the Galileo TUR (Test User Receiver), aimed at validating the Galileo signals, or the GSS (Galileo Sensor Stations), reference monitoring stations integrated in the operational ground segment of Galileo. Other chapters in this book deal with these developments more in detail.

Galileo Receiver Research Through the Framework Programmes

Over the years, the EC has prepared the ground for Galileo through the development of user receivers and applications, or "downstream" activities, through the 6th and 7th Framework Programmes (FP6, FP7).

The 6th Framework Programme of Research and Technological Development was the main tool to finance "downstream" Galileo activities. It covered the whole European Union's financial framework between 2002 and 2006, with a total budget of €110 million for GNSS applications, mission support, test beds and Galileo receivers, establishing the basics for the development of marketable Galileo receivers for consumer and professional user communities. At that time, the EC created the GJU (Galileo Joint Undertaking), with the mission to promote and manage the future exploitation of Galileo, and which was in charge of the FP6 GNSS calls for proposals and projects.

One of the first ever Galileo operational receivers was developed as part of FP6 in the GARDA project, led by Thales Alenia Space Italia and finalized by 2006. It studied Galileo receiver core technologies and developed a software receiver platform and an operational prototype that processed the Galileo E1, E5 and E6 signals as defined at the time. It also set the roadmap for the receiver funding activities that followed, including the segmentation of receivers R&D in *mass market*, *professional* and *safety-of-life*. The GJU followed this segmentation in its R&D funding strategy, allowing a sufficient number of early stage prototype receivers to be developed for all sectors and facilitate the penetration of Galileo technologies in different user markets:

- Mass market receiver projects covered location-based services (LBS) and car navigation users, focusing on E1 signal processing optimization, miniaturization, Assisted-GNSS techniques and high sensitivity.
- Professional receiver projects focused on high-end receivers and areas like AltBOC processing, multi-frequency and high precision through RTK (real time kinematics) and PPP (precise point positioning), or multi-antenna beamforming.
- Safety-of-life receiver activities focused on Galileo integrity services and the development of receiver-specific technologies for safety-critical applications, as antenna, front-ends, interference mitigation techniques, and certification activities. Later, the Galileo safety-of-life was re-profiled, but the effort performed in receiver design and development to comply with the stringent requirements of the aviation and other safety-related communities, as maritime and rail, remains applicable nowadays.

All in all, a significant effort was made during FP6 in Galileo receivers R&D by both the public and the private sectors, as most projects were co-funded by the GJU with about two thirds of total cost. Hundreds of deliverables, studies, IP blocks, simulators and some patents were generated in this process. Here are some of the most significant outputs of these efforts:

- The feasibility and performance of Galileo and GPS-Galileo receivers was demonstrated. This boosted the development of the first 'Galileo-ready' products on the market. For example, by autumn 2006, FP6 participants U-blox,[1] Septentrio[2] and other companies like SiGe Semiconductors[3] appeared to offer the first-ever Galileo-ready GNSS receivers. Figure 11.9 shows some of the Galileo receiver prototypes and products used and developed under FP6.

---

[1] http://www.yournav.com/content/n/225/first-galileo-receiver-upcoming.html.

[2] http://www.gisdevelopment.net/news/viewn.asp?id=GIS:N_suztwvfido.

[3] http://www.newelectronics.co.uk/article/7542/First-Galileo-receiver-launched.aspx.

Fig. 11.9 Example from FP6 projects of Galileo-ready receivers From *left* to *right*: ARTUS prototype, courtesy of IFEN, SWIRLS prototype, courtesy of Septentrio, and GNSS chip used for the GREAT project, courtesy of U-blox

- Galileo main differentiators were studied at receiver level, developing the required competences and mitigating potential implementation risks. At the end of these projects, the feasibility of commercial products incorporating GPS/Galileo L1/E1 receivers for the mass market, triple-frequency wideband receivers (E5, E6, E1) and advanced TCAR (E1, E5, E6) algorithms was insured, and performance improvements due to BOC signals and AltBOC in particular were demonstrated (Consoli et al. 2008).
- A cluster of Galileo simulators and testing facilities not existing before emerged in the EU and was made available to the GNSS community.[4]

Research activities in Galileo receivers were extended into the EC's 7th Framework Programme for the following EU financial framework (2007–2013). As for FP6, this Framework Programme covered principally the GNSS "downstream" activities: applications, receivers, services, etc.

Three calls for proposals have been launched in FP7 (2008, 2009 and 2010), covering the themes 'Applications', 'Receivers', 'Tools' and 'Support'. Concerning FP7 research in receivers, as the basic receiver chain was already developed and tested successfully in FP6 prototypes, FP7's priorities went beyond the basic functionalities of a receiver. The topics proposed so far have been 'Mass Market Receivers', 'Professional Receivers', 'Innovative (multi-technology) Receivers', 'Integrity Receivers' and 'Dual-frequency SBAS receivers'.[5] The overall priorities for this period have been:

- Analysis of Galileo-specific issues in receivers, as wide-band radio frequency front ends and Galileo BOC signal acquisition and tracking.
- Multi-GNSS positioning and synchronisation.
- Hybridisation: research of techniques to combine GNSS with inertial sensors, communication signals and other location information.

---

[4] http://gnss-test-portal.eu/.

[5] A detailed list of FP7 calls, activities, topics and selected projects can be found under http://gsa.europa.eu and more information on FP7 in general can be found under http://cordis.ec.europa.eu.

- Integrity: research on techniques to counterfeit multipath and unintentional interference.
- SBAS upgrade: develop receiver technologies to support the next dual frequency (L1-L5) SBAS equipment generation, including antennas, signal processing and multi-frequency observable PVT modes.
- PRS: RFFE technologies for PRS signals, encryption functionalities and intentional and unintentional interference mitigation.

Here are some non-exhaustive examples of the achievements from FP7 projects related to receivers[6]:

- During the 1st call (2007), devoted to 'mass market receivers', the selected projects proposed to focus their research on low cost, power and small-size dual frequency receivers, including L1-L5 front-end switching methods to reduce power consumption, narrow bandwidth front-end performance, multipath mitigation including non-line of sight multipath, hybridisation of navigation and communication signals in the GRAMMAR project (Della Rosa et al. 2010), or dual frequency and carrier phase navigation hybridised with inertial sensors for precise automotive navigation and spoofing mitigation in the GAMMA-A project.
- In the 2nd call (2008), devoted to professional applications, the awarded projects focused on various aspects, as hybridisation of GNSS with laser and inertial measurements for urban mapping in the ATENEA project (Fernández et al. 2011), the use of differential measurements in the signal tracking correlators in the ART-X project, or the multi-constellation and multi-frequency software radio implementations in the PRECISIO project (Roberts et al. 2011).
- In the 3rd call (2010), devoted to multi-technology platforms, the projects focused on multi-technology localisation aspects as the combination of antenna arrays, visual and inertial sensors (ADVANSYS), or synthetic-aperture antennas for GNSS reference receivers (DISAP), or the translation of high-precision features into mass-market chip and chipset implementations (E-HIMALAYA).

For the years to come, the European Union research programs will continue supporting GNSS through 'Horizon 2020', the framework programme for research and innovation of the European Commission for the period 2014–2020, with a €80 billion total budget.[7]

---

[6] More details on FP7 projects GRAMMAR, GAMMA-A, PRECISIO, ATENEA, ART-X, E-HIMALAYA, ADVANSYS and DISAP can be found under http://www.gsa.europa.eu/r-d/gnss-project-portfolio.

[7] http://ec.europa.eu/programmes/horizon2020/en.

Innovation and Technology Transfer

The ultimate purpose of an investment in R&D is to generate novel products and services that create economic value or increase social welfare by satisfying unmet needs. That is, to innovate. In order to achieve that, technology transfer between universities and industry is vital. This section presents some general and GNSS-specific indicators for EU countries and other regions.

University-Industry Collaboration

Figure 11.10 presents university-industry collaboration in R&D for some EU and non-EU countries. As in previous sections, general metrics are used as a proxy for the GNSS sector. U.S. and U.K. rank highest, followed by other EU and non-EU most developed economies (e.g. Canada, Finland or Germany). Lower industry-university collaboration in some other countries may be compensated by stronger government-university collaboration in support to innovation.

Startups, Acquisitions and Spinoffs

This brief section presents some examples of successful technology startups, acquisitions and spinoffs in the satellite navigation domain over the last years. Mainly from the U.S., some examples startups leading to successful exists in the GNSS domain are SnapTrack, founded in 1995 and acquired by Qualcomm in

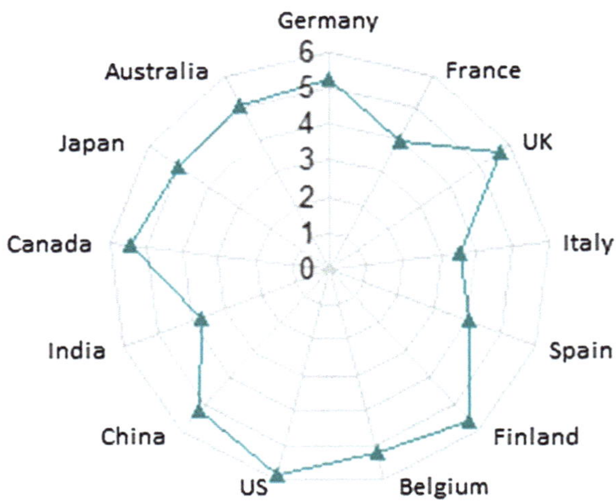

Fig. 11.10 University-industry R&D collaboration

2000, which developed snapshot and assisted technologies, Global Locate, founded in 1999 and acquired by Broadcom in 2007, which developed AGPS and GPS technologies, Nordnav, founded in 2002 and acquired by CSR in 2006, which developed GPS Software Defined Radio (SDR) receivers, and SiRF Technology Inc., founded in 1995 and acquired by CSR in 2009 after becoming a reference in GPS chips. In addition to GPS technologies, digital mapping has become a critical technology for consumer location services, as proven by the acquisition of Teleatlas NV by TomTom (2008) and Navteq by Nokia (2007). In the more recent years, with the boom of smartphone apps, many examples of startups related to location services and applications can be found in Europe, the U.S. and other regions.

Spinoffs, or startups created from a university lab, are good indicators of technology transfer. In Europe, some examples (among many others) of spinoffs that have become small and medium enterprises with a place in the commercial and/or institutional markets are Septentrio Satellite Navigation NV and IfEN GmbH. Septentrio is a professional receiver manufacturer located in Leuven, Belgium, founded in 2000 as a spinoff of the Inter-university Micro Electronics Centre (IMEC). IfEN was founded in 1998 as a spin-off of the Institute of Geodesy and Navigation at the University of the Federal Armed Forces Munich and is providing infrastructure and technical support to the European GNSS programs and commercializing GNSS simulators and receivers.

These are just a few examples of technology transfer through startups in the GNSS user domain. There are many others already and surely more to come in the next years.

Patents

Although only a small percentage of patents generate revenues mechanism, patent filing is an important indicator of innovation, as it is one of the few tangible measures of the outcome of R&D activities. Figure 11.11 presents the number of patents per million habitants in some EU and non-EU countries.

It is difficult to extrapolate these metrics to GNSS, a field where the U.S. has historically been at the forefront thanks to GPS. Figure 11.12 presents a comparison of the volume of patents filed in GNSS receivers for location-based services until 2009. We see a high increase in general in the mid 2000s, with the advent of location services for mobile phones and the incorporation of Assisted GNSS positioning.

Standardization

Standards are a way to open the markets and improve and simplify customer experience. They are also the consolidation of R&D and policy efforts, and in the case of new systems as Galileo, they are the entry point for their use by the user

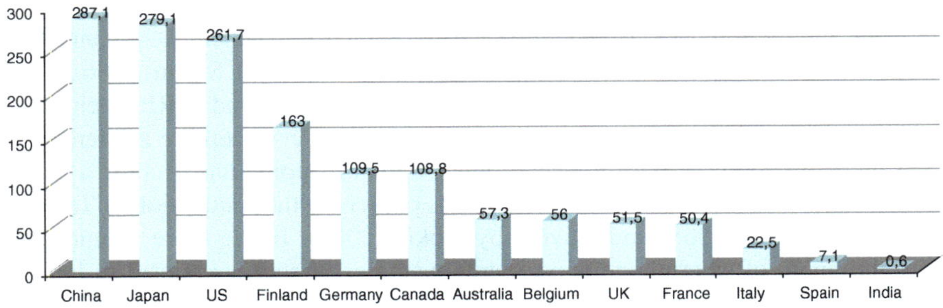

Fig. 11.11 Patents granted in 2009 per million habitants

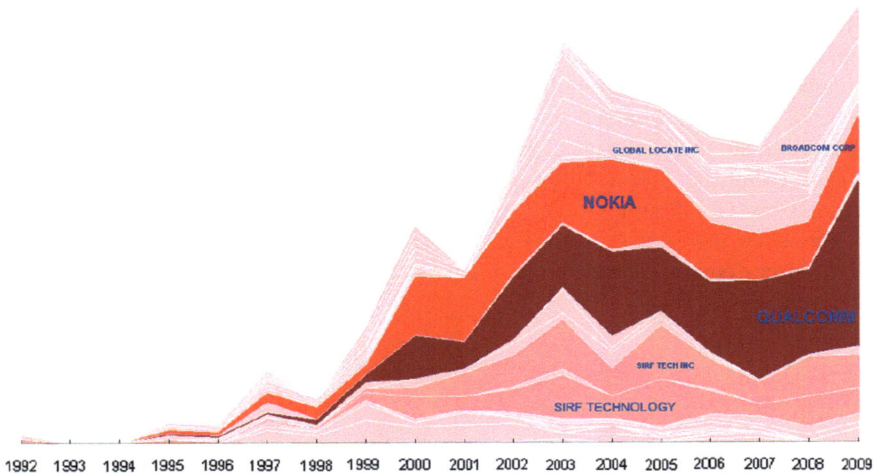

Fig. 11.12 Comparative chart of LBS GNSS receiver patent applications per year (*source* the HIMALAYA project)

communities. Standards are therefore a very relevant aspect of Galileo receiver technologies. Here are some of the main aspects related to standardization of Galileo receivers.

- In mobile communications and LBS, 3GPP has become the main forum of standardization of new technologies, platforms and formats. Galileo was firstly included in 3GPP standards in 2007 through Release 7, proposing a new standard for A-GNSS including information for Galileo satellites and other constellations'.
- Aviation is also a user community very influential regarding GNSS standards, as it requires a high level in standardization and certification due to safety reasons. The International Civil Aviation Organization (ICAO) is the main international body in charge of assuring safe air transport through the approval of SARPs

(Standard And Recommended Practices). In the U.S., the RTCA (Radio Technical Commission for Aeronautics), particularly SC-159 committee, have developed the main Minimum Operational Performance Standards (MOPS) for satellite navigation augmentations. Although aviation standards are currently based on GPS, strategic plans are developed to incorporate other GNSS including Galileo. At European level, the EUROCAE (European Organisation for Civil Aviation Equipment) has been for many years working on the development of Galileo and GPS/Galileo standards for aviation, although the lack of Galileo operational services has slowed down progress in Galileo standardization.
- The maritime community, in addition to founding the principles of navigation over the centuries, has produced some of the ‘de facto’ GPS standards used broadly by consumer and professional users worldwide, as those developed by the NMEA (National Maritime Electronics Association) or the RTCM (Radio Technical Commission for Maritime services), both from the U.S..

Trends and Challenges

Undoubtedly, location technologies are undergoing a quick transformation at all levels. Localisation using navigation satellites has become a commodity for a big part of the world population. Satellite navigation receivers have evolved from professional equipment or handheld devices for open-air leisure activities to miniaturised few euro chipsets integrated massively in smartphones and cars electronic equipment with impressive sensitivity and accuracy. It is difficult to predict how localisation technologies will look like in 10 years from now, especially in fast-moving sectors as the mobile market, but here are some trends based on what is already ongoing.

Multi-GNSS

Since the Selective Availability was turned off in 2000 and the civil community has been able to measure GPS real performance, GPS has significantly improved, reducing satellite orbit and clock estimation errors to usually less than one meter (Van Diggelen 2009) and, at the time this chapter is written, maintaining a constellation of more than 30 operational satellites in orbit. In addition to GPS, GLONASS has replenished most if its constellation[8] and has reached global coverage by around 24 operational satellites in orbit. China, with BeiDou, already provides a regional service over the Asia-Pacific region and will provide a global

[8] http://www.glonass-center.ru/en/GLONASS/.

service in the next few years. Galileo first operational satellites have been launched and provided the first Galileo-only fix and an aggressive launch schedule is foreseeing to deploy an operational constellation in a few years from now. The availability of 100+ ranging sources will allow the selection of those satellites at a good visibility and signal reception conditions, significantly improving accuracy in low-visibility environments.

Navigation and Communications

Thanks to the addition of a communication channel to the receiver, the Assisted GNSS standard has been a major enabler of GNSS performance for the mass market during the last decade, allowing to improve 'cold start' time to first fix (TTFF) from about 1 min to less than 1 s (Van Diggelen 2009), and allowing the improvement of sensitivity that often makes the difference between having a position fix and not having it in urban or indoor environments. New standards developed in the 3GPP and other technological groups, as LPPe (LTE Positioning Protocol Extensions) will permit a higher amount of assistance information. This may enable a higher accuracy through real-time kinematics (RTK) or precise point positioning (PPP) techniques even for consumer users. For that, smartphone receiver antennas, components and signal processing algorithms shall allow stable carrier phase measurement tracking in all environments.

Communications also allow the position computation process to be performed in a server, reducing computing load to the user and permitting the combination of GNSS with other information sources. Thanks to these developments, we can foresee relative or cooperative positioning algorithms or other kinds of location information sharing and combining through *crowdsourcing* processes (e.g. for interference detection).The increased communications combined with the improvement of semiconductors may allow to tag and localize objects and integrate satellite localization technologies as part of the *internet of things*.

Multi-system Integration

Mobile phone location users care about having an instant and moderately accurate position fix, no matter that it comes from a GNSS receiver or other sources. For the mobile sector, GNSS, in spite of its importance, is therefore just one out of many technologies used for location. Mobile phones use *signals of opportunity* from Wifi spots or cellular networks to locate the user, especially in indoor environments where GNSS signals are strongly attenuated.

Hybridisation is already happening also with dead-reckoning systems, as MEMS accelerometers and gyroscopes or maps. In particular, the development of GIS technologies suggests that 3-D and indoor mapping will be soon combined with

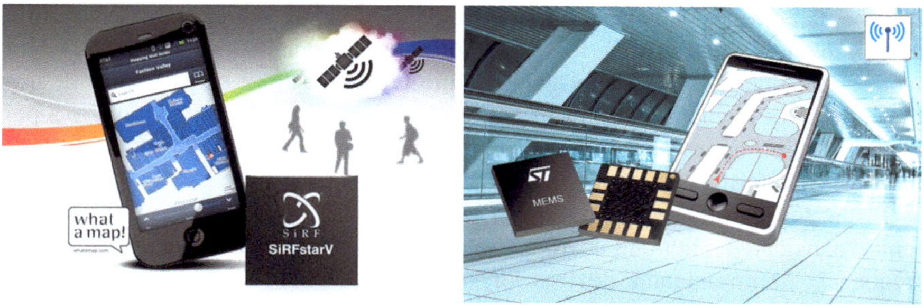

Fig. 11.13 Examples of hybrid chip and indoor applications, courtesy of CSR (*left*) and STMicroelectronics (*right*)

location information and possibly visual sensors to offer new location services (as e. g. georeferenced signatures or location-based billing) and improve the existing ones.

Therefore, indoor accurate and reliable positioning and seamless outdoor/indoor positioning services remain a challenge, possibly unsolvable with GNSS alone, but which GNSS is contributing to solve together with other sources (Fig. 11.13).

Hardware, Processors and Software Defined Radio

The dramatic improvement in semiconductor technologies over the past years has had a huge impact in the development of GNSS receiver's size, power consumption and processing capability. Excluding the antenna component, GNSS chips are orders of magnitude smaller than the smartphones or PNDs they are embedded in. Miniaturisation already allows the development of GPS chipsets in a few square millimetres, and we can foresee this trend to continue in the future, to GPS chips that can be attached virtually everywhere. This includes multiple radio front ends with higher frequency bandwidths and sampling rates. For example, some smartphones as iPhone 4S or iPhone 5 are able to receive GLONASS FDMA signals.

Power consumption and battery life are still an issue, especially for smartphones incorporating higher quality screens and more functionalities, and the GNSS receiver is one of the power-hungriest devices. This is partly due to the high amplification required to process the extremely weak GNSS signals as received on Earth (at around 10^{-16} W) and the continuous signal correlation and tracking. New semiconductors and batteries will allow more autonomous devices that may prolong the debate between an "always on" and a "push to fix" paradigm for consumer location services, or a selection between the two depending on the user application and environment.

Full software-defined radio GNSS receivers, although extremely useful for research purposes (Borre et al. 2007), are still challenging to implement in portable commercial devices. However, CPU and GPU processing power increase according to Moore's law will allow the feasibility of these implementations and the flexibility that can be obtained thanks to them. Future GPS consumer chips may shift from the extreme of having thousands of correlators to speed up the acquisition, to a pure software-based approach where signal acquisition is performed through FFT (Fast Fourier Transform) techniques.

In the long term, cheaper hardware components will allow to incorporate currently considered high-end features to consumer devices, as dual or even triple frequency, multi-antenna devices or ultra-stable CSACs (Chip Scale Atomic Clocks).

Interference, Security and Integrity

Due to the low power with which GNSS signals are received, intentional and unintentional interference is possibly the principal challenge for the world GNSS community, as it has been seen recently with the Lightsquared case in the US,[9] or with the proliferation of GPS jammers, also called *personal privacy devices* (Fig. 11.14), now considered illegal in the U.S. and in many other countries. This, combined with the increasing reliance of the world on GPS, makes authentication of user position, velocity and time a very desirable functionality for GNSS civil users. To achieve that, some GNSS providers as Galileo are studying the provision of authenticable signals for the civil community. Future GNSS signal authentication can be merged with security and augmentation measures at the user receiver (AGC control, Jamming-to-Noise measurements, interference detectors, antenna arrays, inertial sensors, trusted accurate clocks or synchronization protocols, signal processing authentication verifications, measurement redundancy verification, receiver anti-tampering measures, communication authentication) in highly robust and secured receivers offering authenticated location and timing services for civil and commercial purposes.

Together with authentication, position integrity is becoming important for all location users, not only those in the safety-related communities as aviation. Integrity in this context can be understood as the ability of the system to detect and report back to the user when it is receiving wrong position information, and/or to provide a measure of reliability or uncertainty in the position solution. Some examples of applications interested in integrity are emergency calls (E112 or 911), where the position error is bound to a certain amount of meters, liability-related position-based applications, as road tolling, or even indoor location services, where the position fix may not be reported by the software when uncertainty is so high that it may degrade the user experience.

---

[9] http://www.gps.gov/spectrum/lightsquared/.

Fig. 11.14 GPS Jammer, or *'personal privacy device'*

Ultimately, integrity failures from any source need to be treated similarly at the receiver, no matter that they originate from a satellite error, a strong multipath reflection, or a signal interferer, and are therefore part of the same problem.

Given the extreme dependence on GPS of many critical applications, backup navigation systems, both ground-based and satellite-based, have been and are under study, as for example e-Loran or Iridium. This trend will continue in the following years.

GNSS for Air Navigation

Air navigation has been a small in sector terms of market size, accounting only for 0.5 % of the market (GSA 2012), but it has been a very demanding and influential sector driving satellite navigation technologies and infrastructure, and it will still be in the future. In the next years, dual frequency (L1-L5) receivers will be standardized and deployed in aircraft, improving navigation performances thanks mainly to the ionospheric error cancellation. A new standard for L1-L5 SBAS is foreseen to be in place in the following years. For the following generation of avionics, ARAIM (Advanced Receiver Autonomous Integrity Monitoring) architectures will allow the use of Galileo and GLONASS together with GPS for airport precision approaches. They will also distribute the integrity burden between ground monitoring, user equipment and satellites (Blanch et al. 2011). However, certain technical and institutional challenges need to be overcome before the aviation community can benefit from ARAIM flexibility.

In addition to dual frequency standards and new integrity allocation schemes, the FAA (Federal Aviation Administration) in the U.S. has launched an Alternate Position, Navigation and Time (APNT) program to study other navigation systems different than GNSS to increase robustness and reduce reliance on GPS.

Finally, the importance of unmanned aerial vehicles (UAV), or drones, for civilian and military applications is increasing, and, together with other technologies as imaging and inertial sensors and radars, satellite navigation is one of its pillars.

Summary

Over the past years, ICT innovations have brought GPS from bulky and heavy receivers used for hiking or military applications into the pocket of millions of citizens. Currently they are integrating GNSS of with other technologies for the benefit of user location applications. As regards public policy, GNSS vulnerabilities, mainly related to interference, will have to be overcome through appropriate legislation, interference mitigation techniques, and backup location systems. GNSS markets (including Galileo) will continue to grow in the next years. This may imply some growth in R&D investments that, combined with the support to GNSS infrastructure from governments, will allow users to benefit from better-than-ever location and radionavigation services.

References

Blanch J, Walter T, Enge P, Wallner S, Amarillo F, Dellago R, Ioannides R, Pervan B, Fernandez Hernandez I, Belabbas B, Spletter A, Rippl M (2011) A proposal for multi-constellation advanced RAIM for vertical guidance.In: ION GNSS 2011 proceedings, 2011

Borre K, Akos DM, Bertelsen N, Rinder P, Jensen SH (2007) A software-defined GPS and Galileo receiver: a single-frequency approach. Birkhäuser, New York

Consoli A, Gerna D, Göhler E, Piazza F, Euler H-J, Winkel J, Bodenbach M, Falk N, Readman P, Granger R, Simpson S, Lück T (2008) The art of ARTUS–a second-generation Galileo/GPS receiver, InsideGNSS magazine, Spring 2008

Davies A (2008) Education, research and innovation and technology transfer in GNSS—final report, GSA, 2008

Della Rosa F, Hurskainen H, Detratti M, Pérez E, Nurmi J (2010) GRAMMAR: challenges and solutions for multi-constellation mass market user receivers. In: Proceedings ubiquitous positioning, indoor navigation and location-based service (UPINLBS) 2010, Helsinki, Finland, Oct 2010

EU Industrial R&D Investment Scoreboard, Monitoring industrial research (2010) 2010 EU industrial R&D investment scoreboard, joint research centre, Directorate General Research, European Commission, 2010

European GNSS Supervisory Authority (2008) EU sixth research framework programme 2002–2006. EU Publications, Lisbone

European GNSS Supervisory Authority (GSA) (2010) GSA GNSS market report—issue 1, 2010

European GNSS Supervisory Authority (GSA) (2012) GNSS market report, issue 2, 2012

Fernández A, Diez J, de Castro D, Silva PF, Colomina I, Pares E, Dovis F, Friess P, Wis M, Lindenberger J, Fernández I (2011) ATENEA: advanced techniques for deeply integrated GNSS/INS/LIDAR navigation. In: ION GNSS 2011 Proceedings, Portland, Oregon, Sept 2011

Pace S, Frost GP, Lachow I,Frelinger DR, Fossum D, Wassem D, Pinto MM (1995) The global positioning system—assessing national policies, critical technologies institute, RAND, USA
Roberts W, Bavaro M, Dominguez-Tijero E, Legrand F, Vaccaro S, Sage A, Hill C, Fernandez-Hernandez I (2011) PRECISIO—design, development and testing of a multi-constellation, multi-frequency software receiver. In: ION GNSS 2011 Proceedings, Portland, Oregon, Sept 2011
Schwab K (2011) The global competitiveness report 2010–2011, World Economic Forum, 2011
Seco-Granados G (2007) Galileo and R&D at university. Galileo information day, Madrid
Van Diggelen F (2009) A-GPS: assisted GPS, GNSS and SBAS. Artech House, Norwood

Chapter 12
Experimental and Professional Galileo Receivers

Andrew Simsky and Jean-Marie Sleewaegen

Abstract This chapter provides a review of existing experimental and professional Galileo receivers, related application projects and selected design topics, such as generic tracking channels, fast acquisition algorithms and AltBOC tracking. Although at the time of writing Galileo as a GNSS system is not yet fully deployed, quite a number of Galileo receivers have already been created with the help of Galileo simulators, mainly in the context of European projects, the most significant being the Galileo Test User Receiver project (TUR). Some Galileo receivers have also been developed to test the signals of experimental Galileo satellites GIOVE-A/B.

Introduction

Although at the time of writing Galileo as a GNSS system is not yet operational, the design of Galileo receivers is already a remarkably developed field. Availability of Galileo simulators (Boulton et al. 2007; Heinrichs et al. 2007; Irsigler et al. 2009; Artaud et al. 2004) makes the creation of fully functional Galileo receivers possible. The list of known manufacturers of Galileo simulators includes at this moment Spirent (Boulton et al. 2007), Aeroflex, CRS, IFEN (Heinrichs et al. 2007; Irsigler et al. 2009) and NAVIS (Artaud et al. 2004).[1] Most of the Galileo receivers built up to now have been designed within the framework of research projects completely or partially funded by the European Space Agency (ESA) or other European organizations. Development of Galileo receivers by independent researchers or at the universities is still limited, to a great extent due to the high price of Galileo simulators. Galileo receiver design in the private sector is not always covered in

[1] Another chapter of this book contains more information on available Galileo software simulators.

A. Simsky (✉) · J.-M. Sleewaegen
Septentrio Satellite Navigation, Leuven, Belgium
e-mail: simsky@septentrio.com

J. Nurmi et al. (eds.), *GALILEO Positioning Technology*,
Signals and Communication Technology 182, DOI 10.1007/978-94-007-1830-2_12

publications, but is definitely going on quite intensively and has reached a stage of commercialization. Many companies are marketing Galileo-compatible or Galileo-upgradable receivers. *Galileo-upgradable* means that the Galileo functionality is not fully available to users at the point of sale, but an appropriate software upgrade will be made available at the time when Galileo becomes usable (see for example Jia et al. 2009). With some manufacturers, the expected advent of Galileo has already stimulated releases of a new generation of Galileo-ready digital signal processors. These processors are software-configurable ASIC chips capable of tracking all the available ranging signals, not only of GPS and Galileo, but also of the Russian GLONASS and the Chinese Compass. This and other enhancements of the core technology are needed not only to accommodate new code modulations, but also to handle the sheer increase in the number of satellites. The receivers of previous generations were typically designed only for 12–14 dual-frequency GPS tracking channels. New receivers are able to track by far more satellites and signals. With some receivers, the number of available channels nears two hundred. Design of these new *tracking engines* is discussed in the next section. Fast-acquisition units, another new element of Galileo receivers, are described in the last section of this chapter. The tracking of the unique AltBOC signal is also discussed.

In the main part of this chapter we present a historical review of the experimental and professional Galileo receivers known from publications. Significant receiver design effort was stimulated by ESA, which ordered a number of receiver units for the testing and acceptance of Galileo and also reference receivers to be used by the Galileo ground control segment. ESA has also supported the development of special receivers to test the experimental Galileo signals transmitted by GIOVE satellites, and other related activities. In addition to ESA efforts, initiatives to develop Galileo user technologies are managed under the European Commission Framework Programmes; these are described in more details in another chapter of this book.

This chapter was written in May 2011 and reflects the status of projects at that time.

Tracking Engines for Galileo Receivers

When a receiver company begins to design a Galileo receiver, the challenge is not only in the adaptation of receiver algorithms for a new positioning system, but also in the development of a more powerful receiver able to handle a lot more satellites and systems simultaneously without significant increase in power consumption, weight and dimensions. When we say "a Galileo receiver", this means in reality at least a GPS/Galileo receiver, but more typically a multi-GNSS receiver of a new generation capable of using the signals of GLONASS and Compass as well. This challenge calls for a new design of the tracking engine, with a much greater number of tracking channels. The new receivers can be flexibly adapted to the new signals, even those whose modulations are only partly defined. Releases of new tracking engines of this kind have been reported by all major manufacturers of GNSS receivers.

The design of such tracking engines is typically based on a concept of a *generic (or universal) tracking channel*, which can be tuned via SW to track any of a wide variety of signals (Simsky et al. 2005; Lueck et al. 2009; Fortin et al. 2009). As it is not obvious beforehand which combination of GNSS signals might be deemed optimal by tomorrow's user, this kind of engine will be tuned either by a user or by a producer of a specific version of a product to a desired combination of systems and signals.

A typical architecture of a generic channel is presented in Fig. 12.1 (Simsky et al. 2005). The RF input can be any signal of GPS or Galileo. The residual carrier is removed by mixing of the incoming signal with a digitally generated complex carrier. One of the main differences between the current CA-code signal and the new signals of Galileo and GPS is the presence of a pilot component alongside the data-bearing signal. The spreading codes for pilot and data (or memory codes when applicable) are generated by two code generators, which can be configured to produce any spreading code of Galileo/GPS/GLONASS/Compass. Thanks to the coherence between the pilot and data components, a single rate-control unit controls the rate of both code components and also takes care of the BOC modulation. A locally generated signal enters two delay lines. Time-shifted signal replicas created in this way are correlated with the incoming baseband signal, producing all correlation values needed for code and phase tracking, BOC side-peak detection and multipath mitigation. The future receivers will not need indirect tracking schemes (such as codeless or semicodeless techniques currently in use for the encrypted GPS L2 P(Y) signal) because all the codes of future GNSS systems will be tracked directly.

As a side note, the fact that Galileo uses memory codes in the E1 and E6 bands has significant impact on hardware resources used for code generators. Due to the use of memory codes the size of the silicon area for tracking channels increases by about 30 %. The memory codes have been introduced to minimize intra-system interference.

GIOVE Satellites and First Galileo Receivers

Theoretical analyses of Galileo, such as prediction of accuracy and global availability, started from the onset of the Galileo program and were in many ways helpful for the design of receivers. Numerous research papers have been dedicated to the potential benefits of multi-system multi-frequency GNSS for the differential phase processing with fixed ambiguities, so-called RTK. Generally speaking, every work related to Galileo applications, was touching also on the topic of Galileo receivers.

However, the first effort to make actual receivers consisted mostly of hardware design, where several prototype-level boards able to track Galileo signals were built. At that time a fully-blown Galileo simulator was not yet available, and the first Galileo receiver was tested with the use of a custom-made four-channel signal

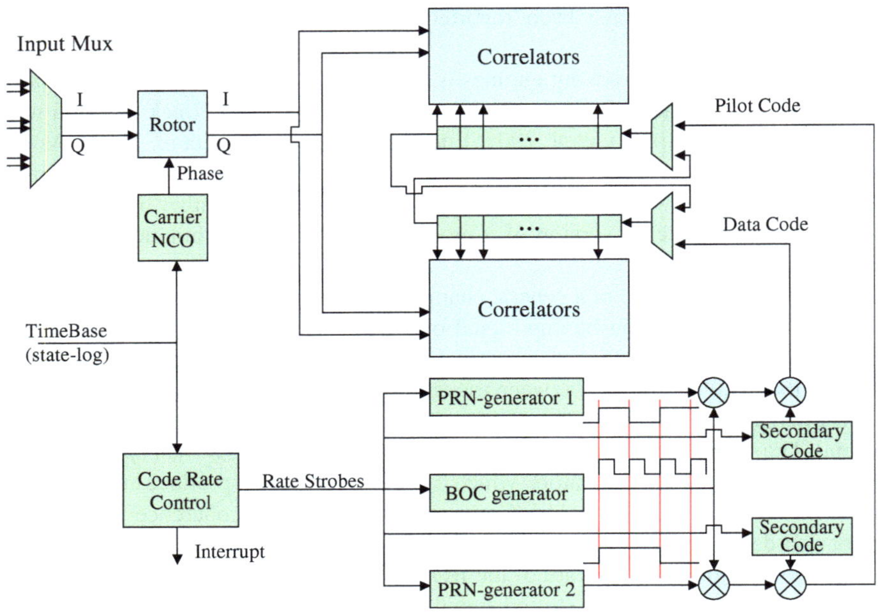

Fig. 12.1 Architecture of a generic tracking channel for a Galileo receiver

generator. This receiver was immediately used for end-to-end verification of the payload of the GIOVE-A satellite, designed to transmit experimental Galileo signals (De Wilde et al. 2004). These "breadboards" evolved later into the Galileo Experimental Test Receiver (GETR), made by Septentrio to ESA's order for the specific purpose of receiving and testing the signals of GIOVE-A and GIOVE-B satellites. The GETR was intended only to generate ranging measurements and to log data—no Galileo-based navigation functionality was foreseen.

The GIOVE-A satellite was launched on Dec 28, 2005 and was transmitting in all the planned Galileo frequency bands. The network of ground stations (so called GESS network Crisci et al. 2007) equipped with GETR receivers was set up by ESA for comprehensive testing of Galileo signals (NovAtel's GTR receivers were added later to some stations, see below). In particular, multipath performance of the new signals was compared to the performance of civilian GPS signals. This research was going on for a few years and concluded that Galileo signals in the same conditions of reception have lower values of multipath noise compared to the GPS L1CA, which is naturally expected due to their higher chipping rate and more advanced structure (Simsky et al. 2006, 2007, 2008a, b, c; Hollreiser et al. 2007). The ranking of Galileo signals according to the value of the average multipath noise was found to be in good agreement with theoretical predictions. Particular attention was paid to the wideband E5AltBOC signal (also referred to as E5 for brevity), which demonstrated record-breaking low-noise performance with code multipath errors of the order of magnitude of 10–20 cm. Exceptionally low code multipath of

E5AltBOC is expected to bring advantages for future differential applications (Fairhurst et al. 2010). Tracking of E5AltBOC was discussed in (Dötterböck et al. 2010; Lestarquit et al. 2008).

The next GIOVE-B satellite, launched on 27 April 2008, was transmitting the MBOC, a modified version of the original BOC(1,1) modulation of the civilian Galileo signal on E1. The multipath resilience of BOC(1,1) was improved by superposition with a higher-frequency spreading code. Expected improvement in multipath performance was confirmed in (Simsky et al. 2008b, c). More importantly, it was shown that with the MBOC all the Galileo signals (except for the outstanding E5AltBOC) would have comparable performance.

The first public demonstration of a receiver successfully computing a Galileo-based position was performed in September 2006 by Septentrio/Spirent on the annual ION-GNSS conference (Fig. 12.2). During the following years other receiver manufacturers (NovAtel, Javad, Topcon, Leica et al.) reported on the development of Galileo-ready professional receivers. Some of these receivers tracked only L1 and E5a signals, a natural choice for future dual-frequency receivers, these two bands being common for Galileo and modernized GPS (E5a band of Galileo is identical to GPS L5). Most of these new receivers were also able to log the signals of GIOVE satellites which transmitted much longer than originally planned. Septentrio's PolaRx3G was used in 2009 to test the reception of GIOVE-A/B signals by the researchers at the University of Delft (Tiberius et al. 2009). Space agencies and independent researchers around the globe set up their own GIOVE logging stations in addition to the GESS network of the ESA. The German Aerospace Center (DLR) reported its own 10-station global COoperative Network for GIOVE Observation (CONGO), which used four types of receivers: Septentrio's GeneRx1 (an analogue of the GETR), NovAtel's Europak 15a, Javad's Delta-G2T/G3TH, Leica's GRX-1200 (Montebruck et al. 2010). The GTR, an all-frequency GIOVE-monitoring receiver developed by NovAtel (Gerein et al. 2009) was used at some GESS stations alongside Septentrio's GETR (Giraud et al. 2009). ESA has provided, on experimental basis, the navigation messages for GIOVE-A/B satellites, so some researchers have computed first GPS/Galileo navigation solution (Bonhoure et al. 2010).

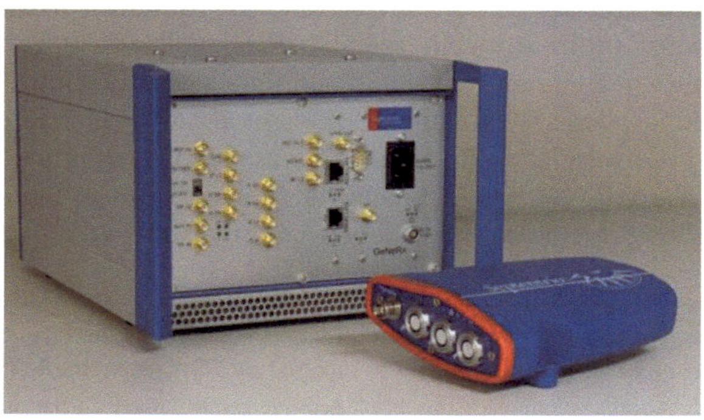

Fig. 12.2 Two GPS/GIOVE/Galileo receivers designed by Septentrio: experimental GETR (*left*) and AsteRx2, a receiver for professional applications

Test User Receivers and Other Galileo Receiver Projects

The Test User Receiver (TUR) project is a large-scale effort organized by ESA. Its purpose is to build a multi-functional GPS/Galileo receiver capable of versatile positioning functionality for the testing of the Galileo system. Flexibility is an important feature of the TUR (Morgan-Owen et al. 2009; Van Bree et al. 2010). The test receiver will be used to verify user requirements for all the Galileo services in a number of pre-defined user environments. For this end, the TUR supports 14 main receiver configurations switchable by user commands. The user environments (e.g. rural, pedestrian, etc.) are characterized by masking conditions and error budgets, which are defined for each combination of signals. The error budgets are given in terms of elevation-dependent User Equivalent Range Errors (UERE) and User Equivalent Range Rate Errors (UERRE). Septentrio has developed two versions of the TUR: TUR-N for the signals and applications of open, commercial and safety-of-life services of Galileo and TUR-P for the Publicly-Regulated Service (PRS) (Davies et al. 2010). At the time of writing, the testing of the TUR is possible only with the use of the simulator.[2] First, the simulator scenarios, which correctly simulate prescribed environments with their respective UEREs, must be created; then the receiver itself must be tested with the use of these scenarios (Morgan-Owen et al. 2009). The TUR is also able to log the signals of the GIOVE satellites. In (Van Bree et al. 2010) the TUR-N was used for the noise assessment of the signals of the GIOVE-B satellite. Elaborate control/analysis software for the TUR has been developed (Romero et al. 2008). Given the criticality of the Test User Receiver for Galileo, ESA has duplicated the TUR contract: another TUR receiver with the same characteristics was independently developed by a consortium led by Thales (Martin et al. 2008a).

As any other global satellite navigation system, Galileo requires for its operation a global network of reference receivers, which send satellite measurements to the Ground Mission Segment, or GMS.[3] The GMS computes and uploads orbit parameters and clock corrections to satellites and monitors the integrity of satellite signals. With Galileo, this important function will be performed by the so-called Ground Receiver Chain, which consists of: two non-PRS reference receivers, one PRS receiver, an antenna, a frequency standard and communication equipment. The non-PRS reference receiver (GRCN), designed by a consortium led by NovAtel, must fulfill stringent specifications: non-stop operation during 20 years, stable group delay, efficient multipath reduction and strong immunity to interference and ionospheric scintillations (Binda et al. 2006; Gerein et al. 2007).

The use of an RF simulator was sufficient to produce a working Galileo receiver. The GIOVE satellites could give a foretaste of the reception of the new signals. However, none of the above was a true simulation of the Galileo signals transmitted

---

[2] The first real Galileo position fix obtained with the four-satellite IOV constellation and with the use of the TUR receivers was reported by ESA in April 2013.

[3] Another chapter of this book provide more details on the architecture of GMS.

by a real constellation and received in real-life environment. An interesting experiment, which comes close to the simulation of the real Galileo constellation, was made in Germany, where the GAlileo TErrestrial test facility (GATE) was built in the German Alps (Heinrichs et al. 2007a, b). GATE includes six ground-based transmitters, located at high points of the mountainous terrain and viewed under the elevation angle of about 15°. These transmitters are not meant to work as pseudolites in the common sense (although they can). GATE operates in the so-called "virtual satellite" mode, in which the signals of a simulated constellation carrying standard Galileo navigation messages are transmitted by the ground transmitters. GATE in this mode works similarly to a usual RF signal simulator, but the signal is transmitted via the air rather than via cables, so that not only receivers, but the whole antenna + receiver combination can be tested with a real vehicle dynamics and multipath. In order to monitor GATE signals, IFEN has built an all-frequency Galileo receiver. Later, in the framework of the ARTUS project, the design of this receiver was developed into a full-fledged prototype Galileo/GPS receiver (Winkel et al. 2007; Lueck et al. 2007). Septentrio's GeneRx1 receiver was used in the GATE environment in the framework of the MAGES project to test helicopter navigation with Galileo signals (Haas and Dixon 2009). Pseudolite-based test environments for maritime applications were also created in the SeaGATE and MARUSE projects with the use of NovAtel's and Septentrio's receivers (Martin et al. 2007, 2008b; Grant et al. 2007).

Two prototype Safety-of-Life GPS/Galileo receivers, compliant with the Galileo Minimum Operational Performance Standards (MOPS), have been reported: a Canadian SOLRX receiver (NovAtel, CRC Canada) (Bates et al. 2007) and a European Galileo receiver for aeronautical applications created by Thales within the framework of the ANASTASIA (Denks et al. 2009) and GIRASOLE projects. The ANASTASIA receiver was used to test the vulnerability of Galileo air navigation to pulsed interference in the E5 band. This interference, caused by range-measuring radar equipment, is a well-known challenge to the use of both E5a of Galileo and L5 of GPS for aerial navigation (Denks et al. 2009; Simsky et al. 2009; Ayaz et al. 2010; Margaria et al. 2009; Steingass et al. 2009). In Simsky et al. (2009) it was shown that the level of interference on the ground is not strong enough to cause real problems to navigation even in the vicinity of big airports. However the intensity of the jamming signal increases with height. In the ANASTASIA project, the impact of interference was assessed with the use of an ingenious hardware setup, which mixed the signals of the Galileo simulator with the real interference signal recorded in flight (Denks et al. 2009). The Galileo receivers will tackle this type of interference with the use of the pulse-blanking technique, which shuts down the reception during the pulse of interference. It has been shown that pulse blanking is effective both in flight and on the ground (Denks et al. 2009; Simsky et al. 2009; Ayaz et al. 2010). Interference vulnerability of Galileo E1-MBOC signal was investigated in Margaria et al. (2009).

Galileo Phase Observable: Pilot or Data?

In each frequency band, Galileo satellites transmit a composite signal, which consists of two components: pilot and data. Because each of the components can be tracked independently, three sorts of measurements can be generated: pilot only, data only, and combined pilot + data. The question is whether Galileo receivers should output observables (ranges, phases, Doppler measurements) for data, for pilot, or for a combination of both? These three sorts of measurements are equivalent in theory, but minor biases dependent upon hardware implementation may be introduced. Therefore it seems desirable to define, at least for high-end receivers, which of the three shall be outputted and used. Here we focus only on carrier-phase measurements, which have mm-level precision. For phase measurements even sub-millimeter-level biases may become non-negligible in high-accuracy differential applications.

Carrier phase measurements of pilot and data signal components differ in two ways:

1. By a nominal carrier shift, either 90° or 180°, depending upon the multiplexing scheme. For E5a, E5b and E5AltBOC signals, this nominal shift is 90° (1/4 cycle). For E1 and E6 signals it equals 180° (1/2 cycle).
2. There is a possibility that they will also differ by small satellite- and receiver-dependent phase shifts caused by hardware effects. For example, on the E1 carrier, the pilot (E1C) and data (E1B) components use different modulations and have slightly different power spectra due to the opposite sign of the BOC (6,1) component of the MBOC modulation. Hence, pilot and data may be subjected to slightly different phase shifts when passing through satellite and receiver filters. On the E5 band, quadrature mismatch between I (data) and Q (pilot) components may cause slightly different phase shifts for I and Q depending upon the details of hardware implementation.

The nominal phase shifts of 90° or 180° are known from the Galileo SIS ICD and are normally irrelevant because they are common to all satellites in tracking and cancel out in double differencing. However, non-nominal hardware biases between pilot and data could be satellite-dependent and hence could have non-negligible impact on high-accuracy phase-based applications if base and rover receivers would track different signal components (e.g., with base tracking the pilot and rover tracking the data). To avoid potential problems, it is advisable to standardize the generation of carrier-phase measurements in professional receivers. Let us look in more detail at the pros and cons of pilot-only tracking.

It is well known that tracking of the pilot component compared to the data component offers quite a few advantages (Ries et al. 2002):

- The probability of cycle slips is lower for the pilot than for the data, and the ability to maintain lock at lower C/N0 values is higher.
- Pilot signals perform better in conditions of high carrier phase dynamics, e.g. during ionospheric scintillations.

- The risk of ½ cycle slips, which may occur with data-bearing signals, does not exist for pilot components.
- Full-ambiguity carrier phase reacquisition after signal outage is much faster with the pilot. Indeed, for the pilot component there is no need to wait for the "preamble" word to fix the full-cycle ambiguity.

However, the tracking of the pilot signal comes at a price of extra hardware complexity. Indeed, the data signal must still be demodulated in order to decode data bits. Therefore, one additional PRN-code generator and one extra punctual correlator are needed to extract the navigation bits from the data signal. Conversely, if only the data signal is tracked, this complexity can be avoided, because the same correlators can be used both for the generation of measurements and for the data demodulation.

Combined pilot + data tracking would require even more correlators, because the whole set of correlators required for tracking (e.g. early, late and punctual correlators) must be duplicated. The advantage of combined pilot + data tracking is that the power of the data and the pilot components is added, which brings about an increase in C/N0 up to 3 dB compared to pilot-only or data-only tracking. Although this seems to be an advantage, it must be emphasized that only the thermal noise is decreased by the added power. Combined pilot + data tracking does not bring any benefit for multipath noise, which is known to be much higher than thermal noise in real-life situations. Moreover, the above-stated 3 dB advantage of combined tracking is only achieved at high C/N0 levels. At lower C/N0 levels, where the increase in power is most desirable, the relative weight of the data component in the pilot + data combination decreases and the gain with respect to pilot-only tracking is reduced. On top of that, for carrier-phase tracking in PLLs at low levels of C/N0, the data component suffers from ½ cycle slips, and mixing it with the pilot component could degrade the signal quality compared to pilot-only tracking. This is an important reason why a combined pilot + data tracking, if ever used, is usually restricted to DLLs (i.e. to code range tracking), while for the PLLs (phase tracking) pilot-only is preferred.

Another potential issue with combined tracking is related to the above-mentioned possibility of small non-nominal hardware biases between pilot and data signals. When pilot and data are combined, their respective hardware biases are also combined in the same way, leading to compound biases, whose values depend on the respective weights of the pilot and the data components, which could be satellite-dependent and C/N0-dependent. Therefore the use of combined pilot + data tracking may give rise to undesirable time-varying satellite- and receiver-dependent biases which may not cancel out completely in double-differencing.

As a final conclusion of this section, generation of pilot-only carrier-phase measurements seems to be the most reasonable choice for future professional Galileo receivers. As a side note, this is one of the reasons why the future GPS L1C signal will transmit 75 % of its power in the pilot channel. At the time of writing, standardization effort is ongoing within several GNSS groups, such as RTCM SC-104 committee and IGS.

AltBOC Tracking

The wide-band E5AltBOC signal brings unrivaled tracking performance in terms of tracking noise and multipath, with pseudorange noise at the centimeter level (Simsky et al. 2006). The E5AltBOC measurements are formed by the processing of the Galileo E5a and E5b bands in a coherent way, i.e. the whole E5 signal (E5a + E5b) must be treated as one signal centered at 1191.795 MHz, the average of E5a and E5b center frequencies. The peculiarity of the baseband AltBOC is that it is complex-valued, i.e. it has real and imaginary components (Sleewaegen et al. 2004; Soellner and Erhard 2003), unlike all other Galileo or GPS baseband signals which are real-valued. If the AltBOC signal is tracked in a usual way using conventional channel structure (Fig. 12.1), the PRN code generators and all related delay lines and correlation logic must be designed to handle complex-valued signals. This would lead to significantly more sophisticated channel architecture.

The research towards novel more efficient algorithms to track the AltBOC signal has been one of the major efforts in the development of professional Galileo receivers (Dötterböck et al. 2010; Lestarquit et al. 2008; Martin et al. 2008a; De Wilde et al. 2009; Dempster and Shivaramaiah 2009). This research led to a remarkable conclusion that despite the peculiar complex structure of the AltBOC signal, it can be tracked in a very efficient way. The AltBOC modulation can be tracked with the use of only two coherent punctual correlators, the first correlating real-valued E5a code and the second correlating real-valued E5b code (Sleewaegen et al. 2004; De Wilde et al. 2009). There is no need to have a pair of early/late correlators.

The Acquisition of Galileo Ranging Signals

The acquisition of GNSS signals is the first step of tracking (tracking of Galileo signals is covered in Chap. 6). It involves correlating the incoming signal with local signal replicas characterized by particular values of frequency shift and code phase. With the traditional approach, receivers sequentially scan through all possible combinations of frequency/code-phase in the search space, until the correlation value would exceed certain predefined threshold. Each pair of code-phase/frequency values defines a *cell*. As an example, commonly used search granularity (cell size) is 0.5 code chips by 500 Hz in code-phase and frequency dimensions respectively. The time spent by the search process in each cell (*dwell time*) determines the acquisition sensitivity of a receiver, i.e., the minimal power of the signal which the receiver is able to detect given a certain probability of missed detection. The dwell time is an important parameter of the acquisition algorithm: long dwell times improve sensitivity, but can lead to unacceptably long overall duration of the acquisition process.

At the *cold* start a receiver has no previous knowledge of the code phase, so the initial acquisition process must scan through the whole length of the code sequence. For Galileo codes a complete scan could take quite long. As an example, for a code length of 10,230 chips (Galileo E5a), with a typical frequency uncertainty of ±5 kHz, and a cell size of 0.5 chips by 500 Hz, the search space would contain more than 400,000 cells. This problem is a known challenge for Galileo, which do not make use of coherent additional short codes, such as a well-known GPS L1-CA (Coarse/Acquisition) code, which was added to a longer L1P-code to facilitate its acquisition. However, the ease of acquisition of the GPS L1 signal with the CA code comes at a price. The shortness of the CA code causes high cross-correlation values between GPS satellites, so-called self-interference: high-power satellite signals can jam weaker ones. In view of this problem, Galileo codes are designed to be much longer in order to have lower cross-correlation values. However, with these longer codes, new acquisition techniques are needed to avoid unacceptable increase in acquisition time.

If a GNSS receiver would use the same number of correlators for Galileo as for GPS, the signal acquisition time with the same sensitivity would be longer for Galileo than for GPS at least by a factor of 10. In Table 12.1 search durations for some GPS, Galileo and GLONASS signals are provided assuming the same level of sensitivity and the same number of correlators for all the listed signals.

In this table, the search duration is directly proportional to the code length, except for the case of Galileo E1. This exception is due to the peculiarity of the modulation used on Galileo E1. With the MBOC modulation, the correlation peaks effectively become narrower, which entails finer search granularity. Hence, although the E1 code is only four times longer than GPS CA code, the acquisition time is longer by a factor of 12 (De Wilde et al. 2006; Shivaramaiah et al. 2008).

With Galileo signals, acquisition engines based on the concept of parallel search will replace traditional acquisition strategies based on the above-described sequential search. Parallel search engines have previously been developed for high-sensitivity acquisition of weak GPS L1-CA signals and also for the direct acquisition of the P-code (Rounds and Norman 2000; Eerola 2000; Betz et al. 2005). The first dedicated Fast Acquisition Unit (FAU) for Galileo signals was reported in 2006 (De Wilde et al. 2006). It combines an N-tap matched filter with an M-tap FFT unit (Fig. 12.3). This is equivalent to the use of several thousand correlators operating in parallel. Such a dedicated acquisition unit is generic and can be used to acquire the

Table 12.1 Duration of search process for the "cold" acquisition of different ranging signals. The duration of the acquisition of the GPS L1-CA code is used as a unit of time

Signal type	PRN code length	Search duration
GPS L1CA	1,023	1
GLONASS L1CA	511	0.5
Galileo E5a	10,230	10
Galileo E5b	10,230	10
Galileo E1	4,092	12

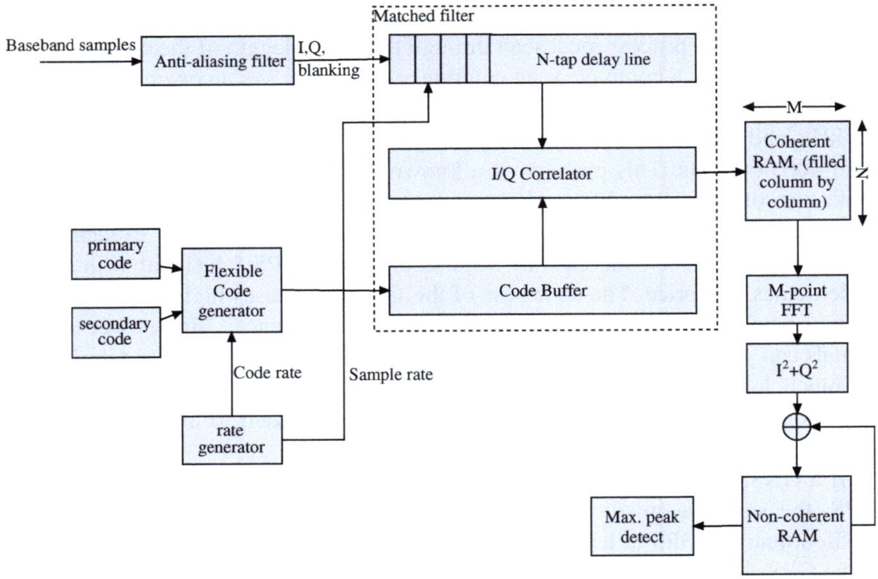

Fig. 12.3 The fast acquisition unit for Galileo signals (De Wilde et al. 2006)

signals of each system: GPS, GLONASS or Galileo. The above-mentioned TUR receiver contains multiple instances of such FAUs.

The ratios of acquisition durations shown in Table 12.1 still hold for the FAU. For example, with the use of the FAU, the GPS L1-CA signal will still be acquired about 12 times faster than with the Galileo E1 code. This means that in terms of the speed of acquisition GPS still has advantage over Galileo. Combined Galileo/GPS receivers typically reach their first fixes using GPS satellites, with the Galileo satellites being acquired later. As the technology evolves, the longer acquisition of Galileo codes with respect to GPS will become less important. For instance, if a future FAU will contain as many correlators as needed to acquire GPS L1-CA in 5 ms, acquisition of the Galileo E1 with the same unit will take 60 ms. With the acquisition durations so short, the difference will remain unnoticed by most users.

Conclusions

Despite the fact that Galileo as a GNSS system is not yet deployed, the satellite navigation industry is ready for its advent. Quite a number of Galileo receivers have already been developed and Galileo-ready receivers are already in the market.

These receivers include new technological elements, such as tracking engines with large number of universal tracking channels and fast acquisition units. They implement the tracking of BOC, MBOC and AltBOC signals. The E5AltBOC signal is particularly promising due to its low multipath noise.

Acknowledgments We are thankful to our colleagues Tom Willems, Wim De Wilde, Frank Boon and Jan Van Hees for discussions and reviewing and to ESA for continuous and dedicated support of Galileo-related activities at Septentrio.

References

Artaud G, De Latour A, Dantepal J, Ries L, Maury N, Denis J-C, Senant E, Bany T (2010) A new GNSS multi constellation simulator: NAVYS. In: Proceedings of ION GNSS 2010, Portland, OR, US, 21–24 Sept 2010, pp 845–857

Ayaz AS, Bauernfeind R, Jang J, Kraemer I, Dötterböck D, Ott B, Pany T, Eissfeller B (2010) Performance evaluation of single antenna interference suppression techniques on Galileo signals using real-time GNSS software receiver. In: Proceedings of ION GNSS 2010, Portland, OR, US, 21–24 Sept 2010, pp 3330–3338

Bates M, Saaied H, Olynik M, Van Wyck B, Gerein N, Blondeau Y, Studenny J (2007) Preliminary results of the Canadian Galileo safety of life receiver (SOLRX). In: Proceedings of ION GNSS 2007, Fort Worth, TX, US, 25–28 Sept 2007, pp 1079–1088

Betz JW, Fite JD, Capozza PT (2005) Getting to M. Direct acquisition of the new military signal, GPS World, April 2005, pp 40-46

Binda S, Pinelli G, Gerein N (2006) Galileo reference receiver. In: Proceedings of ION GNSS 2006, Fort Worth, TX, US, 26–29 Sept 2006, pp 994–1000

Bonhoure B, Boulanger C, Marechal J (2010) Long-term and multi-receiver GPS/GIOVE mixed PVT experimentation. In: Proceedings of ION GNSS 2010, Portland, OR, US, 21–24 Sept 2010, pp 3309–3319

Boulton P, Read A, Wong R (2007) Formal verification testing of Galileo RF constellation simulators. In: Proceedings of ION GNSS 2007, Fort Worth, TX, US, 25–28 Sept 2007, pp 1564–1575

Crisci M, Hollreiser M, Falcone M, Colina M, Giraud J (2007) GIOVE Mission sensor station performance characterization: overview of the results. In: Proceedings of ION GNSS 2007, Fort Worth, TX, US, 25–28 Sept 2007, pp 1028–1041

Davies NC, Black G, Campagne P (2010) Galileo, the public regulated service and PRS/GPS dual mode operations. In: Proceedings of ION GNSS 2010, Portland, OR, US, 21–24 Sept 2010, pp 1863–1868

De Wilde W, Sleewaegen J-M, Van Wassenhove K, Wilms F (2004) A first-of-a-kind Galileo receiver breadboard to demonstrate Galileo tracking algorithms and performances. In: Proceedings of ION GPS 2004, Long Beach, CA, US, 21–24 Sept 2004, pp 2645–2654

De Wilde W, Sleewaegen J-M, Simsky A, Vandewiele C, Peeters E, Grauwen J, Boon F (2006) New fast signal acquisition unit for GPS/Galileo receivers. In: Proceedings of ENC-GNSS 2006, Manchester, UK, 2–6 May 2006, pp 905–908

De Wilde W, Sleewaegen J-M, Seco Granados G (2009) Method and device for demodulating Galileo alternate binary offset carrier (ALT-BOC) signals. US Patent 7486717, Feb 2009

Dempster AG, Shivaramaiah NC (2009) A novel extended tracking range DLL for AltBOC signals. In: Proceedings of the IEEE 70th vehicular technology conference, VTC-FALL-09, Anchorage, Alaska, 20–23 Sept 2009

Denks H, Steingaß A, Hornbostel A, Chopard V (2009) GNSS receiver testing by hardware simulation with measured interference data from flight trials. In: Proceedings of ION GNSS 2009, Savannah, GA, US, 22–25 Sept 2009, pp 1–10

Dötterböck D, Stöber C, Kneissl F, Eissfeller B (2010) Tracking AltBOC with the ipexSR software receiver. In: Proceedings of ION GNSS 2010, Portland, OR, US, 21–24 Sept 2010, pp 1896–1904

Eerola V (2000) Rapid parallel GPS signal acquisition. In: Proceedings of ION GPS 2000, Salt Lake City, UT, US, Sept 2000, pp 810–816

Fairhurst P, Glueckert U, Richter B (2010) The New Leica Viva GNSS receivers. In: Proceedings of ION GNSS 2010, Portland, OR, US, 21–24 Sept 2010, pp 1386–1394

Fortin M-A, Guay J-C, Landry R (2009) Development of a universal GNSS tracking channel. In: Proceedings of ION GNSS 2009, Savannah, GA, US, 22–25 Sept 2009, pp 259–272

Gerein N, Manz A, Pinelli G, Franzoni G, Zin A (2007) Galileo receiver chain: the nonPRS ground reference receiver. In: Proceedings of ION GNSS 2007, Fort Worth, TX, 25–28 Sept 2007, pp 1054–1064

Gerein N, Olynik M (2009) NovAtel's GIOVE monitoring receiver. In: Proceedings of ENC-GNSS 2009, Naples, Italy, 3–6 May 2009

Giraud J, Borrel V, Crisci M (2009) Latest achievements in GIOVE signal and sensor station experimentations. In: Proceedings of ION GNSS 2009, Savannah, GA, US, 22–25 Sept 2009, pp 3025–3036

Grant A, Ward N, Christiansen SE, Dixon Ch, Suskind A (2007) MARUSE: demonstrating the use of maritime Galileo pseudolites. In: Proceedings of ION GNSS 2007, Fort Worth, TX, 25–28 Sept 2007, pp 1923–1930

Haas R, Dixon CS (2009) First MAGES demonstrations: applications of GNSS for emergency scenarios. In: Proceedings of ION GNSS 2009, Savannah, GA, US, 22–25 Sept 2009, pp 1841–1848

Heinrichs G, Irsigler M, Wolf R, Winkel J, Prokoph G (2007) NavX® -NCS: the first Galileo/GPS full RF navigation constellation simulator. In: Proceedings of ION GNSS 2007, Fort Worth, TX, US, 25–28 Sept 2007, pp 1323–1328

Heinrichs G, Loehnert E, Wittmann E, Kaniuth R (2007a) First outdoor positioning results with real Galileo signals by using the German Galileo test and development environment: GATE. In: Proceedings of ION GNSS 2007, 25–28 Sept 2007, Fort Worth, TX, pp 1576–1587

Heinrichs G, Lohnert E, Wittmann E, Kaniuth R(2007b) Opening the GATE. Germany's Galileo test and development environment. Inside GNSS, May/June 2007, pp 45–52

Hollreiser M, Crisci M, Sleewaegen J-M, Giraud J, Simsky A, Mertens D, Burger T, Falcone M (2007) Galileo signal experimentation. GPS World, May 2007, pp 37–44

Irsigler M, Riedl B, Pany T, Wolf R, Heinrichs G (2009) NavX®-NCS: A multi-constellation RF simulator: latest product developments and test applications. In: Proceedings of ION GNSS 2009, Savannah, GA, US, 22–25 Sept 2009, pp 1876–1884

Jia Y, Kenne K, Knight J, Kung C, Williams J (2009) NavCom's SF-3050: a software upgradeable GNSS receiver. In: Proceedings of ION GNSS 2009, Savannah, GA, US, 22–25 Sept 2009, pp 1910–1917

Lestarquit L, Artaud G, Issler J-L (2008) AltBOC for dummies or everything you always wanted to know about AltBOC. In: Proceedings of ION GNSS 2008, Savannah, GA, US, 16–19 Sept 2008, pp 961–970

Lueck T, Winkel J, Bodenbach M, Göhler E, Falk N, Consoli A, Piazza F, Gerna D, Granger R, Reedman P, Simpson S, Euler H-J (2007) Artus: a second generation Galileo/GPS receiver. In: Proceedings of ION GNSS 2007, Fort Worth, TX, US, 25–28 Sept 2007, pp 1008–1016

Lueck T, Winkel J, Bodenbach M (2009) A complex channel structure for generic GNSS signal tracking. In: Proceedings of ION GNSS 2009, Savannah, GA, US, 22–25 Sept 2009, pp 252–258

Margaria D, Savasta S, Dovis F, Motella B (2009) Comparative interference vulnerability assessment of GPS TMBOC and Galileo CBOC signals. In: Proceedings of ION GNSS 2009, Savannah, GA, US, 22–25 Sept 2009, pp 11–21

Martin S, Kuhlen H, Schlotzer S, Schmitz-Peiffer A, Voithenberg MV, Dietz H (2007) SEA GATE: a maritime Galileo testbed in the Port of Rostock. In: Proceedings of ION GNSS 2007, Fort Worth, TX, 25–28 Sept 2007, pp 535–542

Martin N, Guichon H, Revol M, Hollreiser M, de Mateo J (2008a) Architecture of the Galileo TUS receiver for coherent tracking of the AltBOC signal. In: Proceedings of 3rd CNES-ESA workshop on GNSS signals and signal processing, Toulouse, France, 21–22 April 2008

Martin S, Diefenbach S, Fellbach D, Voitenberg MV, Dietz H, Sasum R (2008b) Galileo receiver testing with the GNSS signal generator NSG 5100. In: Proceedings of ION GNSS 2008, Savannah, GA, US, 16–19 Sept 2008, pp 1217–1226

Montebruck O, Hauschild A, Hessels U (2010) Characterization of GPS/GIOVE sensor stations in the CONGO network. GPS solutions. doi:10.1007/s10291-010-0182-8

Morgan-Owen R, de Mateo J, Scarafia S, Hollreiser M, Van den Berg A, Willems T, Pye G, De Wilde W (2009) Galileo test user receiver status, key results and performance. In: Proceedings of ION GNSS 2009, Savannah, GA, US, 22–25 Sept 2009, pp 3242–3251

Ries L, Macabiau C, Nouvel O, Jeandel Q, Vigneau W, Calmettes V, Issler J-L A software receiver for GPS-IIF L5 signal. In: Proceedings of ION GPS 2002, Portland, OR, US, 24–27 Sept 2002, pp 1540–1553

Romero S, Diez J, Fernandez A, Van den Berg A, Simsky M, Van Wassenhove K, Morgan-Owen R, de Mateo C, Hollreiser M (2008) Galileo performance analysis software for test user receivers. In: Proceedings of ION GNSS 2008, Savannah, GA, US, 16–19 Sept 2008, pp 1252–1257

Rounds S, Norman C (2000) Combined parallel and sequential detection for improved GPS acquisition. In: Proceedings of the IAIN world congress, San Diego, CA, US, 26–28 June 2000, pp 368–372

Shivaramaiah N, Dempster A, Rizos C (2008) Exploiting the secondary codes to improve signal acquisition performance in Galileo receivers. In: Proceedings of ION GNSS 2008, Savannah, GA, US, 16–19 Sept 2008, pp 1497–1506

Simsky A, Sleewaegen J-M, De Wilde W, Wilms F (2005) Overview of Septentrio's Galileo receiver development strategy. In: Proceedings of ION GNSS 2005, Long Beach, CA, US, 13–16 Sept 2005, pp 1888–1895

Simsky A, Sleewaegen J-M, Hollreiser M, Crisci M (2006) Performance assessment of Galileo ranging signals transmitted by GSTB-V2 satellites. In: Proceedings of ION GNSS 2006, Fort Worth, TX, US, 26–29 Sept 2006, pp 1547–1559

Simsky A, Mertens D, Sleewaegen J-M, Willems T, Hollreiser M, Crisci M (2007) Multipath and tracking performance of Galileo ranging signals transmitted by GIOVE-A. In: Proceedings of ION GNSS 2007, Fort Worth, TX, US, 25–28 Sept 2007, pp 1596–1609

Simsky A, Mertens D, Sleewaegen J-M, De Wilde W, Hollreiser M, Crisci M (2008a) Experimental results for the multipath performance of Galileo signals transmitted by GIOVE-A satellite. Int J Navig Obs 2008(Article ID 416380):13. doi:10.1155/2008/416380

Simsky A, Mertens D, Sleewaegen J-M, De Wilde W, Hollreiser M, Crisci M (2008b) Multipath and tracking performance of Galileo ranging signals transmitted by GIOVE-B. In: Proceedings of ION GNSS 2008, Savannah, GA, US, 16–19 Sept 2008, pp 1525–1536

Simsky A, Mertens D, Sleewaegen J-M, De Wilde W, Hollreiser M, Crisci M (2008c) MBOC vs. BOC(1, 1) multipath comparison based on GIOVE-B data. Inside GNSS, September/October 2008, pp 36–40

Simsky A, De Wilde W, Willems T, Mertens D, Koitsalu E, Sleewaegen J-M (2009) First field experience with L5 signals: DME interference reality check. In: Proceedings of ION GNSS 2009, Savannah, GA, US, 22–25 Sept 2009, pp 29–37

Sleewaegen J-M, De Wilde W, Hollreiser M (2004) Galileo AltBOC receiver. In: Proceedings of ENC GNSS 2004, Rotterdam, 16–19 May 2004

Soellner M, Erhard P (2003) Comparison of AWGN code tracking accuracy for alternative-BOC, complex-LOC and complex-BOC modulation options in Galileo E5-band. In: Proceedings of the ENC GNSS 2003, Graz, 22–25 April 2003

Steingass A, Hornbostel A, Denks H (2009) Airborne measurements of DME interferers at the European hotspot. In: Proceedings of ION GNSS 2009, Savannah, GA, US, 22–25 Sept 2009, pp 1675–1683

Tiberius CCJM, De Bakker PF, Van Der Marel H, van Bree RJP (2009) Geometry-free analysis of GIOVE-A/B E1–E5a, and GPS L1–L5 measurements. Proceedings of ION GNSS 2009, Savannah, GA, US, 22–25 Sept 2009, pp 2911–2925

Van Bree RJP, Tiberius CCJM, Van den Berg A, Sleewaegen J-M, Morgan-Owen R, Giordano P (2010) Testing the TUR: a performance report on Septentrio's Galileo test user receiver. In: Proceedings of ION GNSS 2010, Portland, OR, US, 21–24 Sept 2010, pp 3320–3329

Winkel J, Luck T, Bodenbach M, Gohler E, Falk N, Wolf H, Kubr H, Minelli L (2007) Galileo payload test system: the receiver subsystem. In: Proceedings of ION GNSS 2007, Fort Worth, TX, 25–28 Sept 2007, pp 509–516

Chapter 13
Software Simulators and Multi-Frequency Test Scenarios for GALILEO

Sarang Thombre and Jari Nurmi

Abstract Global Navigation Satellite System (GNSS) signal simulators are essential in supporting education, research, and also development and testing of advanced positioning receivers. This chapter provides useful insights into the essential components of software simulators of GNSS signals, a brief background study of state-of-the-art in such simulators and finally, some test scenarios for testing performance parameters of multi-frequency GNSS receivers. Section "Introduction" is a brief introduction to the following sections of the chapter. Section "State-of-the-Art Software-Based GNSS Signal Simulators" is dedicated to literature review of state-of-the-art in GNSS software simulators. Thirty-eight example simulators available in the academic and commercial domain were selected and their characteristics compared based on the following parameters: the number of GNSS satellite constellations and signals modeled, the strategy of error modeling, the programming language used, the input and output data formats, and the modeling of the radio frequency front-end (RF FE) effects. Section "Important Components of GNSS Signal Simulators" is dedicated to describing the essential components of a typical GNSS software simulator. Usually, such simulators are modular in design, where each module is responsible for a specific functionality. The different modules are the signal generation module, error sources module, transmission channel module, receiver RF FE module, and finally, the space (satellite constellation) module. Lastly, sect. "Multi-Frequency, Multi-System Receiver Performance Test Scenarios" describes some important receiver parameters and their example test-cases in order to compare the performance of a multi-frequency multi-system receiver over its single-frequency single-system counterpart. It has been observed that in spite of the proliferation of numerous simulators of GNSS signals, there is no single comprehensive reference textbook that explains their baseline theory. This chapter intends to fill this gap.

S. Thombre (✉)
Department of Navigation and Positioning, Finnish Geodetic Institute,
Geodeetinrinne 2, 02430 Masala, Kirkkonummi, Finland
e-mail: sarang.thombre@fgi.fi

J. Nurmi
Tampere University of Technology, Tampere, Finland

J. Nurmi et al. (eds.), *GALILEO Positioning Technology*,
Signals and Communication Technology 182, DOI 10.1007/978-94-007-1830-2_13

Introduction

Future Global Navigation Satellite Systems (GNSS) signals are usually announced to the public before satellites begin transmitting them from the sky. This enables the users to be already equipped with compatible receivers by the time the signals are actually 'turned on live' from satellites. However, the major concern for manufacturers of such innovative-GNSS-signal-compatible receivers is their testing. How do you test whether a receiver is compatible to presently non-existent signals? One possible solution is to create replicas of those signals artificially in laboratories and to test the receivers with them. Devices that locally create such signals are called Simulators or Generators. If the receiver works fine under such 'simulated conditions' (which includes simulated signals and simulated environment) then it is safe to assume that they will work equally satisfactorily under real-life situations. The trick here is to make the simulated conditions as close to reality as possible. For example, if a receiver is intended to be stationed in a dense multipath environment in real-life, while testing in laboratory, the simulated signals should also contain high number of powerful multipath components. If not, the performance of the receiver will be unreliable when used in real-life. In addition to their application as a testing tool, simulators can be useful for education and research by imparting knowledge of the signal characteristics, for example, modulation properties, power profile, frequency spectrum etc. This may also help in locating vulnerabilities in the signal structure and in designing suitable encryption or snooping technologies.

Simulators for GNSS signals are already in existence (GPSWorld 2012) and can be differentiated either as hardware-based simulators and software-based simulators or as academic simulators and commercial simulators. Hardware-based simulators are physical devices containing an actual radio frequency transmitter chain generating physical signals at the output. Such simulators are often bulky and costly. Software-based simulators are built using a software programming language such as Matlab, Simulink, C, C++, etc. They output virtual signals and are mainly used to test receivers also designed using similar software. The focus of this chapter is on software-based simulators and hardware-based simulators are not considered henceforth. Also, for better readability, software-based GNSS signal simulators will be referred to as just 'simulators' throughout the chapter. The other categorization of simulators is based on their origin and possible application. Academic simulators are those which are created primarily for proof-of-concept and information pertaining to them is usually disseminated through research conferences and publications. Commercial simulators are created in industry with the sole purpose of sales and/or in-house testing of GNSS related equipment. Information about them can be gathered through patent filings, white-papers, datasheets, advertisements and in some cases, research publications.

A brief outline of the chapter is as follows: Section "State-of-the-Art Software-Based GNSS Signal Simulators" gives a brief overview of the history of GNSS signal simulators. Thirty-eight references (both commercial and academic) have been compared based on features including, but not limited to, the number of GNSS satellite constellations and signals modeled, the strategy of error modeling, the programming language used, the input and output data formats, and the modeling of the radio frequency front-end (RF FE) effects. These sources are categorized into three generations, based on their year of publication. Generation I is from 2000 to 2003, Generation II is from 2004 to 2005 and Generation III is from 2006 onwards. The results show distinct patterns of evolution from one generation to the next and special features or trends in every generation are discussed. Overall, it was observed that software-based generation of GNSS signals still has considerable potential for innovation especially through the use of C# programming language and in the generation of Russian GLONASS signals. It is hoped that this exercise will help readers form a baseline of state-of-the-art in GNSS signal simulation technology. Section "Important Components of GNSS Signal Simulators" deals with the constituent modules of such simulators. The different modules are the signal generation module, the transmission channel module, the space (satellite constellation) module, the receiver RF FE module and the error signals generation module. Section "Signal Generation Module" describes some important receiver parameters and their example test-cases in order to compare the performance of a multi-frequency multi-system receiver over its single-frequency single-system counterpart.

State-of-the-Art Software-Based GNSS Signal Simulators

Software-based signal simulators for GNSS are a relatively recent invention: pioneered in 1999–2000 (Thombre 2010). In this section, thirty-eight reference sources (from (Corbell 2000) to (Boschen 2010)) are compared to bring out the salient features of such simulators. To place such a considerable number of similar devices in perspective, these sources have been categorized into three generations, based on their year of publication: Generation I (2000–2003), Generation II (2004–2005) and Generation III (2006 onwards). Due to the considerable number of sources, comparison between them was done based on five factors only: GNSS signals simulated, used software platform, error sources modeling, space segment modeling and other factors. Other factors include any special features which would be worth mentioning. Twelve example simulators are compared in easy-to-read tabular form in Table 13.1. It should be noted that, almost all simulators modulate the baseband GNSS signals with an Intermediate Frequency (IF) which are considerably lower than the actual carrier frequencies, since GHz range sampled signal generation is not yet possible with the current software platforms.

GNSS Signals Simulated

American Global Positioning System (GPS) was the first fully operational GNSS and hence the first generation simulators could simulate only GPS L1 signals. All simulators were C/A codes capable, but few were even capable of simulating the

Table 13.1 Comparison of some example simulators from Generation III

	GNSS signals simulated	Software platform	Modeling of error sources	Space segment modeling	Other factors
gLab (Intecs)	Galileo E1, E5, E6, GPS L1, L1C, L2C, L5, EGNOS	ANSI-C	RF FE, filter, ADC quantization, Interference: jammer, intra-system and multipath	N/A	SDR technology, Easy MMI interface, Portability, Tested and validated by ESA, generates log files in text and binary format, possibility to store signals, Fs and IF customizable, 3 types of RF FE and filter effects simulated, Doppler profile, power and system dynamic configurable
GSSF (VEGA)	GPS, Galileo EGNOS	Windows 2,000/XP (.Net) and C++	IONEX and TROPEX files for ionosphere and troposphere errors, clock errors through Clock RINEX files	Keplerian orbital parameters and co-ordinate transformations by reading IGS SP3, YUMA and RINEX files, Inbuilt Walker constellation option	Raw GPS and Galileo data production in RINEX and IGS SP3 format, can exchange data in numerous formats, multi-user and LAN access, also simulates Galileo ground segment, easy visualization of results through inbuilt mapping tool
GDISS (Joo 2007)	GPS L1 C/A, L2C Galileo E1 (B and C), E5a	C++	Internal and external models for ionosphere, troposphere, multipath and clock errors. Filter and quantization effects and AWGN included	Ephemeris data extracted from RINEX files.	Modular, possibility to add external models, signals validated with software GNSS receivers, Raw data generation and digitized IF generation possible, capable of reading RINEX files,

(continued)

Table 13.1 (continued)

	GNSS signals simulated	Software platform	Modeling of error sources	Space segment modeling	Other factors
					GUI for data exchange with user, navigation message also generated
GBTSS (Cosenza 2008)	I/Q format, Galileo E1, E5, E6, CEM, BOC (fs,fc), AltBOC (fs,fc), GPS BPSK	Matlab translated to C	Multipath, ionosphere (based on NeQuick model for Galileo and Klobuchar model for GPS and TEC values), troposphere (Saastamoinen model), Gaussian noise, RF FE effects, interference signals (RADAR, DME/TACAN), Doppler, free space delay, clock offset errors	Orbit propagation module and kinematics module control satellite position, motion and their Doppler frequency effects	Highly modular, entire processing chain from GNSS signal generation to receiver ADC can be modeled
(Lim 2008)	GPS L1 C/A, L5 Galileo E1, E5a	N/A	Clock offsets, timing errors, ionospheric errors can be reconstructed by reading correction parameters from RINEX files. Klobuchar model and modified Hopfield model for iono and tropo errors resptv. AWGN added	RINEX files are read to give the satellite orbit information. Satellite positions are calculated from this information	Multi-channel capability, navigation data generated by reading RINEX files, signals validated by software GNSS receiver, receiver dynamics simulated from Doppler component, RF FE (filter and ADC quantization) effects modeled
Fraunhoffer (Fraunhofer)	Signals are created in	Ported on FPGA	Error sources are read from RINEX files, noise generation based on selected SNR, CW or	Satellite constellation is built from RINEX files	GUI for simulation, analysis, jamming signal simulation and graphical

(continued)

Table 13.1 (continued)

	GNSS signals simulated	Software platform	Modeling of error sources	Space segment modeling	Other factors
	scenarios based on RINEX files		sweep-CW interference, ADC effects		analysis, modular architecture, navigation message is generated, Fs and IF configurable, outputs log messages
CFRS (CFRS)	GPS L1 C/A, P (Y), M, L2C, L5 GLONASS L1, Galileo E1, E5, E6 WAAS, EGNOS	IMPULSE framework	Scintillations, interference and jamming (narrow-band, wide-band, directional), plasma effects, obscuration, multipath effects, ocean scattering, ionospheric effect	Inbuilt satellite orbit data	Offline generation of data file, output can be played back into RF antennas, multiple antenna wavefronts, high-dynamics, specified orbits simulation capability. Schematic-based GUI and simple menu-driven instructions, allows access of various intermediate test points
JGNSS (Ouzeau 2009)	GPS L1 C/A, L5 Galileo E5a	Juzzle = Java + Puzzle	signal propagation errors (delay, attenuation, Doppler), thermal noise, multipath (specular model or statistical model with Rice and or Rayleigh distributions), RF FE effects (down-conversion, filtering, ADC quantization)	Orbitography module implemented for space segment	12 satellite channels simulated, RF FE is simulated, signals validated with software receiver models, GUI for user interaction
EPSS and BaiCES (FAF)	Mostly Galileo	ML-Designer simulation environment	Simulation of multipath effects in urban areas or near airports,	Dedicated satellite orbit model included in library	Capability to read terrain data or city models in three different data formats,

(continued)

Table 13.1 (continued)

	GNSS signals simulated	Software platform	Modeling of error sources	Space segment modeling	Other factors
			other errors contained within specific library models		enables the user to build up a simulation application graphically, by drag and drop models from the model library, new models can be added to library,
GRANADA (Fernández, 2006)	Bit-true Galileo E1, E5, E6, navigation message, BOC modulation, PRN codes	Matlab/Simulink, open source C code	Code Doppler, transmit filter effects, AWGN, multipath, interference, user dynamics, clock errors, atmospheric delays (iono, tropo), tracking errors, ephemeris errors	GPS and Galileo constellation propagator model	Modular and configurable, antenna patterns and RF FE modeled, signals validated with software receiver, GUI to interact with all modules
GRANADA (contd.)	GPS, Galileo, AltBOC	Matlab/Simulink, can be converted into C code	Multipath, interference (Gaussian noise and GPS signals), transmitter filter effects, AWGN, system dynamics (both code and carrier Doppler), UERE in generated measurements—satellite and receiver clock, iono and tropo, ephemeris, receiver tracking and cycle slip errors	Environment and navigation generator has inbuilt capability to simulate GPS and Galileo constellations	Modular and configurable, user friendly GUI, users can embed their own algorithms, two main modules: Bit-true simulator and Environment and navigation simulator (raw data generator), receiver chain modeled, fixed point and floating point arithmetic coding, project also includes a Galileo mono-channel simulator (GMCS)

(continued)

Table 13.1 (continued)

	GNSS signals simulated	Software platform	Modeling of error sources	Space segment modeling	Other factors
MGOS (Dolgansky 2009)	GPS, Galileo GLONASS and COMPASS	N/A	Error source library contains orbital (biases to pseudorange), clock (satellite clock bias arithmetically or by reading SP3 files, receiver clock—RINEX files), atmospheric (ionosphere—Klobuchar model or reading IONEX files, troposphere—Modified Hopfield model), multipath and hardware bias (satellite and receiver) errors	Keplerian orbital parameters and assuming perfect eliptical orbits, integration of satellite motion over time along with perturbations and Earth's gravitational effects, reading orbital data from SP3 files	Pseudorange and carrier phase measurements simulated, GUI and module library to set up scenarios

more complex P codes, M codes or the GPS L2 signal. All simulators in Generation I used random bits to denote navigation data as they neither had the capability to generate actual navigation messages, nor to read downloaded navigation files in Receiver Independent Exchange (RINEX) format. In Generation II, the signals simulated had become more diverse, as Galileo signals (E1 and E5 or even E6), SBAS and the new GPS L5 signal were announced. In Generation III, in addition to all GPS (L1, L2, and L5) and Galileo (E1, E5, E6) signals, simulators had the capability for simulating GLONASS L1, Compass, EGNOS, WAAS and pseudolite signals, and even inertial measurement (IM) data. These simulators could read navigation data from RINEX files, which are readily available on the internet.

Software Platform

The underlying software in which the simulators have been coded has also undergone an evolution from one generation to the next. During Generation I, Matlab was the preferred programming environment, although some simulators contained a few modules of Simulink and C language. In Generation II, C language was more widely used, but always in addition with Matlab. Some of the simulators were later upgraded and were completely coded in C++. Other example programming environments used were Network Simulator 2 (NS2) and VHDL. In Generation III, most of the simulators were encoded using C++, owing to its greater object oriented programming capability. This probably enabled more resource efficient coding, faster generation and processing of data samples and greater modularity for increased flexibility. Some simulators used a graphical user interface (GUI) coded using Microsoft foundation class libraries (MFC). One simulator (Ouzeau 2009) used the Juzzle software platform for coding the simulator. Juzzle is a combination of Java and Puzzle platform and also contains traces of C, Matlab, FORTRAN, Ada, SystemC and Ptolemy. It has been observed that although C# has better memory management and more object oriented programming capability along with Microsoft (MS) .Net support for GUI, it has not yet been used for simulator design. Only one simulator (VEGA) is coded in MS. Net with some modules of C++ and C#, and claims to be compatible with Windows 2000 and Windows XP.

Modeling of Error Sources

As mentioned in sect. "Introduction", simulating the environment is just as essential as simulating the actual signals. In reality, the actual signals are always affected by various error sources during their travel from the satellite to the receiver and also inside both of them. Therefore, the more error sources simulated, the closer the output signal is to reality. In Generation I, many of the simulators had no error modelling, or constants for describing the various error sources, for example using the typical values given in Spilker's clock error model. However, some simulators could simulate a jamming signal environment by creating Frequency Modulated (FM), Amplitude Modulated (AM), Phase Modulated (PM) and frequency swept jamming signals at L1 frequency. One simulator in particular (Dong 2003) had a fairly advanced error modelling scheme consisting of satellite clock, ionosphere, troposphere and Doppler error. Ionosphere delay error was modelled by Spherical Harmonics (SH) and Grid model using Global Ionospheric Maps (GIM) of Total Electron Content (TEC). Troposheric error was simulated using the Hopfield model extended by the Black and Eisner model while a 2 MHz bandlimited additive white gaussian noise (AWGN) was also included.

Starting from Generation II, Multipath was considered the dominant error source and hence almost every simulator modelled it. Up to 7 multipath channels could be simulated, each with a different amplitude, delay and phase using Rayleigh or Rice multipath fading models including near and far echoes. Near echo is simulated using Brenner and Jahn models (Brenner 1998; Jahn 1996) whereas far echo is modelled considering specular reflections using Snell's laws of reflection. Other simulators used empirical models created by observation of multipath errors at a given location over time. Calculation of ionosphere and troposphere errors also becomes more complex (and hence more realistic) in Generation II. Ionosphere error was 'constructed' from correction parameters extracted from RINEX files or from TEC obtained from the global NeQuick model or using vTEC from the more local Ionospheric Exchange (IONEX) files and using the Appleton and Hartree model. All these files are readily available from the internet. Some simulators instead used the Klobuchar model for ionosphere delay modelling. For the troposphere delay modelling, the Hopfield model was generally used, aided by the Black and Eisner model. For clock error modelling, the Winkel model along with Allan Variance was the preferred choice. Some simulators constructed satellite clock errors from correction parameter given in the RINEX file or from clock product files in Standard Product # 3 (SP3) format. Other error modelling included Doppler offset, thermal noise, interference, front-end effects and power loss. It is possible to combine all these errors into a single parameter—User Equivalent Range Error (UERE). This is especially useful when the simulation is conducted over longer intervals or larger geographical areas.

In Generation III, new sources of error included, for example interference signal generation: continuous wave, pulsed, wideband, narrowband, air traffic control signals, tactical air navigation (TACAN) signals and distance measuring equipment signals (DME). The Klobuchar and the NeQuick models are used to generate the ionospheric errors for GPS and Galileo signals respectively. (However, it should be noted that the effect of ionosphere is the same for same frequencies, irrespective of whether they are GPS or Galileo signals. Furthermore, if ionosphere error is simulated with exactly the same model and parameters as the one used in the receiver according to the GPS and Galileo Interface Control Documents (ICD), then the ionosphere error will be completely eliminated within the receiver, which is undesirable for the simulator if the purpose is to assess realistic performances. A better idea would be to introduce some variance or noise in the simulator generated ionosphere error before feeding it to the receiver). The Saastamoinen model is most suitable for modeling tropospheric error. Overall, in Generation III, the environmental error modeling improved greatly, mostly due to the use of 3-dimensional (3D) modeling. In one simulator, ionosphere error modeled using the International Reference Ionosphere Model (IRI) was converted into scintillation and finally timing delay using the Wide Band model (WBMOD) while in another, the Ray Shooting algorithm was used to replicate the local multipath environment.

Space Segment and User Trajectory Modeling

In Generation I, many simulators performed approximate modeling of satellite trajectories using spline smooth fitting or using user input of raw geometric parameters to create satellite dynamics data. Some simulators could read GPS almanac and ephemeris files to recreate satellite position and dynamics as a function of time. In Generation II, simulators had the capability to read ephemeris files (usually in RINEX/SP3 format) or YUMA almanac files for GPS and the Walker constellation parameters for Galileo, to construct entire constellation of satellites. Keplerian orbital parameters were used to calculate satellite co-ordinates as a function of time. The simulator in (Constantinescu 2005) is also unique because it is the only simulator which proposes to simulate the GLONASS constellation. It uses information from Right Ascension of Ascending Node (RAAN) and mean anomaly for this purpose.

User trajectories include static and dynamic scenarios using user input of set of co-ordinates. Typical shapes of user trajectories are along straight line or circle with constant linear or angular velocity. Notable exceptions are (Accord), where complete 3D dynamics of user vehicle are simulated using user input of acceleration, direction and altitude information, and (Constantinescu 2005) which can simulate user trajectory from user inputs, Microsoft Flight Simulator software or X-Plane software. In Generation III, a major advancement was in the ability to read data in any of the co-ordinate systems (for example, WGS-84, ECEF etc.) and to convert from one system to another. Satellite positions, their tracks, azimuth and elevation and velocity were computed as a function of time and their orbits could be tuned considering perturbations and Earth gravitation effects. Additionally, weather data could be combined with satellite orbit data to compute signal strengths from visible satellites.

Other Factors

In Generation I, simulators used flight trajectory data to define the receiver position. Simulators were designed to store large amounts of data over longer simulation periods, or another option was to generate 12 channel Early, Prompt and Late I/Q data using mathematical modelling of GPS. This allowed more data to be generated over smaller simulation periods. In Generation II, simulators modelled radio frequency (RF) front-end effects, especially, analog to digital conversion (ADC) and quantization, filter, antenna gain etc. Users were allowed to define the frequency plan (sampling frequency, intermediate frequency etc.), simulation time, filter and ADC parameters etc. One simulator (Pósfay 2005) integrated an analog to digital converter at its output, offering the capability to convert virtual IF GNSS signals into physical signals. Some simulators would record real GNSS signals from satellites and replay them as scenarios later. In Generation III, innovative features included, for example steering software, indoor positioning capability, use of software defined radio (SDR) design principles, support for different RF FE architectures, flexible/user modifiable software platforms, support for 3D and contour maps, networking/multi-user capabilities (LAN, WAN) among others. Supported input/output data formats expanded to include Tropospheric Exchange (TROPEX) files, log, binary and text formats. Real time processing capabilities were introduced with support for FPGA, VHDL, Windows, and Linux etc. Accurate correction terms were calculated for anomalies such as orbit eccentricities in satellite trajectory or the Sagnac effect (Ashbey 2004). GUIs became more advanced allowing users to modify and redesign the software architecture of the simulators. Simulator code used both floating point and fixed point arithmetic for increased accuracy. For simplicity, some example simulators from Generation III are chosen and a comparison of their parameters is represented in tabular form in Table 13.1.

Important Components of GNSS Signal Simulators

Simulators have a modular design where every module is dedicated to a particular functionality. The different modules are (but not limited to): The signal generation module (including error signals generation), the transmission channel module, the satellite constellation module, and the receiver RF FE module. Figures 13.1 and 13.2 show the block diagram of a typical software-based simulator. The satellite constellation module gives a real-time picture of the geometry of the satellites in the sky. Based on this geometry, intermediate frequency signals from a number of channels in the signal generation module are combined to form one navigation system signal 'bouquet', for example, Galileo E1 or E5a. This combined signal stream is passed through the transmission channel module, which simulates non-linearities of the transmission path through the atmosphere (for example, multipath and additive white Gaussian noise). The RF FE module simulates the effects of the radio frequency front-end of a typical GNSS receiver. It includes filtering, amplification, local oscillator phase noise and ADC quantization effects. All of these blocks are controlled via the underlying software program.

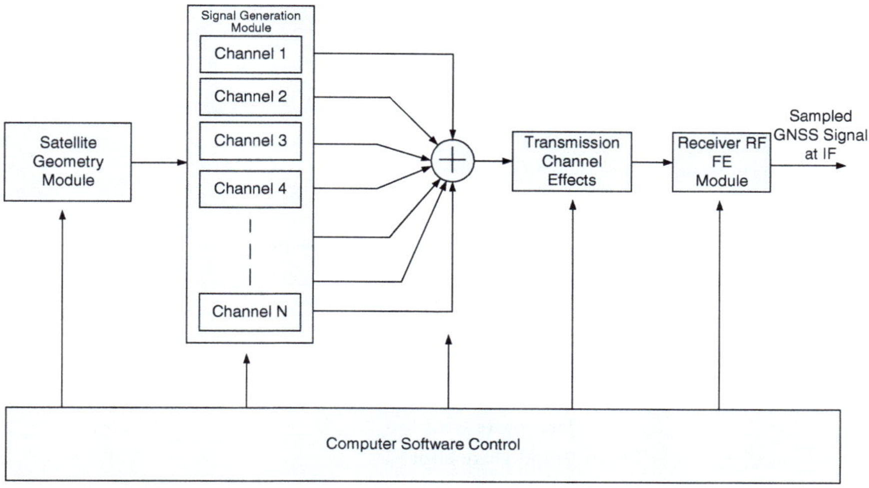

Fig. 13.1 Block diagram of a software-based GNSS signal simulator

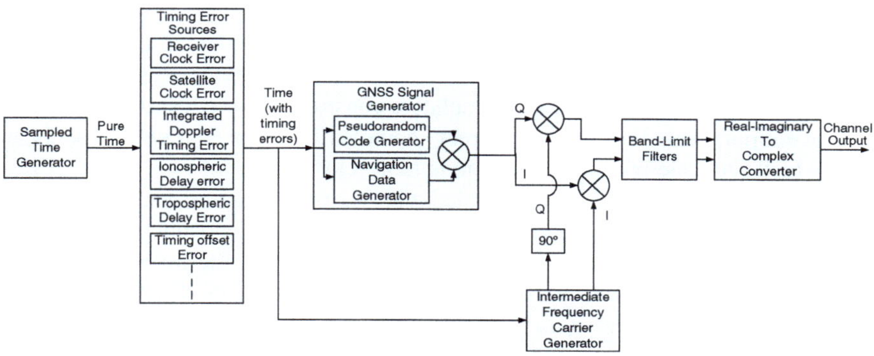

Fig. 13.2 Internal block diagram of one channel of the simulator

Signal Generation Module

The signal generation module is composed of a number of channels. Each channel represents one satellite signal along with its specific error sources. Therefore if, for example, the entire Galileo E1 system has to be simulated, 32 channels would be needed. Figure 13.3 shows the detailed block diagram of one channel within the signal generation module.

The sampled time generator creates time samples from a continuous time source by sampling at a user desired sampling frequency. The sampled time is then fed to the error signal generation block where it is contaminated with various clock and atmospheric timing errors. The most common timing errors and their typical values

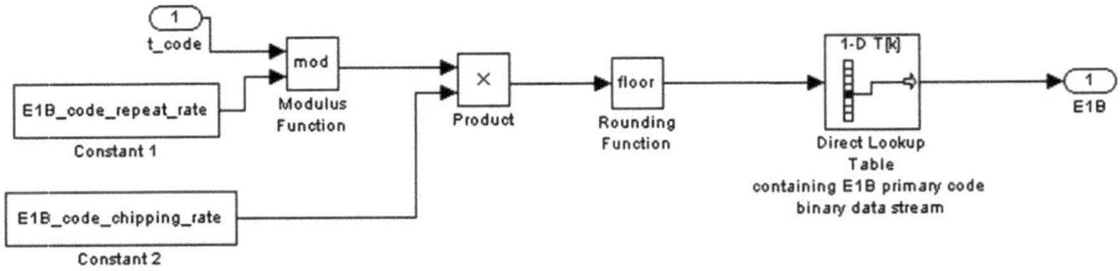

Fig. 13.3 Simulink model for generating Galileo E1B primary PRN code

Table 13.2 Typical values of GNSS signal errors

Parameter	Typical value
Receiver clock error	15 ns
Satellite clock error	10 ns
Ionosphere error	3 m
Troposphere error	1.7 m
Timing signal offset error	few ns
Doppler frequency offset	±12 kHz

are given in Table 13.2. These typical values are generalizations of the average value of these errors over long periods of time and/or distance and may be considered accurate enough for simple simulators constructed for educational or proof-of-concept purposes. For higher accuracy, complex mathematical error models are used which replicate real-world situations more closely.

After the sampled time is contaminated with various timing errors, it is branched into two streams, called t_{code} and $t_{carrier}$. (If the effect of the ionosphere is neglected, then t_{code} and $t_{carrier}$ would be exactly the same. However, the ionosphere has an opposite effect on the code and the carrier portions of the signal. It delays the code, and advances the carrier). t_{code} is used to create the digital components of the overall GNSS signal—the navigation data message, binary offset carriers and primary and secondary pseudorandom noise (PRN) codes. The best solution is to a priori create these signals as a stream of binary data in a software variable. Then, a look-up table approach is used to pick out the correct bit corresponding to the time sample that is currently being handled. This is equivalent to sampling the stored binary data stream at a rate equal to the t_{code} samples. It is also important to take into consideration the bit rate and repetition rate of the code being created. The navigation data and PRN codes are then combined according to the modulation scheme of the desired GNSS system. Figure 13.3 shows an example Simulink model for generating the Galileo E1B primary code. A priori generation of the codes can be done using online sources, for example (Boschen 2010) for GPS L1 and (Galileo-ICD) for Galileo E1B.

The $t_{carrier}$ component is used in the carrier frequency generator to generate the in-phase and in-quadrature (I/Q) components of the analog high frequency carrier signal, according to (13.1) and (13.2).

$$I_{carrier} = A\,cos(2\pi f_{carrier} t_{carrier}) \tag{13.1}$$

$$Q_{carrier} = A\,sin(2\pi f_{carrier} t_{carrier}) \tag{13.2}$$

It is very time consuming to create Carrier samples at GHz frequency range using Matlab and Simulink and hence, it is more convenient to modulate with some IF, for example in the range 2–5 MHz. The I/Q signals are then passed through a band limiting filter whose double-sided bandwidth is dependent on the signal being simulated for example, 2 MHz for GPS L1, 4 MHz for Galileo E1 and 20 MHz for Galileo E5a. Finally, the I/Q components are combined to form a single stream of complex I/Q data, representing one satellite signal.

Receiver Clock Error

The timing offset between the receiver's local clock and the atomic time standard of the GNSS system being used is called the receiver clock timing error. Receivers usually use different types of crystal oscillators, for example temperature compensated (TCXO), oven controlled (OCXO) etc., for timing purposes. These oscillators have a higher rate of drift, but are more affordable than the ultra-stable atomic clocks used in the GNSS satellites. The receiver clock error is a combination of systematic and random errors (Weinbach 2010). The systematic errors are constant time and frequency offset (in terms of parts per million (ppm)). The Allan Variance (AV) (Allan 1966) commonly characterizes the remaining frequency errors. The AV gives the measure of frequency instability of the clock over consecutive samples. Using M samples, AV can be modeled by (13.3) (Stockwell):

$$AV = \frac{1}{2(M-1)} \sum_{i} \left(y(\tau)_{i+1} - y(\tau)_{i}\right)^{2} \tag{13.3}$$

where τ is the time period over which AV is calculated and $y(\tau)_i$ is the ith sample value.

Satellite Clock Error

The time offset between the satellite's local clock and the atomic time standard of the GNSS system under use is called the satellite clock error. Satellites have sophisticated, highly accurate and precise atomic clocks, which drift by an extremely small amount over a long period. The ground control segment continuously monitors this drift and a correction parameter is transmitted through the navigation message. By accessing this parameter through navigation files in various formats such as YUMA, SP3, RINEX etc., the satellite clock error can be modeled in the simulator. The equation for the satellite clock error is given by (13.4) (Farah 2004).

$$dt = a_0 + a_1(t - t_{oc}) + a_2(t - t_{oc})^2 \tag{13.4}$$

where t_{oc} is the reference epoch in seconds, a_0 is the satellite clock time offset in seconds, a_1 is the fractional satellite clock frequency offset and a_2 is the fractional satellite clock frequency drift.

If a simulator is generating the satellite clock errors similar to those corrected in a0, a1 and a2 parameters in (13.4), then the receiver will completely eliminate this error, leading to unrealistically good performances. An alternative option is to use the URA/SISA parameters (EU-US 2012; GEAS 2010) as an input to model the satellite clock errors.

Ionospheric Delay Error

The ionosphere is the outermost layer of the atmosphere lying between 50 km to more than 1,000 km from the Earth's surface. This layer consists of free ions which collide with the electromagnetic waves of the GNSS signals passing through it. This causes the signals to be delayed, thus introducing a time error in the carrier as well as the code part of the signal. The effect of ionosphere is heavily dependent on the carrier frequency and also on the number of free ions in the ionosphere, which is in-turn dependent on the geographic location as well as time of day. A parameter that gives information on the current condition of the ionosphere is the Total Electron Content (TEC). This TEC is specific to a particular location and time of day and can be accessed from the latest IONEX files. The TEC values can be used to model the ionospheric errors. For GPS, usually the Klobuchar model is used to model the ionosphere errors using the 'ION ALPHA' and 'ION BETA' parameters from GPS RINEX files available on the internet. For Galileo signals, the NeQuick model is preferred. The NeQuick model builds on the formulation for the electron density function in the Epstein layer of the ionosphere given by (13.5) (Memarzadeh 2009):

$$n(h) = \frac{4n_{max}}{\left(1 + exp\left(\frac{h-h_{max}}{B}\right)\right)} exp\left(\frac{h - h_{max}}{B}\right) \tag{13.5}$$

where $n(h)$ is the electron density at height h, n_{max} is the peak electron density, h_{max} is the height of peak electron density, and B is called the thickness of the layer. Once the electron density in the ionosphere above the receiver is known, it can be used to calculate the delay and phase errors affecting the various satellite signals passing through it.

Tropospheric Delay Error

The troposphere is the layer of atmosphere up to 20 km from the Earth's surface. This layer also introduces delays in the GNSS signal propagation but unlike the ionosphere, this delay is not frequency dependent. There are many models to simulate the troposphere, however, the most popular is the Modified Hopfield model that in-turn uses a model for the wet and the dry zenith tropospheric delays and a slant delay transformation. The equation for tropospheric error (in meters) using this model is given by (13.6) (Dolgansky 2009):

$$\begin{aligned} &dTropo(P(t), T(t), H(t), El(t)) \\ &\quad = dTropo^{Hopfield}(P(t), T(t), H(t), El(t)) + w \end{aligned} \tag{13.6}$$

where $dTropo^{Hopfield}$ is the slant tropospheric delay (in seconds) generated using the Goad and Goodman tropospheric model, *P(t)* is time variant pressure parameter, *T (t)* is time variant temperature parameter, *H(t)* is time variant humidity parameter, *El (t)* is elevation angle, and w is Gaussian white noise. An alternate equation for tropospheric error (in meters) is given by (13.7) (Farah 2004):

$$T_r^S = 10^{-6} \int_a^b N ds \tag{13.7}$$

where $N = (n - 1) * 106$ is the refractive index along the signal path and a and b define the limits of tropospheric boundary in meters.

Increased accuracy of modeling the atmospheric errors is possible if the obliquity factor is included. However, in that case the signal generation module needs to take into account the satellite geometry and user position information into the error generation (Pullen etal 2009).

Doppler Frequency Offset

Doppler phenomenon is the change in frequency of a signal incident upon a target from a source, one or both of which are in motion. The signal frequency changes at a rate which is dependent on the speed and direction of relative motion between them. In all GNSS systems, the satellites are in constant motion, therefore, there is a distinct possibility that the received signal is never accurately at the stated carrier frequency. The usual range of Doppler offset is about ±12.5 kHz. This includes the satellite relative velocity, the receiver frequency offset, and the receiver velocity (van Diggelen 2009). The code frequency is also affected by the Doppler offset, but since the frequency is much lower the offset will be much smaller, typically in the order of a few Hz.

The Doppler offset has a benefit too. If the receiver is able to accurately identify the amount of this offset, the relative velocity, acceleration and jerk between the receiver and satellite can be determined. By eliminating the constant effect of satellite motion, the speed and direction (heading) of the receiver motion can be accurately calculated. From signal simulator point of view, altering the carrier frequency of every channel within the range stated above, a Doppler offset can be introduced into the simulator output. This is equivalent to simulating receiver motion! Therefore, by a right combination of Doppler frequencies over different intervals of time, a receiver-moving (also called dynamic) scenario can be created.

Transmission Channel Module

The internal block diagram of the transmission channel module is shown in Fig. 13.4. Two errors that are typically part of this module are multipath and additive white Gaussian noise (AWGN). For multipath, three important parameters are the number of multipath components, and the magnitude and the delay (in samples) of each component with respect to the fundamental component. The

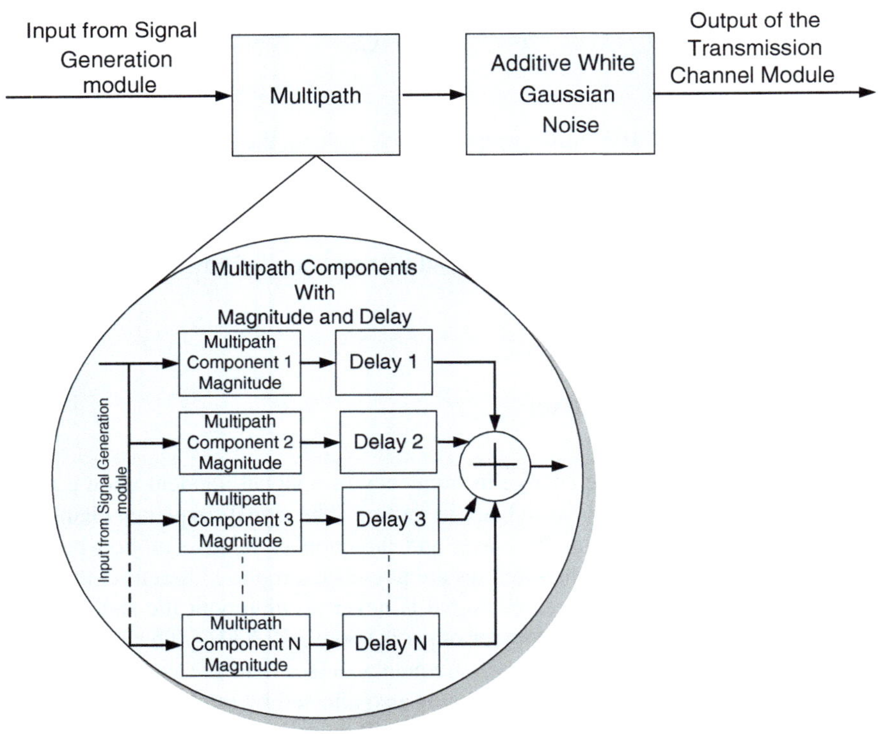

Fig. 13.4 Internal block diagram of transmission channel module

carrier to noise ratio (CNR) defined for every channel determines the level of AWGN that contaminates it. For simplicity, AWGN can be generated as random noise with zero mean and unity variance. From the desired CNR, the maximum noise power can be calculated and using this value, the standard deviation of the random noise can be adjusted to generate enough noise to maintain the desired CNR. The basic function of the transmission channel for the Galileo E1 signal in terms of the sub-carrier can be modeled as shown in (13.8).

$$r_{E1}(t) = \sum_{i=1}^{l} \alpha_i(t) S_{E1}(t - \tau_i) + n \tag{13.8}$$

where $r_{E1}(t)$ is the Galileo E1 received signal after the transmission channel, α_i and τ_i are the complex path coefficient and path delay for the *ith* path respectively, and n is the AWGN.

Radio Frequency Front-End Module

Since the RF FE is the first stage of any communication receiver, it is necessary to model its effects in the simulator. Some software simulators have the capability to model different antenna patterns and power profiles. Typically, only RF filtering effects, amplification, and analog to digital conversion losses are considered.

The RF filter is similar to the band-limited filter used in each channel of the signal generation module, as shown in Fig. 13.2. As before, the filter bandwidth will depend upon the signals being received. The various effects simulated for the filter are band-limiting, insertion loss, passband ripple and group delay. The amplifier includes gain and noise figure effects. Simulating noise figure is equivalent to introducing additional AWGN before the output of the amplifier. ADC quantization causes SNR loss in the received signal. The mathematical expression to simulate this loss in SNR is derived in detail in Shen (2002).

Serna (2010) shows how to include phase noise contamination due to local oscillator non-linearities as a new source of RF FE error. None of the already existing simulators incorporate this feature. The details of the noise generation and ways to add this error into the signal stream are shown in Fig. 13.5. The basic idea is that, the phase noise is a random noise generator which acts like a free running oscillator whose phase noise variance progressively increases with time. Therefore, it is possible to study the effect of different quality oscillators (with varying levels of initial phase noise) over different tracking loop integration times on the quality of baseband processing output. This is shown to be an effective criterion for designers of RF FE local oscillators with regards to the maximum phase noise that can be allowed into the signal stream.

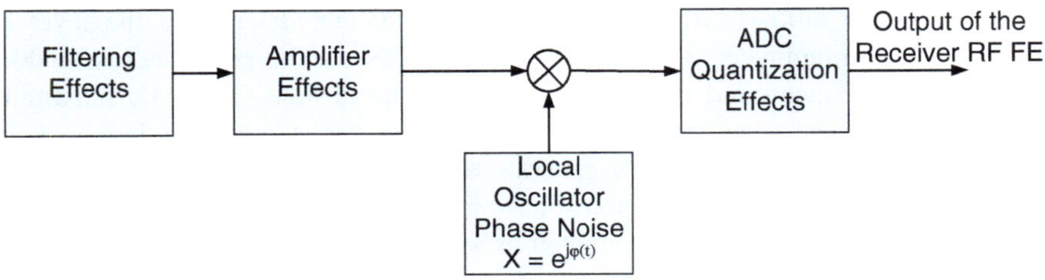

Fig. 13.5 Block diagram of the receiver RF FE module

Satellite Geometry Module

Signals from 32 channels together simulate the entire constellation of one satellite navigation system. The satellite geometry module is responsible for 'modulating' the output of the individual channels according to the current satellite positions and trajectories. If a particular satellite is not visible, that channel output is turned OFF. Conversely, as soon as a satellite peeps over the horizon, its channel is immediately turned ON (albeit with a very low signal strength, which progressively increases if the satellite moves towards the overhead position). Similarly, the signal strength and pseudorange is controlled depending on the elevation angle of the satellite, while the Doppler frequency offset is decided based on the rate of change of pseudorange, which in turn depends upon direction of motion with respect to the simulated receiver position. This is how the satellite geometry module controls the channels within the signal generation module. The next important question is, from where does the satellite geometry module get the current (real-time) picture of the satellite positions and trajectories in the sky? What are its inputs and in what data-format? The answers to these questions have evolved over the years and the level of sophistication and accuracy of modeling the 'real-time constellation picture in space' has improved consistently, as already described in sect. "Galileo Space Segment".

Mathematical Interpretation

This section provides a very simple example of how the background mathematical expressions can be derived. Assuming a Galileo E1 signal generation module, its basic idea is to generate a signal given by (13.9) (GALILEO-ICD 2010).

$$
\begin{aligned}
S_{E1}(t_{code}) \\
&= \left(\alpha sc_{E1-B,a}(t_{code}) + \beta sc_{E1-B,b}(t_{code})\right) \\
&\quad - e_{E1-C}(t_{code})\left(\alpha sc_{E1-C,a}(t_{code}) + \beta sc_{E1-C,b}(t_{code})\right)]
\end{aligned}
\tag{13.9}
$$

Table 13.3 Typical terms included in constructing the overall time parameters

t_{code}	Satellite clock time at the time of transmission (equivalent to the start time of the PRN sequence)
$t_{carrier}$	Time samples for RF carrier generation
$t_{xmit\text{-}true}$	The true time that the signal was transmitted according to Galileo System Time (GST)
δt_{sv}	Satellite clock error
t_{rec}	Receiver clock time
$t_{rec\text{-}true}$	The true time that the signal was received according to GST
δt_{rec}	Receiver clock error
r_{true}	True range between the satellite and the receiver
c	Speed of light = 299792458 m/s
δt_{iono}	Time delay induced by ionosphere
δt_{tropo}	Time delay induced by troposphere
Δf_{dop}	Doppler frequency shift (Hz)
$f_{carrier}$	Galileo E1 carrier frequency = 1575.42 MHz

S_{E1} is the code part of the signal. Modulation on the RF carrier is done according to (13.10).

$$GalileoE1 = A_S S_{E1}(t_{code})cos(2\pi f_{carrier} t_{carrier}) + A_N \qquad (13.10)$$

$e_{E1\text{-}B}$ and $e_{E1\text{-}C}$ are the primary codes for E1B and E1C signals respectively. $sc_{E1\text{-}B}$ and $sc_{E1\text{-}C}$ are the secondary codes or the subcarrier components. α and β are the amplitude components for the subcarriers. A_S and A_N are the signal and noise amplitude components respectively. All of the time variables are actually discrete time (sometimes denoted with t_k). For notational clarity, the subscript 'k' has been omitted. t_{code} and $t_{carrier}$ are the time samples for creating the digital codes and analog RF carrier respectively, as explained in sect. "Bpsk". Table 13.3 defines the typical terms included in constructing the overall time parameters (in seconds).

The SNR determines the relative strength between the signal and the noise, as given in (13.11), (13.12) and (13.13).

$$SNR = 10log_{10}\left(\frac{P_S}{P_N}\right) \qquad (13.11)$$

$$P_S = \frac{A_S^2}{2} \qquad (13.12)$$

$$P_N = \frac{A_N^2}{2} = E\left[Noise^2(t)\right] \qquad (13.13)$$

The relationship between the true transmit time and the satellite's clock time (t_{code}) is given by (13.14).

$$t_{xmit-true} = t_{code} - \delta t_{sv} \tag{13.14}$$

Likewise, the relationship between true receive time and receiver clock time is given by (13.15).

$$t_{rec-true} = t_{rec} - \delta t_{rec} \tag{13.15}$$

It is now possible to set up the relationship between true transmit and receive times, range, and atmospheric delays as in (13.16).

$$t_{rec-true} - t_{xmit-true} = \frac{r_{true}(t)}{c} + \delta t_{iono} + \delta t_{tropo} \tag{13.16}$$

Substituting the true times yields (13.17).

$$t_{rec} - \delta t_{rec} - (t_{code} - \delta t_{sv}) = \frac{r_{true}(t)}{c} + \delta t_{iono} + \delta t_{tropo} \tag{13.17}$$

Solving for t_{code} yields (13.18):

$$t_{code} = t_{rec} - \delta t_{rec} + \delta t_{sv} - \frac{r_{true}(t)}{c} - \delta t_{iono} - \delta t_{tropo} \tag{13.18}$$

Depending on how the simulator is implemented, there are a variety of ways in which the range $r_{true}(t)$ can be represented. It is typical to assume that an initial range and a range-rate (time rate of change of range) is a priori known as in (13.19).

$$r_{true}(t) = r_{true}(t_0) + \int_{t_0}^{t} \dot{r}(t)dt \tag{13.19}$$

The Doppler frequency offset in terms of the carrier frequency is given by (13.20).

$$\Delta f_{dop}(t) = \frac{\dot{r}(t)}{c} f_{carrier} \tag{13.20}$$

Solving for range rate yields (13.21).

$$\dot{r}(t) = \frac{c * \Delta f_{dop}(t)}{f_{carrier}} \tag{13.21}$$

Substituting this result into the equation for $r_{true}(t)$ yields (13.22).

$$
\begin{aligned}
r_{true}(t) &= r_{true}(t_0) + \int_{t_0}^{t} \frac{c * \Delta f_{dop}(t)}{f_{carrier}} \\
&= r_{true}(t_0) + \frac{c}{f_{carrier}} \int_{t_0}^{t} \Delta f_{dop}(t)dt
\end{aligned} \tag{13.22}
$$

Finally, substituting this result into the equation for t_{code} yields (13.23).

$$
\begin{aligned}
t_{code} = t_{rec} - \delta t_{rec} + \delta t_{sv} - \frac{r_{true}(t_0)}{c} - \frac{1}{f_{carrier}} \int_{t_0}^{t} \Delta f_{dop}(t)dt \\
- \delta t_{iono} - \delta t_{tropo}
\end{aligned} \tag{13.23}
$$

The derivation for $t_{carrier}$ is very similar, as shown in (13.24). The only difference is that the sign on the ionosphere error is now opposite.

$$
\begin{aligned}
t_{carrier} = t_{rec} - \delta t_{rec} + \delta t_{sv} - \frac{r_{true}(t_0)}{c} - \frac{1}{f_{carrier}} \int_{t_0}^{t} \Delta f_{dop}(t)dt \\
+ \delta t_{iono} - \delta t_{tropo}
\end{aligned} \tag{13.24}
$$

Multi-Frequency, Multi-System Receiver Performance Test Scenarios

The overall objective of this section is to serve as a guideline for the testing and validation of an example multi-frequency, multi-system GNSS prototype receiver. Using a signal simulator with appropriate test scenarios, a repeatable signal, whose characteristics are fully known, can be presented to the GPS receiver. Figure 13.6 describes the usual and most preferred approach for testing the performance of a GNSS receiver.

For ease of handling a receiver under test (RUT), two example settings can be made in its baseband software: the mode of operation and the mode in which the receiver has to be turned ON. An 8 bit control signal is provided to configure this setting before every test case. Table 13.4 describes the various options and the digital control word corresponding to each option. Shown here are 8 example modes of operation, which require 3 bits to be uniquely represented. However, 5 bits can be assigned in order to accommodate any planned future increase in operating modes. Additionally, there are three modes to turn ON a GNSS receiver, uniquely represented by 2 bits. Therefore, from the 8 bits, only 7 bits are used. The remaining bit is reserved for future use. The mode selection bits are the least significant bits (LSB) of the control word. For example, if receiver has to be set in a

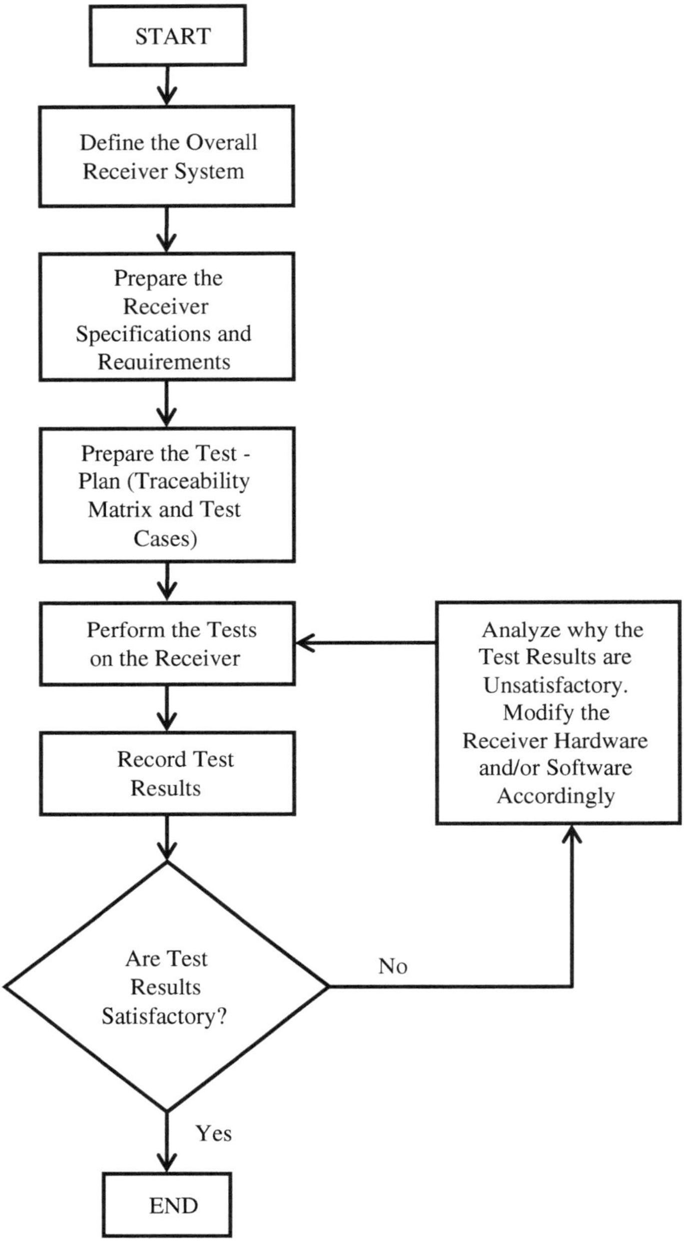

Fig. 13.6 Typical testing procedure for GNSS receivers

Table 13.4 GNSS receiver testing modes

	Example options	Binary representation
Example modes of operation	Position fix = GPS L1 only	00000
	Position fix = Galileo E1 only	00001
	Position fix = GPS L1 and Galileo E1 and no ionospheric correction	00010
	Position fix = GPS L1 + GPS L5 ionospheric correction	00011
	Position fix = Galileo E1 + Galileo E5a ionospheric correction	00100
	Position fix = GPS L1 and Galileo E1 + GPS L5 ionospheric correction (for GPS pseudoranges) + Galileo E5a ionospheric correction (for Galileo pseudoranges)	00101
	Position fix = GPS L1 and Galileo E1 + GPS L5 ionospheric correction (for GPS pseudoranges)	00110
	Position fix = GPS L1 and Galileo E1 + Galileo E5a ionospheric correction (for Galileo pseudoranges)	00111
Turn on modes	Cold start	00
	Warm start	01
	Hot start	10

warm start with position fix using GPS L1 and Galileo E1 (single frequency, dual system) reception with no correction for ionosphere error, the 8 bit control word would be binary "00100010".

Usually, the following minimum tests have to be performed on a GNSS receiver to sufficiently characterize it: time to achieve first position fix (TTFF), acquisition sensitivity, tracking sensitivity, accuracy of position fix, reacquisition time, availability percentage, quality of ionospheric error mitigation, multipath error mitigation and radio frequency interference mitigation, and finally, performance under high dynamics (motion).

Time to First Fix (TTFF) Tests

The amount of time (in seconds) from receiver switch ON to the availability of the first valid location fix is called time-to-first-fix or TTFF. It relates directly to the user desire to have a location fix as soon as possible and hence, TTFF is a critical parameter for testing GNSS receivers under different initial conditions. Receiver initial conditions refer to the amount and quality of data that is present in memory at the time of switching ON. Depending upon the initial conditions, there are three start modes for a typical GNSS receiver: cold start, warm start and hot start.

In a cold start, the receiver has no a-priori information about the approximate time, ephemeris, almanac or last computed position. TTFF in cold start is in the order of 60 seconds under nominal signal power conditions. In warm start, the receiver has knowledge of the approximate time, last computed position, oscillator offset and a valid copy of the almanac is present in the memory. The ephemeris is not yet decoded and hence the time of week information is missing. TTFF is in close to 30 s in case of warm start. In hot start, the receiver has all the data necessary for a position fix already stored in memory. Valid ephemeris and almanac is available for all visible satellites, accurate time, oscillator offset, and last position fix are known. In this case, the TTFF should be around 1 s.

Nominal Cold Start, Warm Start and Hot Start TTFF

The aim of this test is to verify receiver's cold start, warm start and hot start TTFF performances under nominal signal conditions, typically assumed as satellite signal power equal to or greater than −130 dBm.

Low Power Cold Start TTFF

The aim of this test is to verify receiver's cold start TTFF performance under low power input signal conditions, typically assumed as satellite signal power equal to or lower than −142 dBm.

Acquisition Sensitivity in Cold Start

The aim of this test is to verify receiver's acquisition sensitivity. Acquisition sensitivity refers to the lowest satellite signal power level at which the receiver can still perform successful acquisition of this satellite. An example procedure for acquisition sensitivity testing can be as follows; all satellites in the simulator are maintained at the same power level and required numbers of acquisition iterations are performed. In every iteration, acquisition is deemed successful only if the difference between code delay estimated by the acquisition engine and the actual value is less than, for example ±1 chip (300 m) and the Doppler offset difference is less than, for example ±150 Hz. The acquisition engine provides only rough estimates of the delay and Doppler values and hence an additional pull-in module is necessary where more accurate delay and Doppler values can be determined. Next, a percentage of successful acquisitions is computed and overall the acquisition is deemed successful if this percentage is greater than a pre-determined threshold. The input satellite signal power after which the percentage of successful acquisitions falls below the threshold is the acquisition sensitivity of the receiver. Below this power level, we would no longer be sufficiently confident of acquiring satellites.

Accuracy

Positioning accuracy is defined as the magnitude of the distance between the true position of the RUT and the position fix reported by the RUT. This is usually represented in terms of 2-dimensional (2-D) Circular Error Probability (CEP) or 3-D Spherical Error Probability (SEP). A percentile Confidence Level is also assigned to the error probability, for example, 50th percentile CEP is the radius of a circle which encompasses half of the 2-D position fixes. Better the accuracy of the receiver, closer to the true position will be its repetitive position fixes and smaller will be the radius of the circle encompassing half of these points (the true position is usually considered to be the one displayed in the simulator itself).

Tracking Sensitivity

Tracking sensitivity of a receiver is defined as the minimum satellite signal power at which the receiver can successfully continue tracking the satellite. It is important to identify what tracking advantages dual-frequency dual-system operation has over single-frequency single-system operation. Furthermore, when referring to for example, GPS L5/Galileo E5a tracking tests, it is necessary to specify whether this refers to only pilot signal tracking or to combined data + pilot tracking.

An example procedure for tracking sensitivity testing can be as follows; one of the visible satellites is selected and its signal power is initially maintained at nominal level for 30 s to allow the receiver to successfully track this satellite. Next, the power is dropped to a low level and maintained at this level for the next 30 s after which the power is once again raised to the nominal level for the next 30 s. The receiver carrier to noise ratio (CNo) estimator can be used to check whether the receiver continues tracking the satellite throughout this 90 s interval. In the next iteration, the power can be dropped 1 dB lower than in the previous iteration. The power level below which the receiver can no longer compute CNo correctly is the tracking sensitivity of the receiver.

Availability

Availability of GNSS receiver means the percentage of time services of the receiver are useable (Spi02). In other words, percentage of time valid PVT result is computed by the receiver under differing conditions of signal strength. During this test, the receiver is tested under open air good signal conditions, which progressively worsens into a heavy multipath and interference condition. Next the simulator is taken indoors where it is assumed to encounter signal degradation and multipath from the windows. Finally, the receiver is brought back into the open air. User motion is simulated by changing the input signal characteristics after fixed intervals of time. A maximum time for position fix has to be set, up to which if no position fix is made it is regarded that the receiver was not available for this iteration. During availability testing the receiver can be turned OFF and ON repeatedly (unlike continuity testing) or kept ON throughout the test iterations.

Receiver Dynamic Accuracy

Dynamic accuracy tests the performance of the GNSS receiver when in motion. The important parameters of motion are speed, direction, acceleration, jerk and instantaneous position. The RUT should accurately calculate these parameters at speeds up to 120 km/h. (Mitelman 2008) proposes a circular trajectory while (3GPP06) proposes a racetrack trajectory (also called a rectangular trajectory with rounded corners), as shown in Fig. 13.7. Some example values for the distances are: top/bottom lengths = 1,440 m, left/right lengths = 940 m, corner radius = 20 m, start point = left hand top corner, rotation = clockwise, acceleration distance = 500 m, minimum speed = 30 km/h on turn, maximum speed = 120 km/h on straight road. A receiver refresh interval/interval for periodic update of position in a moving scenario has to be fixed. Moving over the selected route 3 times allows a more general idea of RUT performance. It will be convenient if the receiver has capability for mapping the route taken in a dynamic scenario. Alternatively, the performance in this test can also be determined by the ability of the receiver to maintain position fix and periodically update it as its position changes.

Reacquisition Time

Reacquisition time is the time necessary for a receiver to reacquire all visible satellites and make a position fix after a sudden drop in signal power causes all previously tracked satellite signals to be lost. Fast reacquisition time is important for

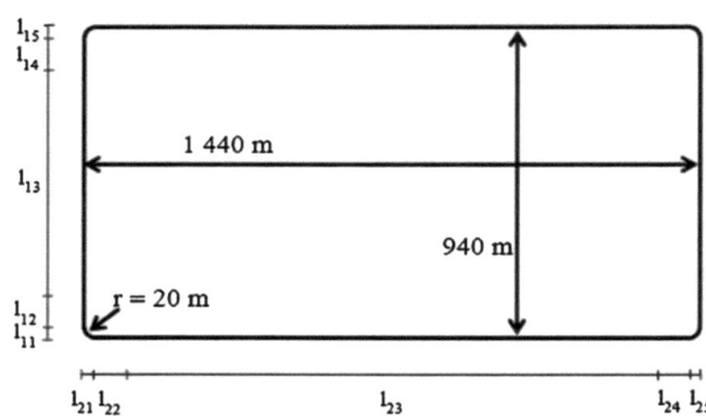

Fig. 13.7 Rectangular (racetrack) trajectory for dynamics testing

in-vehicle navigation systems. Consider a car emerging from a tunnel, in which it has lost all satellite signals. Immediately after the tunnel is a junction at which the driver must make an exit. The navigation system needs to be navigating again quickly in order for it to give the correct instructions as soon as possible after exiting the tunnel (Spi03), (Agi). This test can be simulated by first allowing the receiver to make a stable position fix under nominal power conditions. Then all the visible satellite signals from the simulator should be turned OFF until the receiver loses track on all of them, thus simulating the entry into a tunnel. Finally, all the signals can be simultaneously turned ON at nominal power, thus simulating exit from the tunnel. The amount of time it takes for the receiver to re-obtain stable position fix is the reacquisition time.

Multipath Mitigation

The objective of this test is to determine the ability of the prototype receiver in mitigating the effect of multipath errors. The effect of multipath should be tested for dual-frequency dual-system operation. Signal simulators have inbuilt support for a number of realistic multipath scenarios and models, which can be used for performing this test. Furthermore, Non Line-Of-Sight (NLOS) multipath can be easily generated in a simulator and tested. It is an important error, given that majority of satellite navigation receivers are in smartphones/Personal Navigation Devices (PNDs), and they track and are affected by NLOS multipath components in urban areas.

Radio Frequency Interference (RFI)

This test attempts to establish the ability of the GNSS receiver to operate in the presence of interfering (jamming) signals. There are different possible types of interference signals, as explained in Sect. "Galileo Architecture Overview". The effect of RFI should also be tested under dual-frequency, dual-system operation. The most convenient option is to check whether in presence of RFI of up to −60 dBm, the receiver continues to maintain stable position fix. Testing can be done for interference on a single carrier frequency or on all carrier frequencies processed by the receiver.

Ionosphere Errors

Ionospheric delay errors affect the different frequencies differently. The aim of this test is to check the performance of dual frequency approach in eliminating ionosphere error and improving the accuracy of position fix. For this, the main idea is to test the accuracy in single frequency operation and then comparing with accuracy using dual-frequency operation. The advantage of using dual-frequency can be quantified in terms of meters of improved accuracy. Signal simulators are capable of generating ionosphere errors based on one or more mathematical models, which can be used for this test.

Conclusion

This chapter dealt with software-based signal simulators for GNSS signals and multi-frequency, multi-system GNSS receiver performance testing scenarios. After a brief introduction about the basics of simulators and their use, state-of-the-art software signal simulators were discussed and their properties were compared for some important categories. Software-based simulators have been in existence since the last 10 years and have undergone considerable evolution. This evolution has been described in detail in the background literature study. Next, the essential components and modules of the simulator are discussed, which would enable an interested reader or researcher to have an introduction towards building his/her own simulator. The mathematical background behind the various modules is also presented. In the final section, a set of test scenarios have been presented which will enable the simulator operator to determine the performance of a GNSS receiver based on multi-frequency, multi-system approach. The most common test cases have been presented and literature sources have been recommended to continue a deeper study of these topics for those further interested.

References

(GEAS) (2010) Phase II of the GNSS Evolutionary Architecture Study, Feb 2010. http://www.faa.gov/about/office_org/headquarters_offices/ato/service_units/techops/navservices/gnss/library/documents/media/GEASPhaseII_Final.pdf

3rd Generation Partnership Project (3GPP) (2006–2010) 'TS34.171'; V6.5.0

Abart C, Berglez P, Abwerzger G, Hoffmann-Wellenhof B, Cresens W, Vandeplas T, De Win W (2007) Simulating GNSS constellations—the GAMMA signal generator. In: Proceedings of the 20th international technical meeting of the satellite division of the institute of navigation (ION GNSS 2007), Fort Worth, TX

Agilent Technologies (2008) Typical GPS receiver verification tests using a GPS signal simulator, Application Note, May 2008

Allan W (1966) Statistics of atomic frequency standard. Proc IEEE 54(2):221–230

Ashby N (2004) The sagnac effect in the global positioning system. In: Rizzi G, Ruggiero ML (eds) From the book, Relativity in rotating frames: relativistic physics in rotating reference frames, Kluwer Academic Publishers, ISBN: 1-4020-1805-3

Accord Software and Systems Pvt. Ltd, GPSLAB. http://www.gpslab.us/datasheet/DataSheet-prof.pdf

Boschen (2010) GPS C/A Code Generator, Matlab Central. http://www.mathworks.com/matlabcentral/fileexchange/14670-gps-ca-code-generator&watching=14670. Accessed Jun 2010
Brenner M, Reuter R, Schipper B (1998) GPS landing system multipath evaluation techniques and results. In: Proceedings of the institute of navigation ION GPS-98, Nashville, Tennessee, Sept 1998
Brown A, Gerein N (2001) Advanced GPS hybrid simulator architecture. In: Proceedings of ION 57th annual meeting, Albuquerque, NM, June 2001
Borre K (2009), The E1 Galileo Signal, Lecture at Stanford university. http://waas.stanford.edu/~wwu/papers/gps/PDF/Borre/galileo_sig.pdf
Brown A, Gerein N, Taylor K (2000) Modeling and simulation of GPS using software signal generation and digital signal reconstruction. In: Proceedings of the ION national technical meeting, Jan 2000, Anaheim, CA
Constantinescu C, Landry RJr, Ilie I (2005) Hybrid GPS/Galileo/GLONASS IF software signal generator. In: Proceedings of ION GNSS 18th international technical meeting of the satellite division, 13–16 Sept 2005, Long Beach, CA
Center for Remote Sensing Inc (2009)., Software GNSS Signal Simulator. http://www.cfrsi.com/pdf/Software%20GNSS%20Signal%20Simulator.pdf
Corbell PM, Miller MM (2000) Design and analysis of a matlab based digitized if GPS signal simulator and a simulink based configurable GPS receiver. In: Proceedings of ion GPS 2000, 19–22 Sept 2000, Salt Lake City, UT
Corbell PM, Miller MM (2001) A configurable GPS accumulated I and Q signal component simulator in matlab. In: Proceeding of ION NTM 2001, 22–24 Jan 2001, Long Beach, CA
Cosenza C, Morante Q, Corvo S, Gottifredi F (2008) GNSS bit-true signal simulator-A Test bed for receivers and applications. Satell Commun Navig Syst Signals Commun Technol 3:447–460. doi:10.1007/978-0-387-47524-0_34
DiezFJ, Caramagno A (2006) GRANADA-galileo receiver analysis and design application. In: ESA-GJU Workshop on Tools and Facilities for Galileo Receivers, Mar 2006
Dolgansky A, Szeto A, Bisnath S (2009) Software simulation of multiple global navigation satellite system measurements. In: Proceedings of the IEEE toronto international conference—science and technology for humanity 2009 (TIC-STH 2009) in Toronto Canada
Dong L, Ma C, Lachapelle G (2004) Implementation and verification of a software-based IF GPS signal simulator. In: Proceedings of national technical meeting, institute of navigation, 26–28 Jan 2004, San Diego
EU-US (2012) Cooperation on Satellite Navigation, Working Group C, ARAIM Technical Subgroup, Interim Report, Issue 1.0, 19th, Dec 2012
Farah AMA (2004) GPS/Galileo simulation for reduced dynamic LEO satellite orbit determination. Ph.D. Thesis 2004, University of Nottingham, UK
Fraunhofer Institute for Integrated Circuits IIS (2009), GNSS-SIGNAL-SIMULATOR. http://www.iis.fraunhofer.de/fhg/Images/07_GNSS-Signal_Simulator_neu_tcm278-158408.pdf
Galileo Open Service (Galileo ICD 2010) Signal in space interface control document, (OS SIS ICD, Draft 1), Feb 2010
German Aerospace Center (DLR) (2014), Simulation of satellite navigation signals and channels (SNACKS). http://snacs.sourceforge.net/index.html
Hu C, Tsai M (2007) The implementation of an INS-GNSS software simulator. In: Proceedings of 28th Asian conference on remote sensing (ACRS 2007), Nov 2007, Kuala Lumpur, Malaysia
Intecs (2010), gLAB GNSS Signal Analysis Tool. http://www.intecs.it/pdf/brochure_gLAB_01-02-2010.pdf
Inzerilli T, Lo Forti D, Suraci V (2005) Modeling and simulation of GNSS with NS2. In: Proceedings of 14th IST mobile and wireless communications summit, Dresden, June 2005
Jahn A, Bischl H, Heib G (1996) Channel characterization for spread spectrum satellite communication. In: IEEE 4th international symposium on spread spectrum techniques and applications proceedings, vol 3. 22–25 Sept 1996, pp 1221–1226

Joo I, Lee J, Lee S, Kim J, Lim D, Lee SJ (2007) S/W based IF signal simulator prototyping for L1 C/A, L2C, and E1(B and C). In: Proceedings of The 9th international conference on information and communications security (ICICS 2007), Dec 2007, Zhengzhou, China

Julien O, Zheng B, Dong L, Lachapelle G (2004) A complete software-based if GNSS signal generator for software receiver development. In: Proceedings of ION GNSS 2004, Sept 21–24, Long Beach, CA

Kou ZU, Huang U, Qishan ZZ (2009) Architecture of software-based GNSS signal simulator and implementation of its IF signals generation. J Beijing Univ Aeronaut Astronaut (2009–07). http://en.cnki.com.cn/Article_en/CJFDTOTAL-BJHK200907004.htm

L. Dong (2003), "IF GPS Signal Simulator Development and Verification", Master of Science Thesis, University of Calgary, Canada, November 2003

Lee J, Kim T, Lee S, Kim J (2008) Development of software GNSS signal generator. In: Proceedings of international symposium on GPS/GNSS 2008, Nov 2008, Tokyo, Japan

Lim S, Lim D, Liu M, Moon SW, Park C, Lee SJ (2008) Design of a software-based multi-channel GNSS IF signal generator. In: Proceedings of international conference on control, automation and systems 2008, Oct 2008 in Seoul, Korea

Malik S (2009), Shivaramaiah S Dempster A, Search Engine Trade-offs in FPGA-based GNSS Receiver Designs, University of New South Wales, Sydney, Australia. http://www.gmat.unsw.edu.au/snap/publications/malik_etal2009a.pdf

Marradi L, Franzoni G et al (2006) The GARDA project—building a galileo receiver. November/December, InsideGNSS, pp 40–53

Memarzadeh Y (2009) Ionospheric modeling for precise GNSS applications. Ph.D. Thesis, Delft University of Technology, the Netherlands, Dec 2009. http://www.ncg.knaw.nl/Publicaties/Geodesy/pdf/71Memarzadeh.pdf

Miller MM, Corbell PM, Raquet JF (2000) Design and validation of digitized intermediate frequency GPS signal and receiver software models for developing and comparing advanced gps receiver technologies. In: Proceedings of ION GPS 2000, 19–22 Sept 2000, Salt Lake City, UT

Mitelman A, Nomark P, Reidevall M, Strickland S Apples to apples—standardized testing for high-sensitivity receivers. GPS World, 1 Jan 2008

Nunes A, Ferreira T, Borràs J, Nunes F, Sousa F, Seco G (2004) Signal generator and receiver toolbox for galileo/GPS signals. In: Proceedings of 2nd ESA workshop on satellite navigation, Navitech 2004

National Instruments (2013), The Case for GPS Simulation. http://www.insidegnss.com/special/elib/NI_GPS_Simulation.pdf

NAVSYS Corporation, GPS Signal Simulation Toolbox. http://www.navsys.com/brochures/NAVSYS_Toolbox.pdf

Ouzeau C, Korsakissok J (2009) Performance assessment of a juzzle-based gnss simulator. In: Proceedings of ENC GNSS 2009, May 2009, Naples, Italy

Pósfay A, Pany T, Eissfeller B (2005) First results of a GNSS signal generator using a PC and a digital-to-analog converter. In: Proceedings of ION GNSS 18th international technical meeting of the satellite division, 13–16 Sept 2005, Long Beach, CA

Pullen S, Rife J (2009) Differential GNSS: accuracy and integrity. In: Gleason SD (eds) from the book, GNSS Applications and methods. Gebre-Egziabher, Artech House, ISBN-13: 978-1-59693-329-3

Schubert F, Prieto-Cerdeira R, Steingass A (2008) GNSS software simulation system for realistic high-multipath environments. In: Proceedings of 4th ESA workshop on satellite navigation user equipment technologies (NAVITEC 2008), Dec 2008, Noordwijk, The Netherlands

Spirent Communications (2009), Testing GNSS System Errors, Appln. Note: DAN002 Issue 1–01

Spirent Communications, Fundamental GNSS Receiver Characterisation, Application Note DAN003, Issue 1-01, 2009.

Spirent Communications (2010), Simulation versus Real World Testing—How to undertake controlled testing of your GNSS receiver design. http://www.insidegnss.com/special/elib/Spirent_Simulation_vs_Real_World_Testing.pdf

Serna EP, Thombre S, Valkama M, Lohan S, Syrjälä V, Detratti M, Hurskainen H, Nurmi J (2010) Local oscillator phase noise effects on GNSS code tracking. InsideGNSS, Nov/Dec 2010, pg 52–62

Seynat C, Kealy A, Zhang K (2004) A performance analysis of future global navigation satellite systems. J Global Positioning Syst 3(1–2):232–241

Shen B, Zhang Q (2002) A new method for analyzing the quantization effect of adc in broadband QAM receiver. In: Proceedings of the IEEE 2002 international conference on communications, circuits and systems and west sino expositions, Jun 2002, pp 1262–1266

GPS World (2012) Simulator Special issue, May 2012

Stockwell W, Bias Stability Measurement: Allan Variance", Crossbow Technology, Inc. http://www.xbow.com/pdf/Bias_Stability_Measurement.pdf

Smith AM (2007) Global navigation satellite system signal simulator—an analysis of the effects of the local environment and atmosphere on receiver positioning. Doctor of Philosophy Thesis, University of Bath, UK, Oct 2007

Tan TH (2003) Global positioning system signal simulation, bachelor of electrical engineering (Honours) thesis. The University of Queensland, Australia

Thombre S, Lohan ES, Raquet J, Hurskainen H, Nurmi J (2010) Software-based gnss signal simulators: past, present and possible future. In: Proceedings of the European navigation conference (ENC GNSS 2010), Oct 2010, Braunschweig, Germany

University FAF Munich (2003), GNSS Software Simulation. http://www.ifen.unibw.de/research/gnss_simulator.htm

TeleConsult Austria GmbH, GNSS Multisystem Performance Simulation Environment (GIPSIE). http://www.teleconsult-austria.at/download_public/GIPSIE.pdf

van Diggelen F (2009) A-GPS: Assisted GPS, GNSS, and SBAS", Edn 1, Mar 2009, ISBN-10: 1596933747, ISBN-13: 978-1596933743

VEGA IT GmbH, Galileo System Simulation Facility (GSSF). http://www.gssf.eu/Documents/GSSF%20V2.1%20Fact%20Sheet%20-%20Issue%201.pdf

Weinbach U, Schön S (2010) GNSS receiver clock modeling when using high-precision oscillators and its impact on PPP. J Adv Space Res 47(2):229–238

Chapter 14
An Introduction to EGNOS: The European Geostationary Navigation Overlay System

Javier Ventura-Traveset, Carlos López de Echazarreta, Juan-Pedro Lam and Didier Flament

Abstract Recognizing the strategic importance of satellite navigation and its applications, a European approach was developed in the 90s. The European strategy on satellite navigation was then based on two major steps, what was known as GNSS-1 and GNSS-2, linked respectively to the European Geostationary Navigation Overlay Service (EGNOS) and Galileo infrastructures. The development of the EGNOS system, namely a Satellite Based Augmentation System (SBAS) to GPS, became the first European infrastructure devoted to GNSS. EGNOS, considered by many, as the pioneer and the stumbling block of Galileo, did allow the development of a competent European industry in GNSS and convinced Europe on the importance to achieve full autonomy on GNSS by developing their own GNSS constellation, the Galileo system. EGNOS started to provide Open services on 1st October 2009, targeting mass-market applications for non-safety of life applications. On 2nd March 2011, Safety of Life Service was started, following EC authorization to provide the service. This chapter describes the EGNOS mission, system architecture and services, complemented with some detailed information on specific fundamental aspects such as integrity, ionosphere, signal format and SBAS interoperability with other similar worldwide systems. The chapter also includes some information on EGNOS operations, performance and potential infrastructure evolutions currently under assessment. This chapter is produced for academic purposes only. Official information on the EGNOS actual system status and performances is to be obtained via the European Commission and the EGNOS Operator. Relevant official reference documents and websites (as per May 2013) are provided also throughout this chapter.

J. Ventura-Traveset (✉)
European Space Agency, Madrid, Spain
e-mail: Javier.Ventura-Traveset@esa.int

C. López de Echazarreta · J.-P. Lam · D. Flament
European Space Agency, Toulouse, France

J. Nurmi et al. (eds.), *GALILEO Positioning Technology*,
Signals and Communication Technology 182, DOI 10.1007/978-94-007-1830-2_14

EGNOS Mission and Services

The capabilities of GPS alone, although very adequate for some user communities, present some shortfalls: the lack of a true civilian international control and the limited performances of GPS for some applications. In particular, civil aviation requirements for precision and non-precision approach phases of flight cannot be met by GPS only. Other users, such as marine or land users, may also require some sort of augmentation for improving GPS performances.

The purpose of EGNOS is to implement a system that fulfils a range of user service requirements by means of an overlay augmentation to GPS[1] on the broadcasting through GEO satellites with GPS-like navigation signals containing integrity and differential correction information applicable to the navigation signals of the GPS satellites. EGNOS augments GPS using the L1 (1575.42 MHz) civilian signal band.

EGNOS is the European Satellite Based Augmentation System (SBAS) (see Fig. 14.1). Similarly, other SBAS initiatives are currently operational in USA (the Wide Area Augmentation System—WAAS) and Japan (the Multi-transport Satellite based Augmentation System—MSAS). Three additional SBAS systems are currently under development in India, Russia and China (see Fig. 14.15). All SBAS systems shall comply with the corresponding International Civil Aviation Organisation (ICAO 2004) and Radio Technical Commission for Aeronautics (RTCA 2006) standardisation requirements and should, therefore, be interoperable.

EGNOS is owned by the European Union, with the European Commission being in charge of the overall EGNOS programme management. The European Space Agency (ESA) has led the technical development of the EGNOS system and is now (May 2013), under delegation agreement by the European Commission, the design and procurement agent for system evolutions. EGNOS is today operated by the European Satellite Services Provider (ESSP) SAS under contract with the European GNSS Agency (GSA).

There are essentially three main differentiators of the EGNOS system with respect to GPS, namely

- the provision of integrity positioning with a safety of life quality;
- a better accuracy than GPS; and
- the possibility of establishing a geographical position with reliability.

These differentiators allow EGNOS to address a large multi-modal user community, including civil aviation, maritime and land users; liability services; agriculture; and a large variety of location-based services (EGNOS 2014a). For aviation, the EGNOS service is today (at the time of writing, May 2013) compliant with the Approach with Vertical Guidance (APV-I) requirements, as defined by ICAO Standard and Recommended Practices (SARPs), Annex 10 (ICAO 2004).

[1] EGNOS was originally conceived to augment both GPS and GLONASS constellations, although, finally, only the GPS augmentation has been activated for service provision. For this reason we will speak through this chapter only on EGNOS GPS augmentation function.

Fig. 14.1 EC EGNOS open service (2013) and EGNOS safety of life (2013) definition documents

EGNOS provides three different services[2]:

- The Open Service (OS), freely available to any user;
- The Safety of Life (SoL) Service, that provides the most stringent level of signal-in-space performance to all Safety of Life user communities;
- The Commercial Data Distribution Service (CDDS) for customers who require enhanced performance for commercial and professional use.

Open Service

The main objective of the EGNOS Open Service is to improve the achievable positioning accuracy by correcting several GPS error sources. Notably, the corrections transmitted by EGNOS contribute to mitigate the ranging error sources related to satellite clocks, satellite position and ionospheric effects. The tropospheric

---

[2] The information provided here concerning EGNOS Services is extracted from the official EGNOS service Definition Documents, notably EC-DG Energy and Transport (2013). More detailed information may be obtained from those freely available downloadable documents directly.

Table 14.1 EGNOS minimum accuracy (Source EC EGNOS OS SDD (2013))

	Definition	Value (m)
Horizontal accuracy	Corresponds to a 95 % confidence bound of the bi-dimensional position error in the horizontal local plane for the worst user location	3
Vertical accuracy	Corresponds to a 95 % confidence bound of the unidimensional unsigned position error in the local vertical axis	4

delays are partly corrected through SBAS in the user receiver by a tropospheric delay model based on a priori information. The other error sources (multipath and user receiver contributions) are local effects that cannot be corrected by a global augmentation system.

The Open Service has been available since 1st October 2009, and the corresponding Service Definition Document (SDD), as produced by the European Commission, can be found at EC-DG Energy and Transport (2013). This document describes the EGNOS system and Signal-In-Space (SIS), the performance achieved by the Open Service (OS), and information on the technical and organisational framework at European level for the provision of the OS.

The EGNOS OS minimum Accuracy, as extracted from EC-DG Energy and Transport (2013), is shown in Table 14.1.

Safety of Life Service

The main objective of the EGNOS SoL service is to support civil aviation operations down to Localiser Performance with Vertical guidance (LPV) minima. At this stage, a detailed performance characterisation has been conducted only against the requirements expressed by civil aviation but the EGNOS SoL service might also be used in a wide range of other application domains in the future (e.g. Maritime, Railways, Road…). In order to provide the SoL service, the EGNOS system has been designed so that the EGNOS SIS is compliant to the ICAO SARPs for SBAS (2004).

The EGNOS SoL service is available from the 2nd March 2011, and the corresponding Service Definition Document (SDD), as produced by the European Commission, is available at EC-DG Enterprise and Industry (2013). This document describes the EGNOS system architecture and Signal-In-Space (SIS), the SoL service performance achieved with the current EGNOS System Release and information on the established technical and organisational framework, at European level, for the provision of this service.

One of the major benefits of EGNOS is that it may provide reliable vertical guidance at any airport in Europe, without the need of additional local ground-based infrastructure. General and regional aviation, private/business aviation and helicopter users are all potential EGNOS users. The airport of Pau Pyrenees in

Fig. 14.2 European airports with EGNOS-based approach procedures (existing or forecast) as per May 2013 (*Source* ESSP public and official website)

southern France became in May 2011 the Europe's first airport to use the new EGNOS SoL service for landing using only this highly accurate space navigation signal. Since then, several European airports have defined EGNOS-based approach procedures (see Fig. 14.2). Updated information on those may be obtained through the EGNOS operator official website (ESSP 2014).

The EGNOS SoL minimum service performance characteristics are defined in detail in EC-DG Enterprise and Industry (2013) and summarised in Table 14.2.

Commercial Data Distribution Service

EGNOS CDDS provides authorised customers (e.g. added-value application providers) EGNOS products for commercial distribution. Application Providers will be able to connect to the EGNOS Data Access Service (EDAS), and exploit the

Table 14.2 EGNOS SoL service performance values as declared in EC-DG Enterprise and Industry Safety of Life Service Definition Document

	Parameter	Performance	Comment
Accuracy	Vertical accuracy	4 m (95 % percentile)	Accuracy maps are included in Appendix B of EC-DG Enterprise and Industry.
			Accuracy values at given locations are available at http://www.essp-sas.eu
	Horizontal accuracy	3 m (95 % percentile)	Accuracy maps are included in Appendix B of EC-DG Enterprise and Industry.
			Accuracy values at given locations are available at http://www.essp-sas.eu
Integrity	Integrity risk	2×10^{-7}/approach	
	Time to alert	Less than 6 s	
Availability	Availability	99.9 % for NPA in all the ECAC	See Sects. 6.3.1.3 and 6.3.2.3 of EC-DG Enterprise and Industry (2013)
		99 % for APV-1 in most of ECAC	
Continuity	Continuity	For NPA:	
		$<2.5 \times 10^{-4}$/h in most of ECAC	
		$<2.5 \times 10^{-3}$/h in other areas	
		For APV-1:	
		$<1.0 \times 10^{-4}$/15 s in the core ECAC	
		5.0×10^{-4}/15 s in most ECAC	
		10.0×10^{-3}/15 s in other areas	

Note ECAC stands for European Civil Aviation Conference, defined in (EC-DG Enterprise and Industry 2013)

EGNOS products, offering added-value services to final customers. The main types of data provided by EDAS are:

- Raw GPS, GLONASS[3] and EGNOS GEO observations and navigation data collected by the entire network of RIMSs and Navigation Land Earth Stations (NLES);
- EGNOS augmentation messages, as normally received by users via the EGNOS Geostationary satellites.

[3] GLONASS data may be discontinued in the future.

The EDAS service was formally launched on 26th July 2012 (EC-DG for Enterprise and Industry 2012). The EDAS Service is provided by EGNOS Service Provider (ESSP) under a contract with the GSA. The EDAS Service Definition Document (SDD) presents the characteristics of the services offered to users, including all the necessary information for users to get access to EDAS services (EC-DG for Enterprise and Industry 2012).

Egnos System Architecture

EGNOS has been designed to meet the international SBAS standards and performs the following tasks:

- Collect GPS/GEO signals and data.
- Estimate the integrity data and Wide Area Differential (WAD) corrections for the service area
  - corrections for the orbital and clock parameters of GPS
  - ionospheric vertical delays over the points (IGP) of a predefined two dimensional grid
  - estimations of residual errors: User Differential Range Error (UDRE) and Grid Ionospheric Vertical Error (GIVE) associated to the corrections.
- Transmit to the users, via the GEO satellites, a GPS-like signal, augmented with integrity and correction messages.
- Verify the correctness of these integrity and correction messages.

To accomplish these tasks, the EGNOS architecture is very complex and highly redundant. It can be broken down in three segments as depicted in Fig. 14.3:

- Ground Segment
- Space Segment
- User Segment

EGNOS Functional Overview

The EGNOS system is a widely distributed and redundant system. Data flows from one subsystem to another subsystem have different level of criticality.

The main EGNOS functions are carried out by the Ground Segment through the following (see Fig. 14.4—red part) critical subsystems: the RIMS stations, the Central Processing Facility (CPF) units and the NLESs stations. The Ground Segment is a periodic synchronous and pipelined system. The synchronisation of subsystems located in widely separated geographic places is referred to the GPS

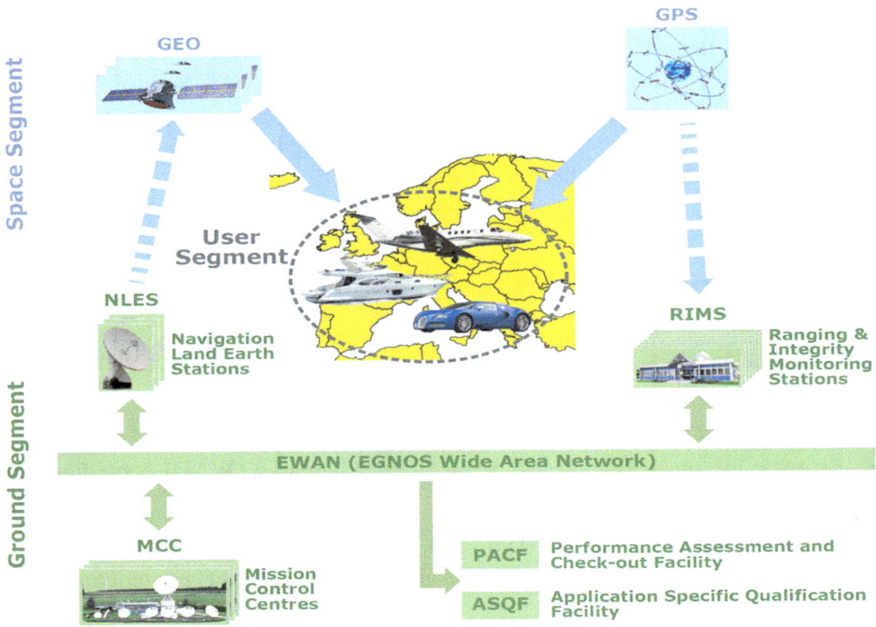

Fig. 14.3 EGNOS system architecture overview (*Source* EC OS SDD (2013))

time by means of GPS receivers providing a one Pulse Per Second (1 PPS) synchronisation pulse to the associated subsystem.

Another feature of the EGNOS Ground Segment is that all Monitoring and Control (M&C) functions performed by humans (non-automatic functions) are centralised and implemented in the CCF subsystem, which is outside of the critical data flow (see Fig. 14.4—green part). Operators on duty in another CCF (hot backup) are ready to take over the system monitoring and control if the master CCF fails. The remaining two CCF (cold backup) can be reactivated if the master one fails.

In red is indicated the real time critical data flow. The subsystems involved in the processing of these critical data are responsible for the achievement of the main EGNOS system performance (accuracy, integrity, continuity, time to alarm, and service coverage).

In green is indicated the non-critical data flow. The subsystems involved in the management of these data perform the Monitoring and Control (M&C) and archive functions.

EGNOS Ground Segment

The Ground Segment is responsible of the main EGNOS functions. It computes precise differential corrections and makes them available to users through a

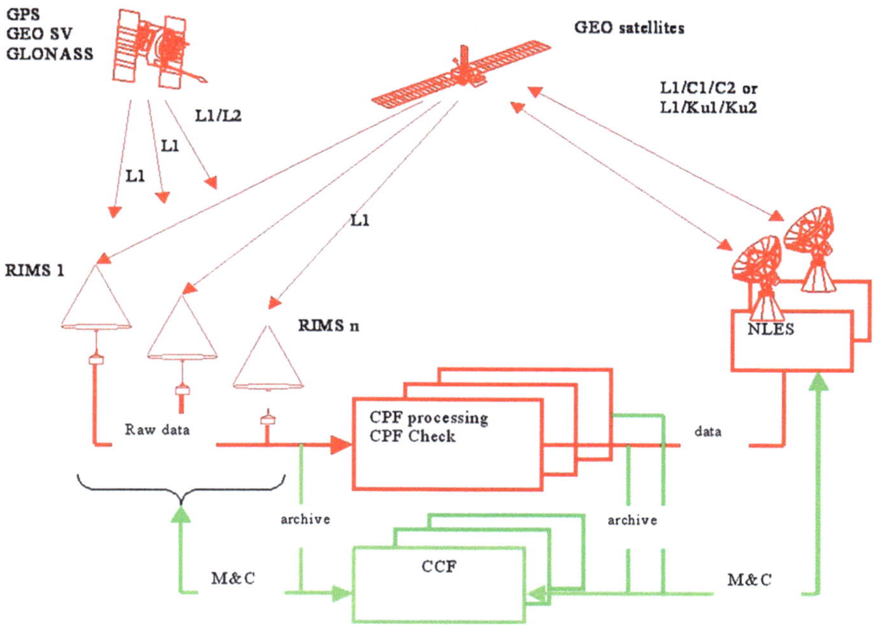

Fig. 14.4 EGNOS data flow (*Source* Chap. 2–2.1 of ESA EGNOS Book (Ventura-Traveset and Flament 2006))

broadcast by the Space Segment. The EGNOS Ground Segment is formed by the following subsystems: the RIMS stations, the CPF and CCF units; the NLES stations and the EGNOS Wide Area Network (EWAN). These subsystems are described here after.

Ranging and Integrity Monitoring Stations

The Ranging and Integrity Monitoring Stations (RIMS) are the ground stations dedicated to the monitoring of the GPS/GEO satellites. RIMS stations may implement two or three independent receiver channels dedicated to different tasks:

- RIMS channel A: it carries out raw measurements on the GPS satellites in view and is connected to the main processing platform (CPF-PS).
- RIMS channel B: it collects the same raw measurements as RIMS A, but in an independent way. It is connected to a parallel processing platform (CPF-CS) dedicated to the verification of the data produced by the main platform.
- RIMS channel C: it monitors and detects rare specific GPS satellites failures named "evil waveform" (EWF) through a multiple correlator implemented in the RIMS receiver. The output of the detection process is then forwarded to the computing platform.

Fig. 14.5 EGNOS RIMS network (*Source* ESSP official website)

Any satellite shall be monitored by a minimum of three RIMS before correction and integrity data can be generated and integrated in the augmentation messages. As the EGNOS performance is strongly relying on the geographical distribution of the RIMS, they have been deployed in some precise location (see Fig. 14.5).

The initial configuration included 34 RIMS stations with the majority of them located in the ECAC region (main service area) in order to ensure an adequate sampling of the ionosphere. Some other RIMS are deployed in remote locations (French Guyana, Canada, South Africa, near the North Pole) to provide a large measurement base to track the geostationary satellites, as well as to have an early visibility of satellites rising from the west and to keep track of satellites going down eastward.

In order to improve the performance of the EGNOS system and enlarge the area where the EGNOS services can be used, an extension of the monitoring network in the southern part of the service area is being achieved with the deployment of additional RIMS. Updated information on the current EGNOS RIMS network may be obtained via the EGNOS operator ESSP official website (2014).

Central Processing Facility

The CPF receives the raw measurements from all connected RIMS stations. The processing module or CPF-Processing Set (CPF-PS) is fed with the RIMS A raw data and generates the Wide Area Differential (WAD) corrections and the integrity data. The verification module or CPF-Check Set (CPF-CS) is fed with RIMS B and C data and carries out independent integrity checks. The EGNOS messages are then sent to the NLES to be broadcast (see Fig. 14.6).

More in detail, the EGNOS computing platform is based on three modules:

- One *Processing Set* module performing:
  - Satellite orbit and time synchronisation determination for all GPS satellites;
  - Ionospheric correction calculation;
  - Constellation and ionosphere integrity surveillance;
  - EGNOS message elaboration;
  - Residual errors (variances) estimation for the produced corrections (UDRE and GIVE).
- Two independent *Check Set* modules. The task of the CPF Check Set is to verify, independently from the CPF Processing Set, the user's positioning integrity and, therefore, to verify the correctness of the messages generated by the CPF Processing Set before and after the broadcast to users who shall be alerted in less than six seconds in case of integrity loss.

There are five identical CPF deployed in the four EGNOS Mission Control Centres (MCC): two in Langen (Germany), one in Torrejon (Spain), one in Swanwick (United Kingdom), and one in Ciampino (Italy). Only three CPFs are

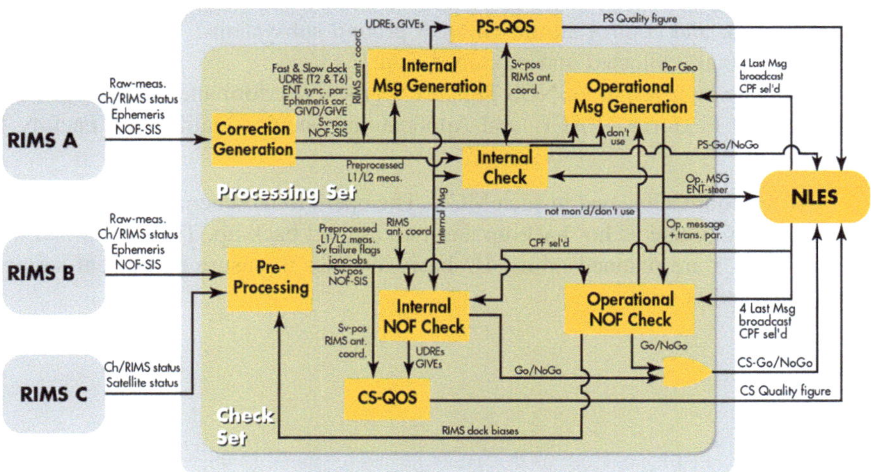

Fig. 14.6 CPF functional architecture (*Source* Chap. 2–2.1 of ESA EGNOS Book (Ventura-Traveset and Flament 2006))

necessary to meet the continuity/reliability requirements. The other two CPFs (including the second CPF installed in Langen) are dedicated to allow integration, validation, verification and qualification activities without introducing service interruptions.

Navigation Land Earth Stations

The NLES stations receive the EGNOS messages from all the CPFs for the upload of the data stream to the geostationary satellites and the generation of the GPS-like signal. For each of the three EGNOS geostationary satellites to upload, two dedicated NLES stations (one active and one as a hot backup) are deployed for data uploading at each of the three EGNOS geostationary satellites. Thus, there are six operational NLES stations in total.

The EGNOS augmentation messages are combined with the PRN codes (assigned to EGNOS satellites) and with additional signals to form a preliminary EGNOS signal synchronized with the L1 frequency before the up-link. The NLES then merges the preliminary EGNOS signals with the up-link signals in C band[4] (the up-link signals contain also other data streams addressed to the satellite). The geostationary satellite, acting as a transponder, filters and shift in frequency the EGNOS signals for the broadcast on the L1 frequency.

Central Control Facility

The Central Control Facility (CCF) is in charge of the monitoring and control of the entire EGNOS system. The main functions are:

- To monitor and control all the EGNOS ground subsystems;
- To archive all collected and produced data;
- To provide interface to EGNOS Support Segment (Performance Assessment and Check-out Platform—PACF- and Application Specific Qualification Facility—ASQF-).

There are four CCFs, one in each MCC. The operational scheme foresees to have one master CCF, one as hot back-up and two as cold back-up. The master and hot back-up MCCs are manned on a 24/7 basis in order to ensure permanent service monitoring and control.

[4] Up-link is in Ku-band for the ESA Artemis EGNOS satellite.

Fig. 14.7 NLES in Torrejon (Spain)

EGNOS Wide Area Network

All EGNOS ground elements are connected through the EGNOS Wide Area Network (EWAN), which uses Multiprotocol Label Switching (MPLS) technology. As EWAN is critical to EGNOS performance, it is permanently monitored and most of the transport links are redundant.

EGNOS Space Segment

The EGNOS Space Segment is composed by three geostationary satellites centred over Europe, which provide redundant data transmission channel to broadcast toward the EGNOS users. Currently, the EGNOS GEO satellites are shown in Table 14.3.

The main criteria followed in the selection of the satellite positions were to improve the measurement geometry and hence the system availability, to provide dual geostationary coverage (minimum) within the core service area and to maximise the visibility angle diversity (and hence minimise the risk of signal blocking).

The geostationary satellites coverage is illustrated in Fig. 14.8 together with the core European Service Area. EGNOS users in the core service area shall be able to track at least two geostationary satellites.

Table 14.3 GEOs used by EGNOS (as per May 2013)

GEO name	PRN number	Orbital slot
INMARSAT 3F2	PRN 120	15.5 W
ARTEMIS	PRN 124	21.5 E
INMARSAT 4F2	PRN 126	25.0 E

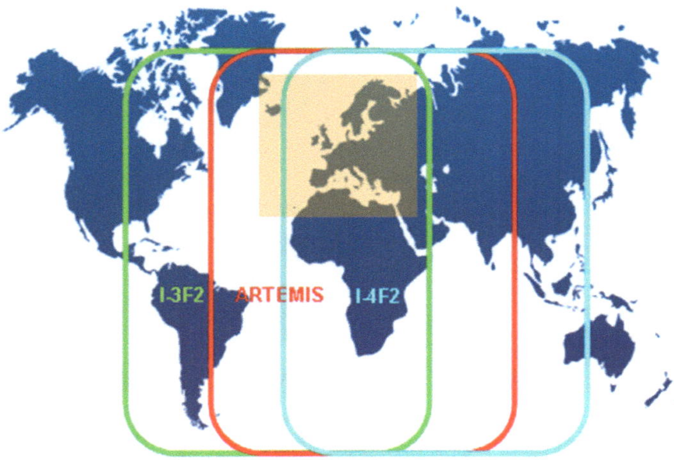

Fig. 14.8 EGNOS satellite coverage (*Source* from ESA EGNOS Book (Ventura-Traveset and Flament 2006))

EGNOS Artemis is planned to be replaced by a new GEO satellite, SES-ASTRA-5 (known also as ASTRA-4B) by mid 2015. INMARSAT 3F2 is planned to be replaced by SES-ASTRA 5-B in 2016.

EGNOS User Segment

Although the EGNOS system has been conceived to meet the stringent requirements of civil aviation users, its unprecedented level of accuracy and reliability, allow the development of a large number of additional applications and EGNOS users. EGNOS users are then today linked to a large number of applications, including: aviation, precision agriculture, maritime, land transportation and location-based services (LBS) (GSA 2014).

For EGNOS Safety of Life applications, EGNOS receivers shall comply with the SBAS Minimum Operational Performance Standards (MOPS) (RTCA 2006), as published by the Radio Technical Commission for Aeronautics (RTCA). Several classes of receivers are defined by the standards depending on the specific intended operations. Details on the EGNOS receivers applicable to aviation are provided in EC-DG Enterprise and Industry (2011).

For non-safety of life applications, a large number of EGNOS-enabled receivers exist on the market, which may be optimized for specific EGNOS applications. Detailed information on EGNOS receivers and a list of available EGNOS receivers in the market may be obtained through the EGNOS Portal of the European GNSS Agency (GSA 2014).

A Specific User Guide for EGNOS Application Developers document has been produced by CNES (French Space Agency) in cooperation with ESA. This guide can be downloaded from CNES (2009).

EGNOS Support Segment

The EGNOS Support segment includes the off-line facilities listed hereafter:

- *PACF* Performance Assessment and Check-out Facility, which provides support to EGNOS management in such area as performance analysis, troubleshooting, operational procedures, as well as upgrade specification and validation and support to maintenance.
- *ASQF* Application Specific Qualification Facility, which provides civil aviation and aeronautical certification authorities with the tools to qualify, validate and certify the different EGNOS applications.

EGNOS Integrity

The EGNOS integrity service should protect users from:

- failures of the monitored GPS satellites (drifting or biased pseudo ranges) by detecting and excluding faulty satellites through the measurement of GPS signals collected by the RIMS network;
- transmission of erroneous or inaccurate differential corrections. These erroneous corrections may in turn be induced from either:
  - undetected failures in the ground segment;
  - processing of reference data corrupted by the noise induced by the measurement and the algorithmic process.

The high-level definition of integrity is provided in SARPs as follows (see ICAO 2004):

> A measure of the trust which can be placed in the correctness of the information supplied by the total system. Integrity includes the ability of a system to provide timely and valid warnings to the user (alerts).

The high-level integrity requirements applicable to SBAS systems, as extracted from ICAO (2004), are summarized in Table 14.4.

Linked to these requirements, the following associated definitions, as extracted from EC-DG Enterprise and Industry (2013), apply:

Table 14.4 SARPs high level integrity requirements on SIS applicable to EGNOS (extracted from SARPs (2004))

Typical operation	Time-to-alert	Integrity Risk	Horizontal alert limit	Vertical alert limit
En-route (oceanic/continental low density)	5 min	10^{-7}/h	4 NM	N/A
En-route (continental)	15 s	10^{-7}/h	2 NM	N/A
En-route, terminal	15 s	10^{-7}/h	1 NM	N/A
Initial approach, intermediate approach, non-precision approach (NPA), departure	10 s	10^{-7}/h	0.3 NM	N/A
Approach operations with vertical guidance (APV-I)	10 s	2×10^{-7} in any approach	40.0 m	50 m

Note NM nautical mile; *N/A* not applicable

- *Integrity risk* the probability that the position error is larger than the alert limit defined for the intended operation and the user is not warned within the time to alert (TTA).
- *Alert Limit* the error tolerance not to be exceeded without issuing an alert. There is a Horizontal Alert Limit (HAL) and a Vertical Alert Limit (VAL) for each operation.
- *Protection levels*
  - The Horizontal Protection Level (HPL) is the radius of a circle in the horizontal plane, with its centre being at the true position, which describes the region which is assured to contain the indicated horizontal position (RTCA 2006).
  - The Vertical Protection Level (VPL) is the half length of a segment on the vertical axis with its centre being at the true position, which describes the region which is assured to contain the indicated vertical position (RTCA 2006).
- *Time To Alert* (TTA) The maximum allowable time elapsed from the onset of the navigation system being out of tolerance until the user equipment enunciates the alert.

The out of tolerance condition is defined in the SARPs in the user position domain (ICAO 2004). Accordingly, formally, a non-integrity event corresponds to the situation obtained when any user Navigation System Error (NSE) in horizontal or vertical dimensions is superior to Horizontal or Vertical Alert Limit (HAL or VAL), *while not providing timely and valid warnings to the user.* This situation (NSE > HAL or NSE > VAL) is often referenced as "Hazardously Misleading Information (HMI)" case.

ICAO SARPs (2004) includes the following complementary definition applicable to the SBAS system designers:

> Given any valid combination of active data, the probability of an out-of-tolerance condition for longer than 5.2 consecutive seconds shall be less than 2×10^{-7} during any approach, assuming a user with zero latency. An out-of-tolerance condition is defined as a horizontal

error exceeding the HPLSBAS or a vertical error exceeding the VPLSBAS (as defined in B.3.5.5.6 in RTCA (2006)).

The Horizontal and Vertical Protection Level (HPL and VPL) are elaborated within the user receiver (ICAO 2004) at each epoch by combining ground transmitted parameters, aircraft parameters and geometry of the user with respect to satellites used in the position calculation. Then, an out-of-tolerance condition will be produced when (NSE > HPL or NSE > VPL). This is often referenced as "Misleading Information (MI)" case.

Since this definition implies the knowledge of the NSE, a standard user may obviously not apply this out of tolerance test to raise a flag in case of non-integrity event. The test to be done at user level to check the correctness of transmitted data is defined in ICAO (2004):

> The receiver shall compute and apply horizontal and vertical protection levels defined in B.3.5.5.6 of RTCA (2006).
>
> … If the computed HPL exceeds the Horizontal Alert Limit (HAL) for a particular operation, SBAS integrity is not adequate to support that operation. The same is true for precision approach and APV operations, if the VPL exceeds the vertical alert limit (VAL).

This test (HPL > HAL or VPL > VAL), which is implemented at each epoch, allows to declare the SBAS "system unavailable" for a given level of operation since in this case the probability of an MI (and HMI) event is high. Note that xPL and xAL (x stands either H or V) are now known by the user.

If a SBAS is SARPs compliant as defined in Sect. 2.2 of ICAO (2004), then a user applying the above test will be protected to the required level.

The three above discussed integrity tests (HMI, MI and system unavailable) appear more explicitly in Fig. 14.9.

Another practical representation of these different cases is obtained through a 2D plot of the Vertical Position Error (VPE) against the VPL where each pixel corresponds to a measurement epoch as in Fig. 14.10. This is usually known in Europe as the Stanford diagram. As an example, Fig. 14.10 illustrates the trade-off between integrity and availability (Stanford diagram) as obtained through EGNOS real measurements at ESA in Toulouse (France), on 23rd April 2012. The diagonal traces the limit between the safe operation of the system (left side) and the unsafe conditions (right side).

The EGNOS System is shown to be safe in the nominal test conditions of Fig. 14.10, with an availability of APV-1 of 100 % for this specific test period.

When focusing on integrity, an interesting modification of the all in view Stanford diagram is the *Stanford-ESA Integrity diagram* (Ventura-Traveset and Flament 2006), where for a specific user location and period interval, every computed sample represents the worst possible geometrical integrity situation for that user at that location. Thus, if no integrity risk violation is observed with the Stanford-ESA Integrity diagram computation at a given location and for a given period of time, we may certainly conclude that for all possible GPS satellite geometries and SBAS information that could potentially be used in that location

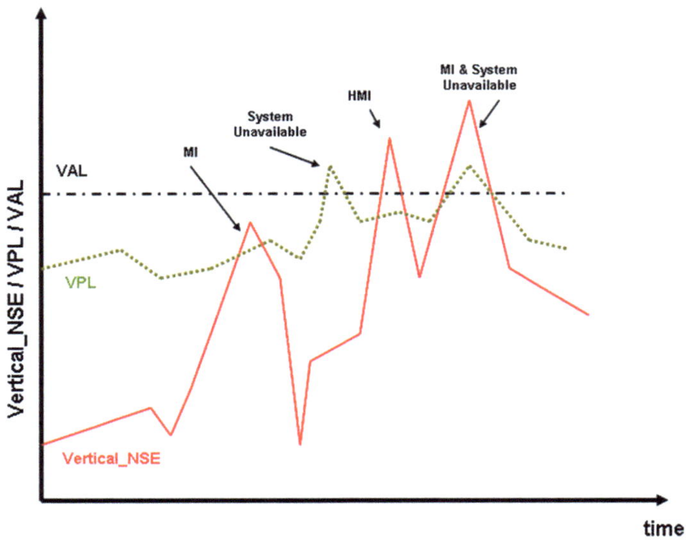

Fig. 14.9 Example of the different non-integrity definitions and tests (*Source* ESA EGNOS Book (Ventura-Traveset and Flament 2006))

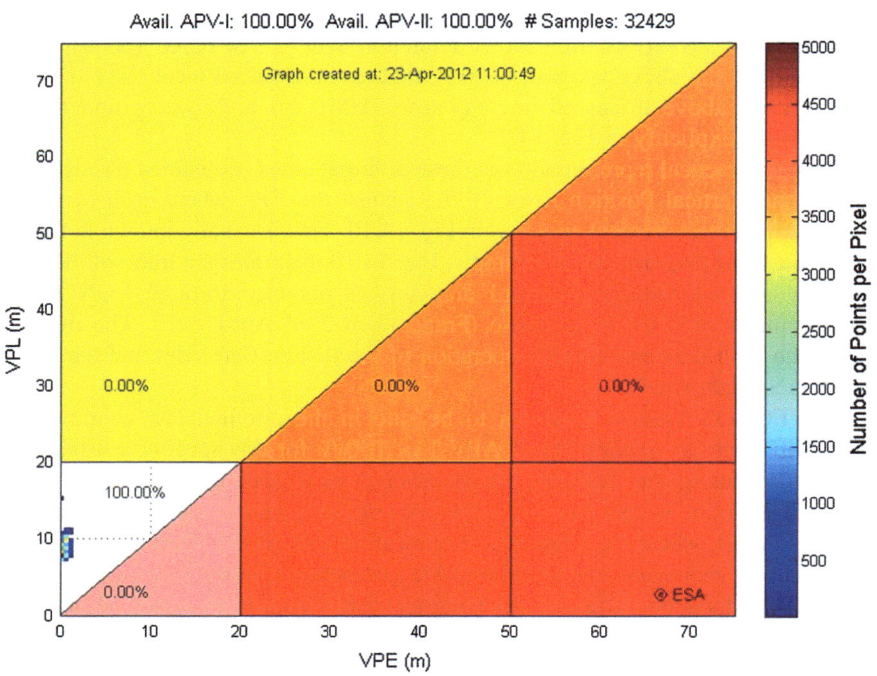

Fig. 14.10 EGNOS measured Stanford diagram for APV (23rd April 2012, Toulouse)

and for that period, the system was safe (please refer to Chap. 1–1.5 in Ventura-Traveset and Flament (2006) for more details).

For a more complete elaboration on EGNOS Integrity and on the way Protection Levels are computed, the reader is referred to Ventura-Traveset and Flament (2006).

Ionosphere for SBAS Systems

For GPS single-frequency user receivers in good visibility environments, the largest remaining satellite positioning error source is linked to the ionosphere delay. Thus, the provision of ionospheric corrections and of associated integrity bounds is a major role for SBAS systems. Given the importance of this error component in overall EGNOS performances, we discuss here briefly about the origin of this error and the EGNOS correction strategy to cope with that. Most of the information provided here is extracted from the ESA EGNOS Book (Ventura-Traveset and Flament 2007), for which the reader is referred for more details.

The ionosphere extends from a bottom height of about 50 km out to about 2,000 km. The extension of the upper ionosphere along the Earth's closed magnetic field lines is known as the plasmasphere, which reaches all the way to the orbital height of GPS. The charged particles are created by the solar radiation and their density shows a strong diurnal variation (Davies 1990). In addition, the Sun undergoes a cyclic variation with an average period of 11 years (SIDC 2011) (Fig. 14.11).

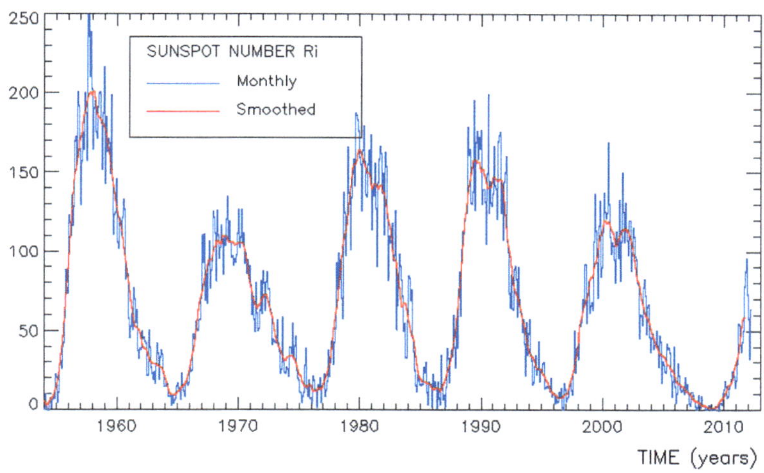

Fig. 14.11 Monthly averaged Sunspot numbers from 1950 until 2012 (*Source* Solar Influence Data Analysis Center (2011))

Ionospheric Propagation Effects on GNSS Signals

The ionosphere is a dispersive medium, which means that the group delay of a wave travelling though depends on the wavelength (Hofmann-Wellenhof et al. 2001).

The group-delay expressed in meters is given in Eq. (14.1):

$$\Delta s = \frac{40.3 * TEC}{f^2} \text{ [m]} \tag{14.1}$$

where $TEC = \int Ne\ ds$ is the Total Electron Content and f is the carrier frequency in [Hz].

The delays at frequencies used by GPS (at 1575.42 MHz) can reach up to 25 m in the vertical (ITU-R 2005).

Because of the dispersive nature, dual frequency receivers should be able to correct for the first order component of the ionospheric delay by estimating the Total Electron Content (TEC) from the differential delay seen between the two signals. Single frequency receivers GPS users, instead, use model estimation (the Klobuchar model (1986)) of the actual state of the ionosphere to make a correction. In Satellite-Based Augmentation Systems, the correction of the ionospheric delay is done much more accurately by transmitting a grid of vertical ionospheric delay values. This is explained in the following section.

The Ionospheric Correction in EGNOS

In EGNOS, the RIMS stations observe all visible GPS satellites and assess potential error sources that could affect the navigation accuracy of the users inside the service region. For the estimation of the ionosphere, the receivers make use of both GPS frequencies, L1 and L2.

Assuming that both signals have propagated along the same path and have therefore encountered the same Total Electron Content, the differential delay between L1 and L2 can be used to establish this TEC value. In the EGNOS CPF, all the slant TEC values from all the RIMS are then converted to a grid of vertical TEC values (at a defined height of 350 km) whereby the typical grid point spacing is 5° in longitude and latitude (in the Polar regions the grid is thinned out to 10°). The vertical TEC value is then converted to vertical delay at the L1 frequency, characterised by a value known as the Grid-point-Ionospheric Vertical Delay (GIVD). In addition, the CPF calculates also an error bound for these delay values, which depends on the quality of the observation (how many links contribute to the grid-point value), the observed gradients and other factors. This error bound is characterised by the so known, Grid-point-Ionospheric Vertical Error Bounding (GIVE). GIVD and GIVE values are being transmitted to the user at a typical update rate of about 5 min.

The user estimates the ionospheric delay on each satellite link as follows: First the location of the "pierce-point", the point where the received signal crosses the 350 km imaginary thin-shell, is established. Then, using the GIVDs of the adjacent grid points, the receiver performs an interpolation to calculate the ionospheric vertical delay at the pierce point. Finally, using an obliquity function, the receiver calculates the slant ionospheric delay. The associated errors are calculated in a similar manner and the resulting error is used for deriving the integrity of the solution.

One of the key challenges of EGNOS qualification and certification has been to demonstrate that the system is robust and provides the adequate performances (notably integrity and availability) against the worst possible ionospheric conditions that the user may face on the EGNOS ECAC service area.

EGNOS SiS and EGNOS Messages

SBAS Signal in Space (SiS) and SBAS associated Messages are specified in detail in ICAO (2004) and RTCA (2006).

EGNOS Signal RF Characteristics

The EGNOS GEO satellites transmit right-hand circularly polarised (RHCP) signals in the L band at 1575.42 MHz (L1). The broadcast signal is a combination of a 1023-chip PRN navigation code of the GPS family and a 250 bits per second navigation data message carrying the corrections and integrity data elaborated by the EGNOS ground segment (EC-DG Enterprise and Industry 2011).

EGNOS SIS is such that, at all unobstructed locations near ground level from which the satellite is observed at an elevation angle of 5° or higher, the level of the received RF signal at the output of a 3 dBi linearly polarised antenna is within the range of −161 to −153 dBW for all antenna orientations orthogonal to the direction of propagation (EC-DG Enterprise and Industry 2011).

Signal Data Structure

SBAS interacts with the user via the Signal in Space. The way the SBAS delivers to the user the aforementioned corrections and integrity data as well as some ancillary information (timing, degradation parameters, etc.) is through messages encoded in the signal. These messages are sent each second with a data rate of 250 bits.

Table 14.5 SBAS message format (components)

Position in message	Name	Purpose
0–7	Preamble	Assure frame synchronisation
8–13	Message type identifier	Define the type of message
14–225	Data field	GIC/WAD information
226–249	Parity information	Redundancy and error checking

The raw navigation message of the SBAS contains 500 bits. These raw data are ½ convolutional encoded with a Forward Error Correction (FEC) code, which means that 250 bits of information are available every second at user level.

The 250-bit message has different parts, including an 8-bit preamble and 24 ancillary bits to include redundancy and error checking within the message.

Table 14.5 and Fig. 14.12 summarise the message format. Bit 0 is considered the most significant bit, i.e. the bit that is transmitted and received first.

A brief explanation of the different Data Fields is presented hereafter:

- *Preamble*. It is a unique 24–bit field, distributed over three successive words. The preamble sequence (in three consecutive words) is 01010011 (83)—10011010 (154)—11000110 (198). It is assured that the start of the preamble is synchronous with a 6-second GPS sub-frame. Thus, the preamble allows the receiver to achieve frame synchronisation.
- *Message Type Identifier*. It is a 6-bit field, which permits up to 64 different messages (identifiers 0–63). Depending on the message type, the information included in the Data field has different meaning. Even though 64 type messages are available, only 20 are currently defined.
- *Data Field*. It contains different corrections and integrity information that depends on the type of message.
- *Parity Information*. The 24 bits of the end of the message (CRC parity bits) provide protection against burst and random error (for more information on parity bits algorithm generator refers to RTCA (2006)).

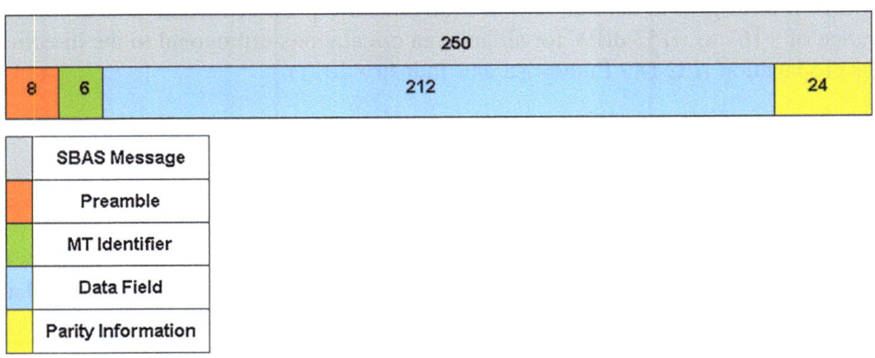

Fig. 14.12 SBAS Message format (lengths in bits) (*Source* ESA EGNOS Book (Ventura-Traveset and Flament 2006))

Messages are interrelated using the Issue of Data parameters (IOD), which are present in the message data. Also satellite messages are related with satellite navigation services ephemeris via the issues of data.

The sequence of transmission of the messages is not fixed and is responsibility of the SBAS Service Provider (each second the provider decides which message is to be sent). There are requirements in MOPS concerning the refresh time for each type of information and about the alarm conditions (problems with one or more satellite information or with ionospheric information) (RTCA 2006). Under an alert condition, the SBAS System must repeat the message with the alert information three more times (i.e. four messages in 4 s in total), so to ensure adequate reception at user level.

EGNOS Message Types

SBAS messages have a 6-bit message type identifier, which informs the receiver about the information the message holds. Due to the limited size of the type identifier (6 bits), 64 Message Types (MT) are possible, from which, 20 MT are currently defined and 44 MT reserved for future messages (RTCA 2006).

In the case of EGNOS, today, it may transmit up to 18 different messages types. The following table, extracted from EGNOS Official SoL SDD Document (EC-DG Enterprise and Industry 2011), summarises the current EGNOS message types and the contained informations are shown in Table 14.6.

Table 14.6 EGNOS SIS transmitted message types (Source EC EGNOS SoL SDD (2013))

Message type	Purpose
0	Discard any ranging, corrections and integrity data from that PRN signal. Used also during system testing
1	Indicates the slots for GPS and GEO satellites provided data
2–5	Range corrections and accuracy
6	Accuracy for all satellites in one message
7	Information about the degradation of the fast term corrections
9	EGNOS satellites orbite information (ephemeris)
10	Information about the degradation of the long term corrections
12	Parameters for synchronisation of SBAS network time with UTC
17	GEO almanacs
18	Indicates for which geographical point ionospheric correction data is provided
24	Fast-term error corrections for up to six satellites and long-term satellite error correction for one satellite in one message
25	Corrections for satellite ephemeris and clock errors for up to two satellites
26	Vertical delays/accuracy at given geographical points
27	Defines the geographic region of the service
63	Filler message if no other message is available

Detailed information on the content of the listed MT and their specific use by EGNOS users is provided at ICAO SARPs (2004) and RTCA SBAS MOPS (2006).

EGNOS Operations

On the 30th of September of 2009, the European Commission contracted the Operation of EGNOS and the SoL service Provision for a period of 51 months to a consortium composed of:

- The European Satellite Services Provider (ESSP), acting as the Consortium Prime Contractor. The ESSP was founded in 2001 as an European Economic Interest Grouping (EEIG), and evolved to a Limited Liability Company in 2008, in order to prepare for the provision of the SoL service.
  The shareholders of the ESSP are the Air Navigation Service Providers of France (DSNA), Germany (DFS), Italy (ENAV), Portugal (NAV-EP), Spain (AENA), Switzerland (Skyguide) and UK (NATS) (ESSP 2010).
  The ESSP has been involved in the operation of EGNOS since its foundation. In July 2005, ESA assigned the contract for the EGNOS Initial Operations Phase, which resulted in the qualification of the ESSP to operate EGNOS in April 2009. Since the launch of the Service Provision contract, the ESSP has been preparing its certification as an Air Navigation Service provider (ANSP). This certification was achieved in July 2010 (ESSP 2010).
  After a competitive dialog process, in June 2013 the ESSP signed with GSA a contract for the provision of EGNOS services until 2021.
- The companies in charge of the operation of the MCCs:
  - AENA, for the MCC of Torrejón.
  - DFS, for the MCC of Langen
  - ENAV, for the MCC of Ciampino
  - NATS, for the MCC of Swanwick.
- The hosting entities in change of the maintenance of each of the NLES and RIMS sites.

EGNOS System Operation

The Operation of EGNOS as implemented by the ESSP is organized through the following functions:

- Control Function:
  This function is mainly carried out by the MCC Operators through the CCF MMI. It consists of commanding remotely the different EGNOS assets to ensure that all the subsystems remain in nominal conditions.

The function is shared by the four MCCs (in Langen, Swanwick, Ciampino and Torrejón) in a rotation scheme. The Master MCC, in charge of the Control of EGNOS, is staffed in a 24/7 shift. In addition, there is always a Back-up MCC ready to assume the Master role in case of any issue in the MSR MCC.

• Engineering Function:
This function is mainly carried out by the ESSP, through two engineering facilities:
  - The System Operation Unit (SOU):
    Located in Toulouse, France, the SOU is in charge of all the engineering activities associated to ensure the day-to-day life of the System: they schedule the planned operations with the hosting entities, they manage the observations on the system behaviour, they coordinate the investigation of anomalies, and they perform the isolation of HW failures.
  - The Service Provision Unit (SPU)
    Located in Madrid, Spain, the SPU is in charge of the assessment of the service provided to user communities, in particular the level of performance achieved by the SoL service Maintenance Function.
    The execution of this function is the result of the close cooperation of the operators in several facilities.

The corrective maintenance cycle can be summarized as follows:

1. The failure is detected by the Master MCC Operator, who verifies that the faulty asset has been removed from the computation chain. Then, the Operator executes some first-level recovery procedures with the support of the hosting entity of the affected site. If the failure persists, it is reported to the maintenance team of the SOU.
2. After retrieving the information needed to characterize the failure, the SOU maintenance team isolates the equipment to be replaced and organizes the shipment of a spare from the Central Warehouse to the hosting entity.
3. Once the spare arrives on-site, the hosting entity replaces the faulty equipment. This is supervised by the SOU and the Master MCC. When the activity on-site is concluded, the MCC reintroduces the asset into the computation chain.
4. The previous three steps take usually around three working days for EGNOS assets in Europe. EGNOS has been designed with the sufficient level of redundancy to ensure that this time to recover a failure is compatible with the SoL service levels.
5. The faulty equipment is shipped to the manufacturer for repair. Once it is repaired, it will be stored in the Central Warehouse.

The preventive maintenance is organized in a similar way. The major difference is that in this case the trigger of the maintenance activity is not a failure in the HW, but an activity planned in the operations schedule. The management of preventive maintenance is carried out at the SOU.

- Safety Function:
  In order to provide a SoL service, the ESSP has enforced a strong Safety System within its organization and within the organization of their subcontractors. The core of the safety activities is carried out in the ESSP Headquarters in Toulouse, although safety responsibles are nominated at all the EGNOS facilities.

EGNOS Performances

In June 2006, EGNOS started broadcasting in MT0/2 mode,[5] meaning that users all over Europe could start profiting from GPS augmentation for non-SoL applications. Since then, EGNOS SiS has been available almost 100 % of the time, with only a few minor interruptions during the test phase.

The EGNOS version running when the European Commission declared the SoL service was the V2.2-ext. This version was qualified in 2008 and then deployed by the ESSP. The test phase to remove the MT0/2 and start the SoL mode was carried out during the second half of 2010. Finally, on the 2nd of March 2011 the EC declared the SoL service available to users.

As an indication of the performance achieved with V2.2-ext, the following figure shows the average APV-I availability measured during January 2011.

At this time, EGNOS signal was broadcast through PRN 120 (INMARSAT 3F2) and PRN 126 (INMARSAT 4F2), and a network of 35 RIMS was used.

It is noted that in January 2011 MT0/2 had already been removed and the System was broadcasting in SoL mode.

Next figure confirms the level of performance expected from V2.2-ext: 99 % of availability of APV-I service in most of the land-masses of Central and Western European Civil Aviation Conference (ECAC), with a continuity risk in the order of $10^{-4}/15$ s. This level of performance has been considered as sufficient by EUROCONTROL and other European ANSPs to allow the implementation of approach procedures based on EGNOS (Fig. 14.13).

For the reader's reference, a very complete summary on EGNOS System performances for the yearly period from the 1st of April 2013 to the 31st of March 2014, as produced by the ESSP EGNOS official operator, is available in (EGNOS Service Provision Yearly Report 2013–2014). EGNOS monthly performance reports are also regularly produced by the ESSP and available via ESSP (2014).

---

[5] EGNOS SIS has three modes of operation: (1) Test Mode: This mode is defined by the broadcasting of the MT0 (Message Type 0) in the signal. It indicates that the system is on test and is not usable. (2) Non-Safety-Of-Life Mode: This mode still indicates that the system is on test, but allows receivers to compute EGNOS corrections, as long as the signal is only used for non-safety-of-life applications. The mode consists of overlapping a message type 2 on the frame reserved for the message type 0 (mode known as MT 0/2). (3) Safety-Of-Life Mode: EGNOS SIS may be used for Safety of Life applications. In this case, no message Type 0 is broadcast and message type 2 is broadcast normally (ESSP 2010).

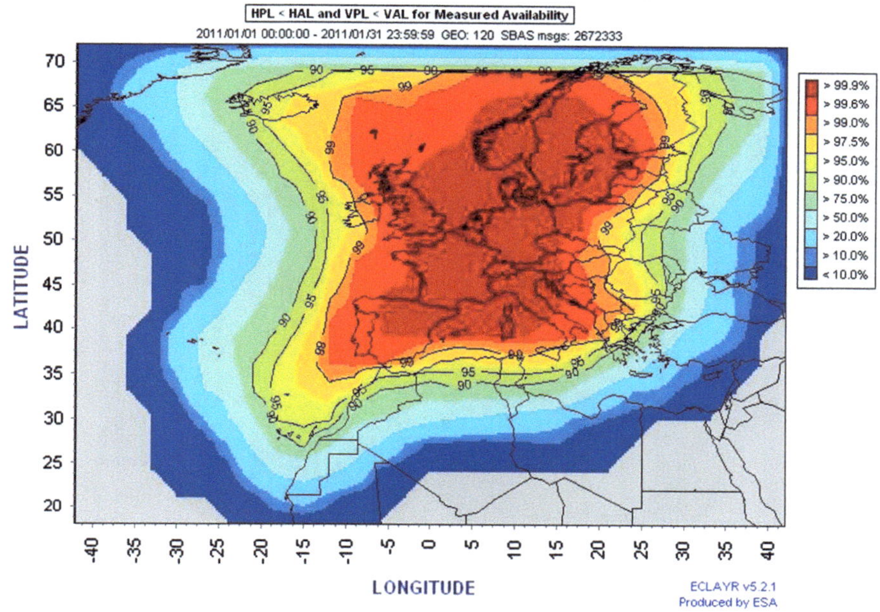

Fig. 14.13 APV-I Availability in January 2011 with V2.2-ext (PRN 120)

In parallel to the V2.2-ext lifecycle, ESA has continued working to improve the performance offered by EGNOS. In fact, EGNOS programme has been organised to ensure a progressive evolution of the System. The feasibility and definition phases of a new EGNOS version are already launched before the previous release is already qualified. This allows reducing the lead time between successive EGNOS versions, speeding the implementation of System evolutions.

The following releases have been produced since V2.2-ext as a result of this "continuous-evolution approach":

Version V2.3.1p. This version included an extension of the RIMS network to the South of ECAC (new RIMS of La Palma in Spain, Athens in Greece and Alexandria in Egypt), plus an optimization of the algorithms in charge of the ionosphere corrections. The deployment of this version was carried out by ESSP during February 2012.

Version V2.3.1i. This version improved the system robustness to the ionosphere activity, in response to the increasing solar activity expected at the peak of the solar cycle (foreseen around 2013-2014). The deployment of this version was done beginning August 2012. V2.3.1i was broadcast through EGNOS GEOS PRN 120 and PRN 126.

Version V2.3.2. This release included an extension of the coverage area, thanks to the addition of 2 new RIMS (Agadir in Morocco and Abu-Simbel in Egypt), the update of the EWAN technology (from frame relay to MPLS) and a further optimization of the iono algorithms. The deployment of this version was done in

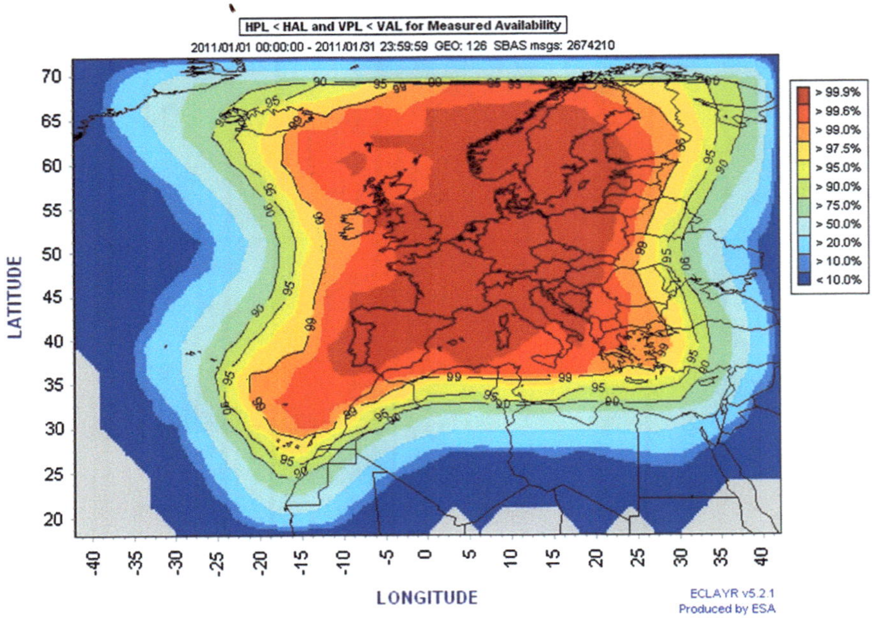

Fig. 14.14 APV-I Availability in January 2011 with V2.3.1p (PRN 126)

October 2013, and is the current release at the time of writing (July 2014). V2.3.2 is broadcast through EGNOS GEOS PRN 120 and PRN 126.

Figure 14.14 shows the average APV-I availability performance achieved with V2.3.1p, as measured during January 2011 with PRN-126 (Test Platform). This period is the same as the one reported with V2.2-ext in Fig. 14.13 to ease the comparison.

As it can be observed, the improvement in the coverage area of EGNOS achieved with V2.3.1p with regard to V2.2-ext was significant.

In the meantime, a new release is already under preparation:

- V2.4.1 M. This release introduces the modernisation of the NLES to prepare them to the new generation of Geostationary satellites (L1 and L5), the replacement of ARTEMIS, PRN 124, by ASTRA-4B, and the resolution of the Obsolescence of the CCF. The deployment is expected by mid 2015.

For the actual dates on the deployment of these releases, the reader is directed to the official ESSP EGNOS operator information (ESSP 2014).

Other SBAS Systems and Interoperability

In addition to EGNOS, there are other SBAS systems in operations or under development. The Wide Area Augmentation System (WAAS) in USA was formally

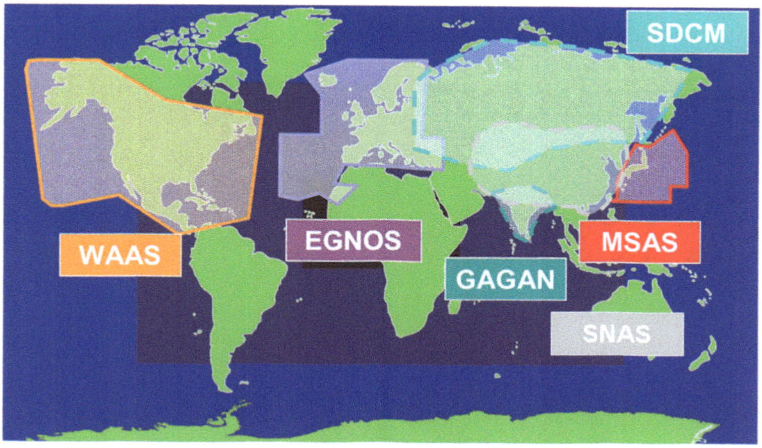

Fig. 14.15 Existing and under development SBAS Systems

commissioned in July 2003 (FAA 2010). In Japan, the Multi-functional transport satellite (MTSAT) Satellite-based Augmentation System (MSAS) provides certified SBAS services since September 2007. Three other SBAS systems are currently under development: in India the GPS Aided Geo Augmented Navigation (GAGAN 2012); in Russia the System of Differential Correction and Monitoring (SDCM 2013); and in China the Satellite Navigation Augmentation System (SNAS). Figure 14.15 indicates the expected coverage areas of those existing and planned SBAS systems.

Although all SBAS are currently defined as regional systems, they are all designed against the same ICAO and RTCA standardisation requirements (see ICAO 2004; RTCA 2006); what has been known as interoperable systems. This interoperability ensures that the same user receiver may work in any of the SBAS service areas. This fundamental point was well understood from the very beginning by the SBAS designers, which decided to create the SBAS Interoperability Working Group (IWG). The group addresses commonly any possible different interpretation of the SARPS ICAO and RTCA documents, exchanges technical lessons learned on the systems development and agrees on the set of common assumptions against which different SBAS performances could be measured. The IWG has been meeting in an uninterrupted way since 1997 and is now integrating the new SBAS emerging systems.

Assuming all SBAS systems become finally operational, there are three additional potential SBAS evolutions that could lead to achieving a true seamless worldwide global SBAS navigation system. First of all, existing SBAS systems have plans to extend their service areas to neighbouring regions with the provision of additional reference stations; secondly, it could be expected that SBAS system do evolve to augment, in addition to L1, the new L5 Aeronautical Radio Navigation Service (ARNS) protected band from modernised GPS-IIF and GPS-III satellites;

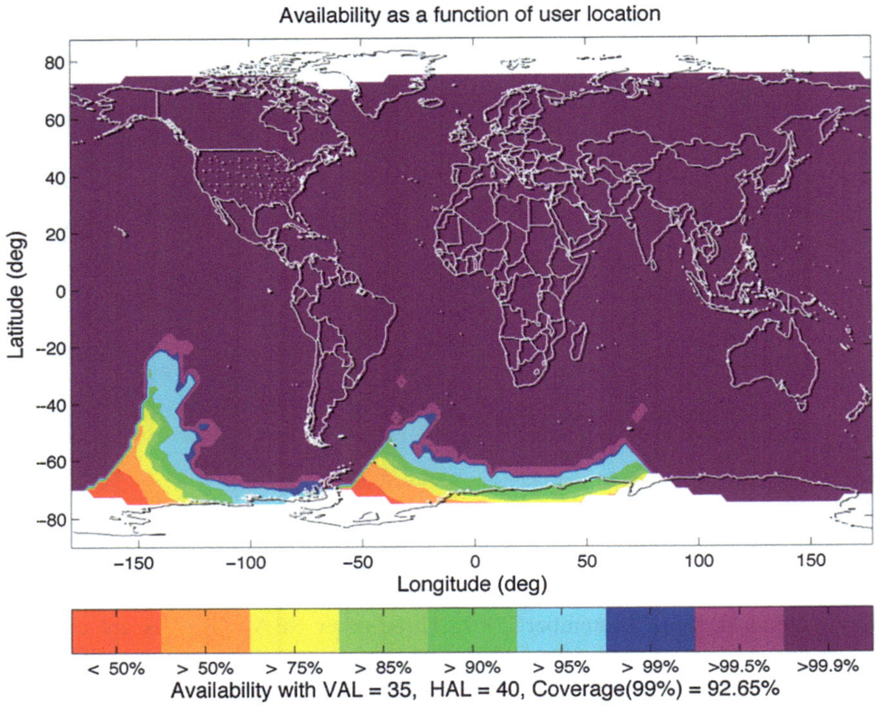

Fig. 14.16 LPV-200 availability maps assuming extended SBAS systems, dual frequency and dual constellation systems (*Source* Walter et al. 2010)

finally, it could also be conceived that, similar to GPS, SBAS systems do also include in the future the possibility of augmenting Galileo satellites. This could then result on a dual constellation, dual frequency and global SBAS system.

Walter et al. have analysed in (2010) this potential final evolution scenario in terms of LPV-200 service provision. Figure 14.16 shows the results, where it is assumed that all identified SBAS system are operational, additional stations added in the southern hemisphere and the augmentation of GPS and Galileo dual frequency constellations. As we may see, the whole Earth is covered with LPV-200 provision with an availability above 99.9 %, with the only exceptions of the very North and South latitudes (beyond 70° North and 70° South), where GEO SBAS augmentation messages are not reachable. Further, this limitation could potentially be resolved with the provision of non-GEO broadcasting such as MEO or IGSO SBAS broadcasting, as proposed in Airbus (2009).

EGNOS Potential Evolutions

The SBAS and their associated ground infrastructure will exist and will be operated at least until 2028–2030 because, as a key mission objective, the current augmentation services will be maintained for those users equipped with mono frequency (L1) SBAS/GPS receivers until end of life of their equipment (typically 15 years). Yet, over this period, the SBAS will have to evolve for several reasons which are the following:

- Intrinsic obsolescence of some of the EGNOS infrastructure elements;
- Evolution induced by an evolving GNSS environment (e.g. GPS, WAAS, GALILEO and other new constellations);
- Need to pursue the improvement of system operability, and/or the reduction of operation costs;
- Opportunity of new services for the various communities.

These different potential drivers of EGNOS evolution are currently under assessment by ESA and the EC, together with the elaboration of a suitable infrastructure evolution strategy, which could become the basis of EGNOS new generation (EGNOS V3). Some of these evolutions drivers are briefly discussed here after.

Evolutions Induced by GPS Signal Changes

With regard to signal evolution, EGNOS presently uses two of the GPS legacy signals:

- L1 C/A signal: no EGNOS evolution is necessary with respect to this signal, as there is no US Government plan known today to change it.
- L2 P(Y) signal: for this signal, not used by the EGNOS user but necessary for the EGNOS System to monitor the ionosphere, the US Government only commits to maintain its characteristics until 2020, when the second and third civil signals (L2C and L5) are planned to be broadcast from a minimum of 24 GPS satellites. After that date, the characteristics of the L2 P(Y) signals transmitted by any or all GPS satellites broadcasting two or more civil-coded signals may change without further notice and may preclude the use of this signal.

EGNOS will be adapted to cope with L2P(Y) suppression by 2020, so as to ensure the legacy service for mono frequency users beyond that date.

Evolutions Induced by Interoperability with WAAS

WAAS is certified in operation since 2003 and supports precision approaches with vertical guidance up to 250 feet above the runway. With the implementation of short term upgrades until 2013 the WAAS LPV performance will be enhanced down to 200 feet decision height, the so called LPV200 operation. Following the introduction of WAAS LPV-200, the Federal Aviation Administration (FAA) plans to start introducing an operational dual frequency augmentation service in the L1/L5 bands by 2020 (while maintaining the L1 only legacy service until 2028). This new service will be based on a new user receiver standard to be built by 2015.

Maintaining interoperability with WAAS, in terms of performances and functionalities, could lead EGNOS to also evolve towards an LPV200 dual frequency augmentation service by 2020, while still maintaining the L1 only legacy service until at least 2028. It has to be noticed that the dual frequency EGNOS service could also allow to provide extended LPV200 service over large regions like Africa with a minimum number of RIMS to deploy.

Opportunities with New Constellations: Multi SBAS Dual Frequency Multi Constellation

New constellations broadcasting highly interoperable signals with GPS in L1/L5, like GALILEO, are now being deployed which will become operational in the coming years. When combined together with SBAS/GPS in a global navigation solution, these new constellations could offer new opportunities, and new services. Potential evolutions of EGNOS in this context are currently under assessment by ESA and the EC. An important point to be taken into account is that even if the SBAS systems evolve in this direction, thanks to the backward compatibility of SBAS standard legacy services for L1 only users will be maintained in each SBAS service area.

The performance that the multi SBAS dual frequency multi constellation concept could offer to equipped users has already been discussed and it is illustrated in Fig. 14.16.

Potential Evolutions Linked to Operations

A new EGNOS design will also be an opportunity for investigating in depth operations concept enhancements, such as:

- Further automation of operations;
- Optimisation of built-in redundancies in the system architecture;

- Optimisation/simplification of HW diversity requirements, when compatible with Reliability Availability, Maintainability and Safety (RAMS) constraints, so as to facilitate the management of spares and organization of logistics, as well as to reduce the costs associated to maintenance;
- Streamlining of integration, qualification and deployment processes, with minimised impacts on operations;
- Reduction of re-qualification costs and schedule for new releases
- Synergies of EGNOS Operation with Galileo GS Operation;
- Adaptation of performance monitoring to user communities different from civil aviation ones.

EGNOS V3 Studies

The evolutions discussed above and the possible adaptation of EGNOS, what has been known as EGNOS V3, are currently (May 2013) under the definition phase (extended phase A an phase B) by ESA, in close coordination with the European Commission, in charge of the overall EGNOS programme management. These studies will provide additional elements which will support programmatic decisions on the evolution path of EGNOS.

Learning EGNOS

ESA has developed tools and educational material for students and engineers interested in learning on SBAS systems in general or more specifically on EGNOS (2012). Within the academic context of this book, it is considered of interest to make available here this information:

- *SISNeTlab* The SISNeTlab tool is an EMS (EGNOS Message Server)-based application, which provides users with hands-on experience of the EGNOS information messages broadcast. Free download at ESA (2002).
- *SBAS TeACHER* The SBAS TeACHER allows decoding and encoding SBAS by means of user-friendly interface components. Free download at ESA (2000).
- *SBAS Message Generator* (SBAS MeNTOR) An extension of the SBAS TeACHER tool. Where TeACHER enables users to modify contents of SBAS messages in visual way but one at a time, the SBAS MeNTOR enables users to define the desired values and states of the whole SBAS and let the software generate all the specific messages of the selected period. Free download at ESA (2008).
- *SBAS Simulator* This simulator gives approximate EGNOS performance based on simplified macro models for ECAC region (Free download at ESA (2010).

- *The EGNOS Book* "EGNOS—A Cornerstone of Galileo", is a valuable resource for SBAS students, teachers and other GNSS professionals. The EGNOS book presents a complete view of the EGNOS mission, system, architecture, performance and applications. Some of the information presented in this chapter has been extracted from this book. More information at Ventura-Traveset and Flament (2006).
- *NAVIPEDIA* (http://www.navipedia.org) NAVIPEDIA is an initiative of ESA aiming at becoming the reference for GNSS general knowledge on the Internet (ESA NAVIPEDIA 2012). It is conceived as a collaborative GNSS encyclopaedia with the objective to foster the transfer of knowledge in this field. EGNOS has a dedicated category in Navipedia.

Summary

EGNOS is the European Satellite Based Augmentation System (SBAS) designed to comply against the corresponding ICAO International Standard and Recommended Practices (SARPs 2004). EGNOS provides an augmentation to GPS based on the broadcasting through GEO satellites of GPS-like navigation signals containing integrity and differential correction information applicable to the navigation signals of the GPS satellites. EGNOS augments GPS using the L1 (1575.42 MHz) civilian signal function.

The first generation of the EGNOS system was developed by the European Space Agency and handed over to the European Commission on 1st April 2009. EGNOS is owned by the European Union, with the European Commission being in charge of the overall EGNOS programme management and ESA being in charge of the design and procurement agent for system evolutions. EGNOS Open Service were formally open on 1st October 2009 and on 2nd March 2011 Safety of Life services did start, culminating an effort of 15 years of work by European industry, European civil aviations, Eurocontrol, the European Commission and the European Space Agency.

The performances of EGNOS are excellent, allowing a large multi-modal user community that includes Civil Aviation, Maritime and Land users; liability services; agriculture; and a large variety of location-based services.

EGNOS is part of a multi-modal inter-regional SBAS service. Other SBAS systems are currently operational in USA (the Wide Area Augmentation System—WAAS) and Japan (the Multi-transport Satellite based Augmentation System—MSAS). Three additional SBAS systems are currently under development in India, Russia and China.

EGNOS potential evolutions will be closely harmonised with ICAO standards and the evolutions of the other existing SBAS systems. In this sense, it is envisaged that long-term SBAS could include dual constellation and dual frequency augmentation services, which together with the new emerging SBAS systems could lead to achieving a true seamless world-wide global SBAS navigation system.

Acknowledgments The success of the EGNOS System development and operations is the result of a collective work of many engineers and staffs from several European organizations, including European space industry, Eurocontrol, the European Commission, the European GNSS Agency, National Space Agencies, National Civil Aviations, the ESSP-SAS EGNOS services provider and the European Space Agency, which the authors wish to acknowledge here.

Disclaimer

This chapter is produced for academic purposes only. The characteristics of the service offered to users by EGNOS are provided in the European Commission EGNOS Service Definition Documents. EGNOS service provision is subject to specific disclaimers of liability as noted in EC-DG Energy and Transport (2013). Official information on EGNOS operational status and performances is provided by the EGNOS Operator, ESSP (2014). A specific Helpdesk service is available through the EGNOS operator by-email (egnos-helpdesk@essp-sas.eu) or by phone for any question concerning EGNOS use ESSP (2014).

References

Administration (2008) (ref. "Results from a GPS Timing Criticality Assessment James Carroll, DOT/RITA Volpe Center April 2008")

Airbus (2009) Satellite landing systems—presentation at RTCA February 2009

CNES (2009) User guide for EGNOS application developers. http://www.cnes.fr/web/CNES-fr/7877-st-guide-egnos.php

Davies K (1990) Ionospheric radio, Peter Peregrinus Ltd, ISBN:0 86341 186X

EC-DG Energy and Transport (2009) Potential EGNOS extension scenarios, EC-D (2009) 68925

EC-DG Energy and Transport (2011) The European GNSS programmes: EGNOS and Galileo, February 2011

EC-DG Energy and Transport (2013) EGNOS service definition document—open service, EGN-SDD OS V2.0 2013

EC-DG Enterprise and Industry (2013) EGNOS safety of life service definition document, EGN-SDD SoL, V2.0, 2013

EC-DG for Enterprise and Industry (2012) EGNOS data access service—service definition document, July 2012

EGNOS Service Provision Yearly Report (April 2011–March 2012) ESSP Reference, ESSP-DRD-6673P, June 2012. http://www.essp-sas.eu/printed_documents

EGNOS Tools for Educational Research Factsheet (2012) European Space Agency. http://download.esa.int/docs/Navigation/Factsheet_EGNOS_for_education.pdf

ESA EGNOS (2012). http://www.esa.int/esaNA/egnos.html

ESA NAVIPEDIA (2012). http://www.navipedia.org

ESA EGNOS for Professionals (2007). http://www.egnos-pro.esa.int/index.html

EUROCONTROL (2008a) EUROCONTROL policy on GNSS for navigation applications in the civil aviation domain, SCG/8/AP1028/04/2008

EUROCONTROL (2008b) EUROCONTROL policy on GNSS—ION September 2008

European GNSS Agency (GSA) (2012) EGNOS Data Access Service information web http://www.gsa.europa.eu/go/egnos/edas

European GNSS Agency (GSA) (2014) EGNOS. http://egnos-portal.gsa.europa.eu/business-support/commercial-information

European GNSS Agency (GSA) (2014) EGNOS case studies. http://egnos-portal.gsa.europa.eu/discover-egnos/about-egnos/case-studies

European Satellite Services Provider (ESSP) (2014). http://www.essp-sas.eu/
European Space Agency (ESA) (2000) SBAS Teacher learning software: http://www.egnos-pro.esa.int/sbasteacher
European Space Agency (ESA) (2002) Sisnetlab learning software: http://www.egnos-pro.esa.int/sisnetlab
European Space Agency (ESA) (2008) SBAS Mentor learning software: http://www.egnos-pro.esa.int/sbasmentor
European Space Agency (ESA) (2010) SBAS Simulator learning software: http://www.egnos-pro.esa.int/sbassimulator
European Space Agency News (2011) Europe's first EGNOS airport to guide down giant Beluga aircraft. http://www.esa.int/esaNA/SEMTV1ASJMG_index_0.html. Accessed March 2011
FAA (2010) Navigation services—wide area augmentation system (WAAS). http://www.faa.gov/about/office_org/headquarters_offices/ato/service_units/techops/navservices/gnss/waas/
GSA Presentation (2011) Pushing for European GNSS market results 24 February 2011—Administrative Board Meeting 28
Hofmann-Wellenhof B, Lichtenegger H, Collins J (2001) GPS theory and practice, 5th edn. Springer, New York, ISBN:3-211-83534-2
ICAO (2004) Standard and recommended practices (SARPs), Annex 10, vol 1, (Amendment 79 of 25th November 2004)
ISRO GAGAN (2012). http://www.isro.org/scripts/futureprogramme.aspx#Satellite
ITU-R Rec. P.531-8 (2005) Ionospheric propagation data and prediction methods required for the design of satellite services and systems, Geneva
Klobuchar JA (1986) Design and characteristics of the GPS ionospheric time delay algorithm for single frequency users. In: Proceedings of the PLANS-86 conference. Las Vegas, Nevada, pp 280–286
RTCA (2006) Minimum operational performance standards for global positioning system/wide area augmentation system airborne equipment, RTCA-DO 229 D, December, 2006
SDCM (2013). http://www.sdcm.ru/smglo/staticpages?version=eng&site=extern&title=about
Solar Influence Data Analysis Center, SIDC (2011) Royal observatorium of Belgium. http://sidc.oma.be/sunspot-index-graphics/sidc_graphics.php
Thales Aerospace (2008) Thales Aerospace at ENC GNSS 2008 through their presentation "GNSS in commercial aircraft avionics, questions and perspectives"
Ventura-Traveset J, Flament D (2006) The EGNOS book: EGNOS—a cornerstone of Galileo, ESA Publications. ISBN:9789290924531/9290924535, http://www.egnos-pro.esa.int/education/book.html
Ventura-Traveset J, Flament D (2006) The EGNOS book: EGNOS—a cornerstone of Galileo, Chapter 1–1.6 (p 73), ESA Publications. ISBN: 9789290924531/9290924535, http://www.egnos-pro.esa.int/education/book.html. [IONOEXPERTTEAM] EGNOS Book, Iono Expert Team Chapter
Walter T, Blanch J, Enge P (2010) Coverage improvement for dual frequency SBAS. In: Proceedings of the 2010 international technical meeting of the Institute of Navigation, January 2010. San Diego, CA, pp 344–353

Chapter 15
Issues in Implementing the Galileo Safety-of-Life Service

Boubeker Belabbas

Abstract To understand the issues of implementing the Galileo Safety-of-Life service, it is necessary to explain the underlying motivation and the technical challenges induced by the targeted performance.

Introduction

To understand the issues of implementing the Galileo Safety-of-Life service, it is necessary to explain the underlying motivation and the technical challenges induced by the targeted performance. To illustrate the explanations and the keywords used in this chapter, the aviation application has been chosen. This application demands specific and exacting safety requirements in terms of accuracy, integrity, continuity and availability of satellite navigation systems depending on the phase of flight. Standards development groups like the Radio Technical Commission for Aeronautics—Special Committee 104 (RTCA SC-104) in the US and the European Organisation for Civil Aviation Equipment—Working Group 62 (EUROCAE WG62) in Europe have contributed to establish Minimum Operational Performance Standards for aviation equipment using satellite based navigation systems like the Global Positioning System (GPS), the Wide Area Augmentation System (WAAS), the European Geostationary Navigation Overlay Service (EGNOS) and the European satellite navigation system Galileo. Many experts from industry, civil aviation authorities, navigation service providers, academia, and research institutes participate in these groups and provide technical and operational expertise. The other reason that motivates the choice of aviation is the high level of integrity that must be proven to exist under severe conditions before aviation applications of GNSS can be certified.

B. Belabbas (✉)
German Aerospace Center, Bavaria, Germany
e-mail: boubeker.belabbas@dlr.de

J. Nurmi et al. (eds.), *GALILEO Positioning Technology*,
Signals and Communication Technology 182, DOI 10.1007/978-94-007-1830-2_15

A large number of books and articles detail the concept of integrity for satellite based navigation systems as it is the key metric for Safety-of-Life services (Parkinson and Spilker 1996; Parkinson and Jr Spilker 1996; Misra and Enge 2006; Kaplan and Hegarty 2006) and especially (RTCA 2006).

This chapter is organized as follows: First a definition of the performance metrics associated with a safety of life service is given, then a brief description of the Galileo Safety-of-Life service and in particular the original Galileo integrity concept are presented. Based on the requirements for civil aviation taken as an example, the technical issues for implementation is described. In a last section a possible alternative to the Galileo integrity concept is presented. And finally a conclusion and future considerations closes the chapter.

Definition of the Galileo Safety-of-Life Service and Integrity Parameters

In the definition phase of Galileo, along with the open service providing free of charge positioning, navigation and timing service for everyday applications, additional services like the Galileo Commercial Service (CS), the Galileo Public Regulated Service (PRS) and the Galileo Safety-of-Life service (SoL) have been proposed which provide guarantees of performance.

-What is meant by "Safety-of-Life" service for a navigation system?

Obviously, this should be understood as a navigation service for safety critical applications. Or, which is equivalent, a navigation service that, if not fulfilling one or more “safety” requirements, may result in an unsafe situation. In this definition, some questions naturally arise: What is meant by Safety-of-Life service (or “safety requirements”) not being fulfilled?

Ideally, a navigation system should provide a user equipped with a receiver his or her position over time, and this position should be as close as possible to the true one. “As close as possible” means that the deviation from the true position should not exceed a certain margin in a probabilistic sense. One of the challenges for GNSS is that it is hard to find another system that can be used as a reference because GNSS usually achieves the best performance compared to almost all other positioning systems. It is so accurate that geodesists have adopted it to measure reference points on Earth with an accuracy of few millimeters.

For aviation, a much larger error (e.g., 1 – 2 meters) would be fully acceptable for even the most stringent phases of flight. In fact, the challenge is not accuracy but the need to provide guaranty that the instantaneous error is below a margin (alert limit) with a probability of 0.9999999. Past events have shown that, although the average error can be kept small, there are events that could introduce persistent errors of many hundreds of meters in vertical direction, such as excessive drift of a satellite clock. This anomaly can occur (and become potentially unsafe) before the ground monitor detects it and flags the satellite as unhealthy.

As a simple definition of the Galileo Safety-of-Life service, we consider the one that can be found in the European Space Agency website http://www.esa.int/Our_Activities/Navigation/The_future_-_Galileo/Galileo_services.

> The *Safety-of-Life Service* (SoL) improves the open service performance through the provision of timely warnings to the user when it fails to meet certain margins of accuracy (integrity). It is envisaged that a service guarantee will be provided for this service.

The choice of this definition is motivated by its clarity and its completeness. Clarity, because every term can be well understood, and completeness, because the information one could expect from such a service is not more than what is proposed. Sepcifically, we just want to be warned when something is going wrong with the service, and we expect some responsibility to be taken by the service provider. Therefore, we expect this service to be guaranteed.

The aim of this section is to explain the underlying aspects implied in this definition. This will help the reader understand the complexity of implementing such a service.

If we examine this definition in detail, it suggests that the system providing the open service (positioning for typical users) should be augmented by an additional system monitoring this service. This monitoring system should detect if an error in position is larger than a given threshold (Alert Limit) with a specified probability, which is known as probability of misleading information, or P_{MI}. The probability of failing to detect an unacceptable error is known as the probability of missed detection, or P_{MD}, which may differ from P_{MI} depending on the assumed probability of anomalies that can lead to large errors. In addition, users located in a given service area should be warned if such an error occurs within a maximum allowed time to alert. These requirements must be fulfilled for all possible user locations in the service area and for all possible GNSS satellite geometries.

What is not explicitly required but must also be fulfilled is that the positioning service and the service providing timely warnings should work nominally (provide positions with errors below the alert limit without warning) during a sliding period of time with a probability of not meeting this condition that is less than a separately-specified continuity risk requirement. A nominal error distribution requirement must also be fulfilled. This is usually called the "accuracy" requirement and, for aviation, is typically defined by a 95% error threshold, or roughly at two standard deviations of a zero-mean Gaussian distribution.

A service is declared available for a given phase of flight when all the requirements defined above are fulfilled. The availability (in terms of a percentage over time) should be higher than a given service availability requirement (e.g., a minimum of 99 % for precision approach under low visibility conditions).

The Galileo Integrity Concept

Galileo, as opposed to GPS, was foreseen to provide along with satellite navigation data additional integrity information in the form of:

- Signal-in-Space Accuracy (SISA), which is the standard deviation of Gaussian overbound of the Signal in Space Error (SISE) for a given required integrity risk.
- Signal-in-Space Monitoring Accuracy (SISMA), which is the Gaussian overbound of the difference between the true SISE and the estimated SISE (eSISE) from the Galileo Sensor Stations.
- An integrity flag for a given satellite, which is set if a satellite fulfills its integrity requirements. This notion will be further defined later.

Errors impacting user position are unknown and, except for receiver thermal noise, have non-Gaussian distributions in general. These errors are generally handled as random variables. As their distributions are unknown, a concept of Gaussian overbounding has been proposed in (Kaplan and Hegarty 2006).

Two levels of overbounding can be defined. First,

if f_{er} represents the probability density function (pdf) of a residual position error in one direction, f_{ov} is an overbounding pdf if:

$$\int_{-\infty}^{-AL} f_{er}(x)dx \leq \int_{-\infty}^{-AL} f_{ov}(x)dx \quad \text{and} \quad \int_{AL}^{+\infty} f_{er}(x)dx \leq \int_{AL}^{+\infty} f_{ov}(x)dx$$

In this expression AL is a given alert limit (the maximum allowable position domain error for a given application). This is generally part of the performance requirements.

A second and equivalent definition can be given using the cumulative density function (cdf). If F_{er} represents the cdf of residual position error, F_{ov} is a cdf overbound of the error if:

$$F_{er}(-AL) \leq F_{ov}(-AL) \quad and \quad F_{er}(AL) \geq F_{ov}(AL)$$

The equivalence of these 2 definitions is straight-forward because

$$F_{er}(x) = \int_{-\infty}^{x} f_{er}(x)dx \quad and \quad F_{ov}(x) = \int_{-\infty}^{x} f_{ov}(x)dx$$

A Gaussian overbound is the case where f_{ov} is the normal density function.

This notion of Gaussian overbounding, which corresponds to Gaussian tail overbounding assumes that the underlying error distribution is stationary and that the second moment of the underlying error distribution exists.

To simplify the problem, the errors impacting GNSS pseudorange measurements are generally assumed to be of six kinds: satellite orbit error, satellite clock error, residual ionosphere error, residual troposphere error, multipath error, and receiver noise error. A more complete list of error sources can be found in Kovach (2000). These errors are independent by nature, and only their sum can be observed by a single frequency receiver (assuming a known reference position). Generally, true position is estimated using post-processed carrier phase measurements (as commonly used to evaluate flight trial results). Individual error components are also estimated in post-processing using a network of stations and multi-frequency combinations as offered for example by the International GNSS (Global Navigation Satellite System) Service (IGS) (an example of individual error analysis using IGS products is shown in Belabbas et al. (2005)). The Galileo Safety-of-Life service was aimed to provide timely warnings when an error larger than a predefined threshold arises by using a network of sensor stations and multi frequency combinations. It was expected that this service would also provide a prediction of its availability.

The Galileo constellation service provider would have provided a monitoring of errors occurring in the space or in the ground segment. Three different types of information were planned to be provided by each satellite, which are the integrity flag, the SISA, and the SISMA as listed in the beginning of this paragraph. The SISMA is the standard deviation of the Gaussian distribution overbounding the distribution of the difference between SISE and eSISE. This can be represented mathematically in the following way:

$$SISMA = \sqrt{E^{F_{ov}}\left[(SISE - eSISE)^2\right]}$$

where $E^{F_{ov}}$ is the expectation with respect to the overbounding distribution F_{ov}

Each satellite should be observed by at least 4 ground reference stations at any given time in order to determine its position and to estimate the signal in space error by comparing this position to the position generated from the navigation message broadcast by the satellite. The SISA is generated by considering the overbound of the error distribution projected at the worst user location within a predefined Galileo service area. The worst user location is determined based on the user location with the highest position dilution of precision factor.

The Orbit Determination and Time Synchronization (ODTS) process for integrity differs from the ODTS for navigation in the sense that it has to be done on a real-time basis. This constrains the architecture of the sensor station network, as each satellite must be always visible by at least 4 ground reference stations. Indeed, to update the three components within a 6-second time to alert worldwide, a dense network of stations is necessary with distributed uplink stations to upload potential integrity information for each satellite at all times.

This is what is done at continental level by the Space Based Augmentation System (SBAS) using geostationary satellites to broadcast the integrity message.

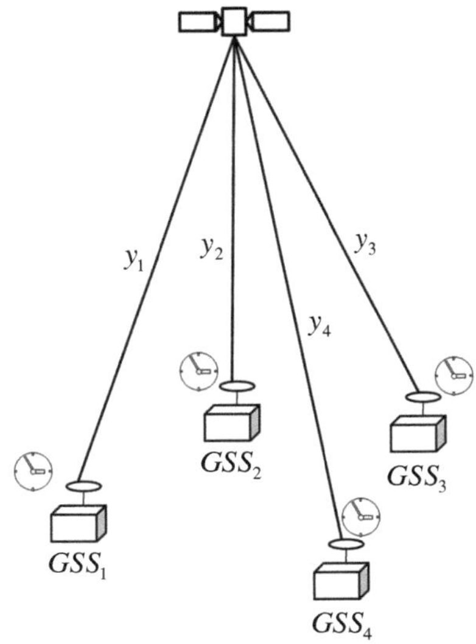

Fig. 15.1 Satellite position determination using synchronized Galileo Sensor Stations (locations not necessarily representative)

Thanks to highly precise satellite orbit propagation models, it is possible to calculate the estimated SISE at the system level with an acceptable level of accuracy. The estimated SISE for a given satellite is determined within a certain confidence level which depends, among other things, on the location of the ground stations with respect to the satellite (see Fig. 15.1).

In this example, we examine the case of 4 Galileo Sensor Stations (GSS), but usually we take all stations in view of the given satellite. Let's call y_i the pseudorange observation of the satellite under consideration at GSS station i. Because all stations are synchronized to Galileo System Time (GST), it is possible to estimate the position of the observed satellite using an inverted GNSS technique; i.e., the satellite plays the role of the user and the stations the role of the satellites.

At least 4 stations are necessary to provide 3-D position and the satellite clock offset to GST. In order to calculate the precise orbit of the satellite, it is common to use a Kalman filter with a very precise motion prediction model of the satellite. The measurement/observation equations can be written as follows:

$$\begin{pmatrix} y_1 \\ y_2 \\ y_3 \\ y_4 \end{pmatrix} = \mathbf{G} \begin{pmatrix} x_1 \\ x_2 \\ x_3 \\ b_{clk} \end{pmatrix}$$

With **G** being the matrix containing the unit vector from station to satellite and a fourth column of ones for the projection of the ranges to the satellite clock offset.

To determine the position of the satellite and its clock bias, it suffices to invert the preceding equation:

$$\begin{pmatrix} x_1 \\ x_2 \\ x_3 \\ b_{clk} \end{pmatrix} = \mathbf{G}^{-1} \begin{pmatrix} y_1 \\ y_2 \\ y_3 \\ y_4 \end{pmatrix}$$

If the number of ground stations involved exceeds four, we would apply the pseudo-inverse method that minimizes the least square error:

$$\begin{pmatrix} x_1 \\ x_2 \\ x_3 \\ b_{clk} \end{pmatrix} = \left(\mathbf{G}^T\mathbf{W}\mathbf{G}\right)^{-1}\mathbf{G}^T\mathbf{W} \begin{pmatrix} y_1 \\ \vdots \\ y_k \end{pmatrix} \quad \text{for } k \geq 4$$

In this expression, **W** represents a weighting matrix, which is usually taken to be the inverse covariance matrix of the pseudorange errors.

The resulting filtered position solution of the satellite and the precise satellite clock estimation is then used as a reference for estimating the SISE.

One unique aspect of the integrity concept is that the user is calculating the integrity risk depending on the maximum allowed error (alert limit) given an integrity risk requirement. The calculated integrity risk should lie below the integrity risk requirement to declare the system available for the given application/ operation.

The following user equation determines the integrity risk or probability of misleading information (P_{MI}) given a required Vertical Alert Limit (VAL) and Horizontal Alert Limit (HAL) (extracted from Oehler et al. 2005).

$$P_{MI}^{user} = f(SISA, SISMA, P_{Sat}, \mathbf{G}, \mathbf{VAL}, \mathbf{HAL}, \sigma_{\mathbf{L}})$$

With:

P_{Sat} being the probability of satellite failure,

G being the geometry matrix introduced earlier (unit vectors from satellites to user with an additional column of ones for the unknown receiver clock bias).

VAL and *HAL* being the vertical and horizontal alert limits, respectively, corresponding to a particular user operation,

σ_L being the standard deviation of the local range error overbound (errors not considered in the SISA but entering into the global user error budget like residual tropospheric error, multipath and receiver noise error). This standard deviation is usually satellite elevation dependent.

Integrity risk is calculated as the weighted sum of the combination of all faulted and fault free modes. A fault free mode or hypothesis is defined as the typical behavior of the system under nominal performance conditions. A faulted mode or hypothesis is defined as a condition where one or more pseudoranges are affected by an abnormal condition. Most faulted mode can be modeled as a deviation or bias in the affected pseudorange (in addition to the presence of nominal error sources).

It was assumed that only one faulty satellite was considered in a faulted mode at any given time. As explained above, the error for this satellite is usually modeled as a non-centered Gaussian distribution, i.e. the fault is modelled as a bias in the measurement.

We invite the interested reader to consult (Misra and Enge 2006) for detailed explanations.

This concept of integrity (Oehler et al. 2005) differs from the integrity concept based on protection levels as adopted in the aviation community (RTCA 2006). The protection level approach used in aviation assumes a fixed allocation between vertical and horizontal integrity risk. For example, SBAS allocates 98 % of the total risk to the vertical direction and 2 % to the horizontal. In the case of Galileo, no allocation per direction is assumed. Galileo is aiming to serve different safety-of-life applications and not focus its service on the aviation community only.

For example, in maritime, the along track and cross track components of 2-D horizontal positioning are the critical ones. The directions having the most stringent requirements should receive the largest amount of integrity risk budget.

In order to satisfy the needs of all possible safety-of-life applications, it is left to the user communities to define the appropriate risk allocations based on the requirements and the critical directions relevant to each application.

The system should fulfill the following requirements:

The acronyms used in Table 15.1 are defined as follows:

TTA: Time To Alert: The maximum period of time between the occurrence of an unacceptable error in the system and the time when the user is alerted about the abnormal situation (or the unacceptable error is removed),

HAL/VAL: Horizontal Alarm Limit and Vertical Alarm Limit. These are the maximum errors allowed, in 2-D horizontal and 1-D vertical position directions, for a given application during a given phase of operation.

Table 15.1 System integrity requirements (Oehler et al. 2005)

	Level A requirements	Level B requirements	Level C requirements
Integrity risk	2.0×10^{-7} in any 150 s	1.0×10^{-7}/1 h	1.0×10^{-5}/3 h
Continuity risk	0.8×10^{-6} in any 15 s	1.0×10^{-4} to 1.0×10^{-8}/1 h	3.0×10^{-4}/3 h
Availability of service	99.5 %	99.5 %	99.5 %
TTA	6 s	10 s	10 s
HAL/VAL (NA= Not Applicable)	40/20 m	556 m/NA	25 m/NA
Dual Frequency E5a + L1 or E5b + L1	Yes	Yes	Yes
Single Frequency L1 or E5b	No	Yes	Yes
Coverage	Global	Global	Global

These requirements should be applied along with any additional user-specific requirements to determine if a given user can trust the Galileo Safety-of-Life service or not. For the majority of applications, these system requirements are sufficient. However, for some aviation phases of flight (such as Category I precision approach guidance to a 200 ft decision height), it is necessary to either define a new column with more stringent requirements than Level A or it is necessary to target a less stringent phase of flight. Another challenging aspect is the necessity to provide global coverage with a required availability of both integrity and continuity of up to 99.5 % for non-precision approach.

User Requirements: Example for Aviation

In the approach segment of aviation applications, GNSS requirements are driven by the minimum decision height, which can be defined as the altitude above the ground at which the pilot should be able to see the runway and thus not rely on GNSS-based positioning. The following figure gives an illustration of these requirements and, in the left part of the drawing, the corresponding GNSS augmentation that could enable the indicated phase of flight.

The acronyms of Fig. 15.2 are explained as follows:

GPS-Baro: Global Positioning System using a barometric altimeter for vertical guidance,

SBAS: Space Based Augmentation System, such as WAAS in the US and EGNOS in Europe,

ARAIM: Advanced Receiver Autonomous Integrity Monitoring (using Multiple Hypothesis Solution Separation as defined in Blanch et al. (2007), Pervan et al. (1998)).

ISM: Integrity Support Message (for ARAIM to achieve LPV-200: Localizer with Vertical guidance for a 200 ft Decision Height),

DH: Decision Height: altitude above the ground during the approach phase at which the pilot should see the runway and (a) decide to proceed with the final approach and landing, or (b) execute a go-around if he or she cannot see the runway

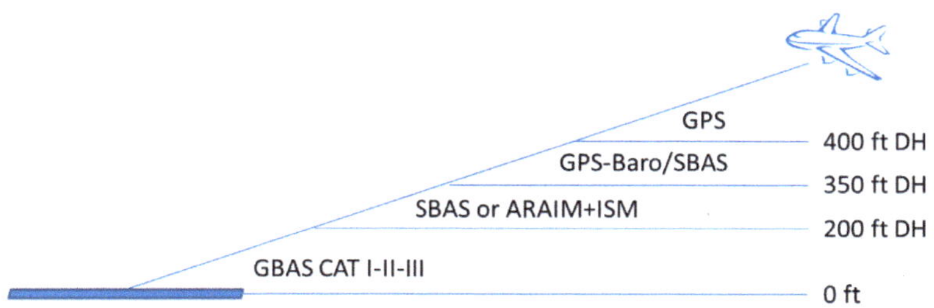

Fig. 15.2 Navigation solution as a function of the decision height for the approach phase (derived from ICAO (1996))

or estimates, after evaluating the situation, that the position or the speed of the aircraft are outside the parameters of a safe final approach and landing.

GBAS: Ground Based Augmentation System also called the Local Area Augmentation System in the US.

CAT I-II-III: Category I-II-III of precision approach under low visibility conditions. CAT I has a minimum decision height of 200 ft, CAT II: 100 ft and CAT III: from 50 to 0 ft.

Note: CAT III must be conducted using an automatic landing system certified for this type of approach and landing.

Table 15.2 summarizes the system requirements that GNSS-based navigation used on aircraft should fulfill.

WAAS (Wide Area Augmentation System) in the US and EGNOS (European Geostationary Navigation Overlay Service) can almost achieve Category I precision approach. The limitation to achieving CAT I is the low vertical alert limit of 10-15 meters. Therefore a new procedure has been proposed (localizer performance with vertical guidance down to a 200 ft decision height, in short, LPV-200) which differ from Category I precision approach by setting the vertical alert limit to 35 m and a so called Effective Monitoring Threshold of 15 m. At the time of writing, some details of this requirement are still under discussion.

At present, the objective for aviation is that SBAS (Space Based Augmentation Systems) like WAAS, EGNOS, the Indian GAGAN (GPS Aided Geo Augmented

Table 15.2 navigation requirements for aviation (ICAO 1996)

Typical operation	Horizontal alert limit	Vertical alert limit	Integrity risk	TTA	Continuity risk	Availability
En-route	7.4 km (4 NM)	N/A	1×10^{-7}/h	5 min	1×10^{-4}/h to 1×10^{-8}/h	0.99–0.99999
En-route, Terminal	1.85 km (1 NM)	N/A	1×10^{-7}/h	15 s	1×10^{-4}/h to 1×10^{-8}/h	0.99–0.99999
Initial approach, Intermediate approach, Non-precision approach, Departure	556 m (0.3 NM)	N/A	1×10^{-7}/h	10 s	1×10^{-4}/h to 1×10^{-8}/h	0.99–0.99999
Approach operations with vertical guidance (APV-I)	40 m (130 ft)	50 m (164 ft)	2×10^{-7} per approach	10 s	8×10^{-6} in any 15 s	0.99–0.99999
Approach operations with vertical guidance (APV-II)	40 m (130 ft)	20 m (66 ft)	2×10^{-7} per approach	6 s	8×10^{-6} in any 15 s	0.99–0.99999
Category I precision approach	40 m (130 ft)	15–10 m (50–33 ft)	2×10^{-7} per approach	6 s	8×10^{-6} in any 15 s	0.99–0.99999

Navigation), and the Japanese MSAS (Multi-functional Satellite Augmentation System) aim to achieve LPV-200. The Safety-of-Life service of Galileo must be redefined to take into account this goal.

For Category I, II and III precision approach, Ground Based Augmentation System (GBAS) is the baseline accepted by the major aviation research programmes (NextGen in the United States and SESAR in Europe). Several versions of Category II and III GBAS requirements exist but are still undergoing development and validation testing.

Three options are left for the Galileo Safety of Life service in the long term: Either it can achieve better performance than SBAS (keeping in mind that SBAS is expected to support dual-frequency service in the future), it could fill in the gap in requirements left by the other augmentation services and backup to these augmentation services within their service volumes or it could be used to improve the Galileo core system performance.

Service Architecture and Technical Issues for Implementation

As already mentioned, the provision of guaranteed error overbounds at a stringent integrity risk requirements and short times to alert require a dense network of stations to achieve real time orbit determination and time synchronization.

One of the remaining problems with the Galileo Safety-of-Life service definition is lack of knowledge about the magnitude of SISA and SISMA. The overbounding aspect is of high interest. It took WAAS many years of measurements to establish a realistic and robust Gaussian overbounds of its ranging errors. In a conservative way, it is necessary to inflate the SISA/SISMA in order to cover rare events that may violate integrity requirements if not detected within the time to alert. A total of 40 GSS are foreseen to be installed worldwide to enable the Galileo Open Service (OS). This is probably not sufficient to enable a Safety of Life service unless the system requirements in terms of integrity and continuity are reconsidered. To guarantee a time to alert of 6 s, it is necessary to deploy a network of uplink stations enabling at each instant the uplink of integrity flags to any satellite in the Galileo constellation. In other words, any location in space potentially occupied by Galileo satellites must be visible to at least one uplink station at all times.

Originally, the Galileo integrity concept didn't foresee additional integrity monitoring at user level. This constrains the architecture significantly. The current approach is to consider a user level integrity monitoring algorithm with system support for the signal in space performance. An "Advanced" Receiver Autonomous Integrity Monitoring (ARAIM) algorithm has been suggested. This algorithm, based on multiple hypothesis solution separation techniques (see Blanch et al. 2007; Pervan et al. 1998)), has the possibility to evaluate the overall performance of the system by calculating the position-domain impacts of many different failure scenarios using redundant information. This algorithm in order to achieve the stringent

LPV 200 requirements worldwide requires an Integrity Support Message (ISM). The role of ISM is to provide information with guaranteed integrity to users that would not otherwise be available, such as ARAIM parameter inputs (i.e., a priori probability of unflaged satellite failure during a given exposure time, SISA, SISE, maximal and nominal biases...). Integrity flags for each satellite are not required within the stringent 6-s time to alert because ARAIM is present to cover faults that appear suddenly. ARAIM performance is sensitive to a high satellite failure probability P_{sat} (if P_{sat} becomes large, the probability of independent failures of multiple satellites becomes significant within the integrity risk budget and cannot be neglected). Also, a large SISA may reduce the availability of the integrity service (due to large protection levels). Constellation wide failures, like for example errors in the implementation of the Earth Orientation Parameters, are responsible of an increase of the constellation failure probability P_{const}, which may also reduce service availability.

The content and format of the ISM to support ARAIM remains in the definition phase. Two different approaches are for the moment under consideration: the long and short latency architectures. The long latency architecture considers a message update frequency on the order of weeks to months, while the short latency architecture considers an update rate of minutes to hours (as of this writing, the details of these architectures are still under investigation).

The Galileo system will take advantage of ARAIM by integrating it as part of its service portfolio. This offers a unique opportunity for Galileo to redesign (with acceptable cost) the original Galileo Safety-of-Life service as a support to ARAIM to provide, in combination with GPS and eventually GLONASS and other emerging constellations, a global LPV-200 service.

An Incremental Threat Analysis Scheme for Galileo

The provision of guaranteed error overbounds with stringent integrity risk requirements is a particular challenge for a constellation like Galileo that has not yet reached its full operational capability.

For the certification of Galileo for aviation applications, it is necessary to characterize the system in detail using observations.

It took almost 10 years for WAAS to be certified for aviation applications on the 10th of July 2010 after almost 20 years of GPS availability and about 7 years of WAAS availability (following the achievement of WAAS Initial Operational Capability in 2003).

Galileo will take advantage of the long heritage of GPS, and it is possible to reduce the length of time needed to get the system qualified for aviation use. Nevertheless, some specific aspects not necessarily of a technical nature will bring additional challenges, like the reliability issues posed by a system produced and operated by a consortium of European entities (in contrast to GPS, which is exclusively operated by the US Department of Defense).

A minimum observation time is not avoidable if we want to characterize the behavior of the Galileo constellation and establish a consequent threat analysis to make the system operational for Safety-of-Life service.

A first step could use the planned GSS configuration to see what level of performance could be reached. The measured level of SISA/SISMA over time would determine if it is necessary to extend the monitoring facilities or not. An important challenge is the need for worldwide coverage. Even if 4 sensor stations are sufficient to determine the position of a given satellite, it is necessary to use more than 4 pseudorange measurements to increase the accuracy of the estimated SISE and therefore reduce the level of SISMA.

But all these performance parameters can only be adjusted when the system has been sufficiently characterized. For that, a "step zero" would consist of starting to observe the system during its deployment phase and gradually update the key threat parameters together with their statistical confidence levels.

One approach would be to define a confidence interval for each parameter of the threat space as a function of spatial and temporal observations of the system. Intuitively, the larger the observation sample size, the smaller the confidence intervals of the inferred statistical parameters.

The ISM parameters together with their confidence intervals must achieve minimum requirements to declare the constellation to be qualified for safety of life applications. These minimum requirements need to be defined, and the method to be developed will have to follow some standards coming from the International Civil Aviation Organization for aviation applications for example.

Conclusion and Future Considerations

The Galileo integrity concept, which is the basis for the proposed Galileo Safety-of-Life service, has been defined. The assumptions adopted in this concept are acceptable so far. The work done to establish the mathematical foundation of the concept is well established. Analysis based on simulations has proven that, under the hypotheses made by these studies, the resulting performance is satisfactory.

But the decisions to implement this service cannot only be based on simulations. A good example is the development of WAAS. Additional monitors and architecture modifications were implemented due to unexpected events, such as the appearance of severe ionosphere gradients introducing unacceptable undetected ranging errors. These events had to occur and be noticed before they could be understood as threats, which led to countermeasures being developed.

In the same way, although Galileo has many similarities to GPS, it is a different system for which a new threat analysis must be conducted based on the lessons learned from GPS and its own history of measurements. Continuous monitoring of Galileo performance and potential threats will motivate changes to the service architecture if needed while providing data to international teams and justifying increased confidence in the system over time.

It took more than 10 years for GPS (from the full operational capability of GPS in April 1995 to the certification of WAAS for LPV-200 operations) to reach the level of confidence necessary for civil aviation.

Another point that needs to be investigated is the combination of an Integrity Support Message (ISM) with Advanced Receiver Autonomous Integrity Monitoring at the user level. This is a very promising technique because the information to be broadcast to users (in the ISM) does not need to be updated very frequently. This should reduce the density demands on the Galileo sensor stations and therefore the costs of deployment of the Galileo Safety-of-Life service.

References

Parkinson BW, Jr Spilker JJ (1996) Global positioning system: theory and applications, vol 1. American Institute of Aeronautics & Astronautics, Washington

Parkinson BW, Jr Spilker JJ (1996) Global positioning system: theory and applications, vol 2. American Institute of Aeronautics & Astronautics, Washington

Misra P, Enge P (2006) Global positioning system, signals, measurements, and performance. Ganga-Jamuna Press, Massachusetts

Kaplan ED, Hegarty CJ (2006) Understanding GPS, principles and applications. Artech House, Boston

RTCA SC104 (2006) DO-229D: minimum operational performance standards for global positioning system/wide area augmentation system airborne equipment, RTCA

Kovach K (2000) New user equivalent range error budget for the modernized navstar global positioning system GPS. In: Proceedings of the institute of navigation ION national technical meeting

Belabbas B, Petitprez F, Hornbostel A (2005) UERE analysis for static single frequency positioning using data of IGS stations. In: Proceeding of the ION NTM, San Diego

Oehler V, Luongo F, Trautenberg HL, Boyero J-P, Krueger J, Rang T (2005) The galileo integrity concept and performance. In: Proceedings of ENC GNSS

International Civil Aviation Organization (ICAO) (1996) aeronautical telecommunication annex 10 to the convention on international civil aviation, vol 1. International Standards and Recommended Practices

Blanch J, Ene A, Walter T, Enge P (2007) An optimized multiple hypothesis raim algorithm for vertical guidance. In: Proceedings of the institute of navigation ION GNSS

Pervan B, Pullen S, Christie J (1998) A Multiple Hypothesis Approach to Satellite Navigation Integrity. Navigation 45(1) p. 61-71

Chapter 16
GIMS: Galileo Interference Management System

Philippe Morgand and Anne Ferreol

Abstract This chapter focuses on Galileo Interference management. First the concept of LAAS is described, then GIMS is presented in detail.

Introduction: LAAS and GIMS

The concept of LAAS is described hereafter in order to introduce the architecture and equipment on which the GIMS is based.

A LAA System is a standard approach used to improve the performances of a positioning system. It is used as aircraft landing system and it is also called GBAS by the ICAO.

The LAAS is mainly based on the differential correction of the GPS signal (a similar system can be envisaged for the Galileo signals) obtained by local receivers which send them to a control centre usually located at the airport. This information is then sent to the users through using a radio channel who can exploit it to correct the received GPS data.

Figure 16.1 shows the typical architecture of the LAAS.

This system is based on the following main equipments:

- *PSEUDOLITES* (blue cubes in Fig. 16.1): devices whose position is known, emulating additional Galileo satellites; they transmit Galileo signals, as any other satellite belonging to the Galileo constellation, which are received directly by the final user;
- *Differential Augmentation Terminals* (*DAT*) (yellow cube in Fig. 16.1): devices comparing their actual position with the position computed through receiving the Galileo signals; they extract the differential corrections and send them to the

An erratum to this chapter is available at DOI 10.1007/978-94-007-1830-2_18

P. Morgand (✉) · A. Ferreol
Thales Communications, Gennevilliers, France
e-mail: philippe.morgand@fr.thalesgroup.com

J. Nurmi et al. (eds.), *GALILEO Positioning Technology*,
Signals and Communication Technology 182, DOI 10.1007/978-94-007-1830-2_16

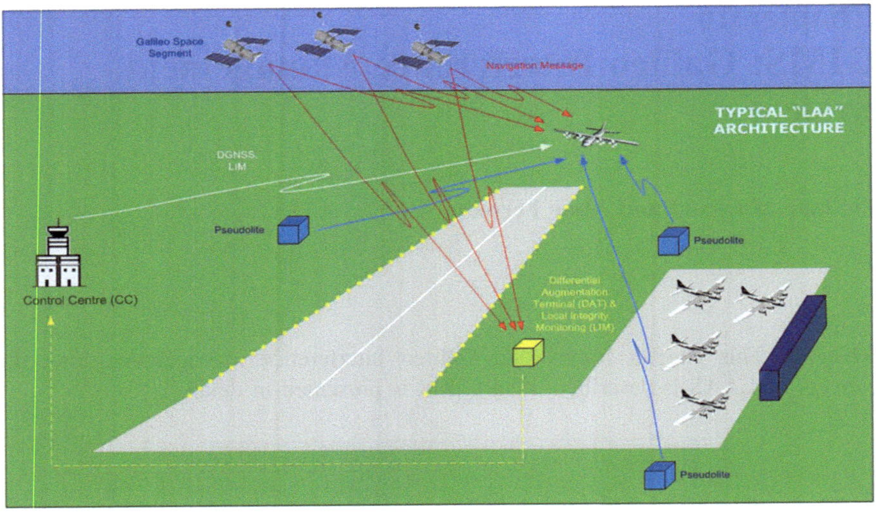

Fig. 16.1 Typical LAAS architecture

Control Centre (CC) of the airport; finally from the CC the differential corrections are sent to the GRs which use them to improve its own position estimation;

- *Local Integrity Monitoring* (*LIM*) (yellow cube in Fig. 16.1): devices monitoring the correctness of the Galileo signal through receiving the Galileo signals themselves; the integrity information is sent to the CC and exploited to obtain a better accuracy excluding a given satellite from the position computation if its contribution degrades the accuracy;

This equipment can be analysed taking into account their vulnerability to potential interferences as reported in the following points:

- *PSEUDOLITES*: it is transmitting equipment which typically doesn't need a mitigation strategy being their performances not depending on the presence of disturbs coming from the environment;
- *DAT*: these types of devices calculate the differential corrections so they must include a GR which can be disturbed by the presence of any interference; therefore the GR part of the DAT can exploit any available mitigation strategy aimed at minimising the effect of these interferences;
- *LIM*: these types of devices monitor the correctness of the Galileo signal so they must include a GR which can be disturbed by the interferences present in the environment; therefore they can take advantages from a suitable mitigation strategy;

Summarising it is possible to state that the GIMS can help the DAT and the LIM devices to improve their performances in presence of interfering sources.

The concept of GIMS is depicted below in order to complete the architecture proposed for the target LAAS including the GIMS structure.

The GIMS is the strategy proposed by the MAGIC Project to face the problems caused by the presence of interfering signals. It is able to manage three main functions:

- Interference Detection and Classification: to discover the presence of interfering signals affecting the managed area and to determine their main characteristics;
- Interference Isolation: once the presence of the interference has been detected, to find their exact position;
- Interference Mitigation: to provide to the GR all the parameters necessary to put in force a mitigation strategy (which is not part of the GIMS but it is studied in the MAGIC project to be implemented inside the GR);

Figure 16.2 shows the typical architecture of the LAAS and the GIMS put side by side.

The GIMS is based on the following main equipments:

- *Interference Observation Site* (*IOS*): devices able to detect the presence of interfering signals and to provide information on them like frequency, bandwidth, AOA and so on;
- *Interference Control Centre* (*ICC*): control centre which receives the information from each IOS and extracts the position of each interfering signal associating it with its main characteristics. Furthermore it also provides: (I) A Signal Alert to warn the area about the interference presence; (II) The Notification to the Surveillance Authority; (III) The Mitigation Strategy to each GR (including those belonging to the LAA) belonging to the controlled area;

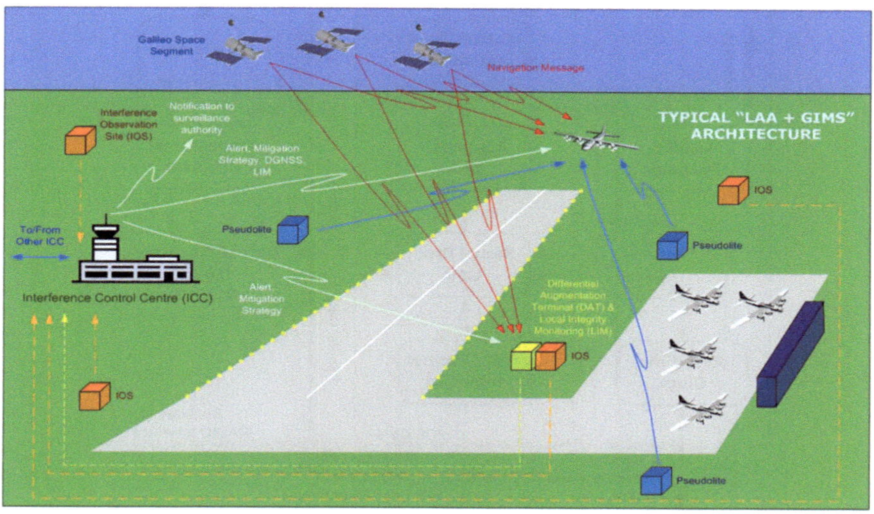

Fig. 16.2 Typical LAAS with GIMS architecture

GIMS Elements

Three main elements composed the GIMS such as Interference Observation Site, Interference Control Centre and a Galileo receiver unit upgraded with mitigation capabilities.

IOS: Interference Observation System

Figure 16.3 shows the block diagram of the IOS.

The IOS can be divided into three main sections:

- *IOS Core*: it includes the processing for detecting/classifying and isolating the interference;
- *SYNCHRONIZATION Interface*: it includes the interfaces necessary for the synchronization of the IOSs in case of TDOA-AOA approach;
- *Auxiliary RADIO Interface*: it includes the radio interface between the IOS and ICC to be used where a cable interconnection is not possible;

The next paragraphs describe each part in more depth.

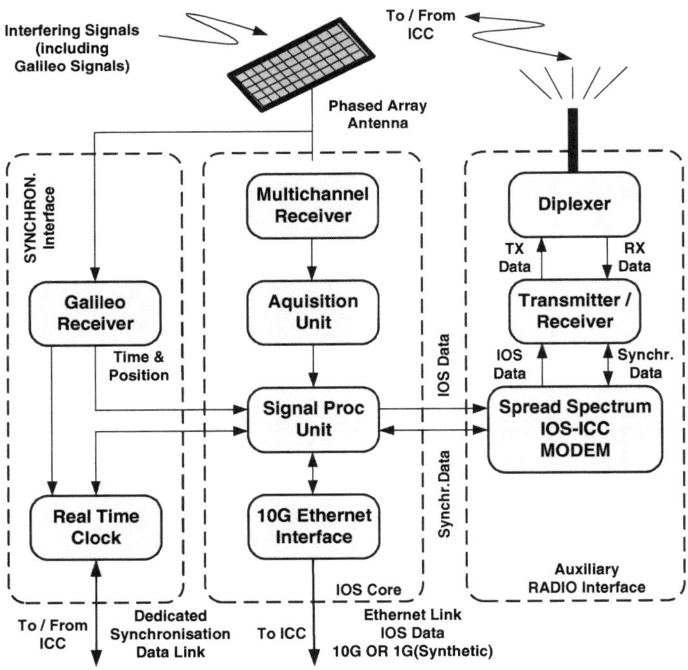

Fig. 16.3 IOS block diagram: (10G Ethernet only useful for AOA-TDOA localisation techniques)

IOS Core

The main functional blocks of the IOS Core are:

- A Phased Array Antenna: it is required for the Isolation processing when based on AOA technique (AOA-AOA and AOA-TDOA approaches); this antenna should be able to receive signals coming from all the three Galileo bandwidth (L1, E5, E6) therefore it should be a wideband antenna covering a frequency range from about 1,100 MHz to about 1,700 MHz;
- A Multi-channel RF Receiver: it is necessary to receive the elements of the phased array antenna; it should also include a triple down converter to allow processing the three Galileo bands at the same time; a scaled version of this stage can foresee only a single programmable down-converter whose central frequency can be changed in order to process the three Galileo bands in different times;
- An Acquisition Unit: required to perform the Analogue to Digital Conversion at 200 MHz of each channel; in case of triple down converter it should be able to convert a triple number of channels;
- A Signal Processing Unit implementing the Detection algorithms and the first stage of the Isolation algorithms (designed using real time HW based on FPGA); this unit is also used to manage the peripherals of the IOS, that is the several interfaces to the external world;
- The 10 Gbps Ethernet Interface to transfer data toward the ICC; a scaled version using 1 Gbps Ethernet can be used with the penalty of having not a continuous monitoring in case of TDOA technique;

SYNCHRONIZATION Interface

This interface includes two main functional blocks:

- a Real Time Clock to provide the time-stamp to the IOS; it can be set through using a GR or through an external connection to the ICC which provides the necessary time triggers;
- a GR to obtain both the estimation of the time and the coordinates of the IOS;

Both the peripherals are managed by the Signal Processing Unit which uses the time-stamp to label the signals for the TDOA analysis (performed in the ICC).

The synchronisation through using a GR provides an accuracy of about 100 ns. This accuracy is available obviously in absence of interferences. A local (inside the IOS) mitigation strategy can be applied provided that the duration of the interference is enough to allow detection, classification and isolation (the overall computing flow must be executed from IOS to ICC and return).

The synchronisation through a dedicated data link connecting the IOS and the ICC is also available which can be used to exploit the available synchronisation technique (unidirectional synchronisation, round-trip synchronisation, reference broadcasting, etc....). This approach can be used in case of Galileo signal unavailability or to improve the time precision.

Auxiliary RADIO Interface

The main functional blocks of the Auxiliary RADIO Interface are:

- a spread spectrum MODEM able to provide a base band modulated signal including in the data fields both the information of the interfering signal (ISO Data) and the Synchronisation Data;
- an RF Transmitter/Receiver which up converts the base band signal (and down convert the IF signal) in order to feed the diplexer (and to accept the diplexer output);
- a Diplexer necessary because the link is bidirectional;
- an Antenna;

This RADIO Interface is useful to completely replace the cable interconnection (both for IOS Data and Time Synchronisation) in case of unavailability of the cabled network.

ICC

Figure 16.4 shows a functional block diagram of the ICC.

Many blocks have been foreseen to implement all the functions. In particular the ICC architecture can be ideally divided into the following sections:

- *ICC Core*: it includes the capabilities required to manage the entire ICC;
- *Dedicated Interfaces*: it includes all the interfaces necessary for communicating with each element of the GIMS architecture;
- *GIMS RADIO Interface*: it includes the radio interface between the IOS and GRs;
- *Auxiliary RADIO Interface*: it includes the radio interface between the IOS and ICC to be used where a cable interconnection is not possible;

Each section is described in more depth in the following paragraphs.

ICC Core

It is composed of two main functional blocks:

- The Signal Processing Unit used to elaborate the data coming from the IOSs. In case of AOA-AOA technique it performs only the last step of the processing (the synthesis stage of the isolation algorithm), in case of AOA-TDOA technique it has also to perform the cross-correlation among the wideband signals coming from the IOSs;

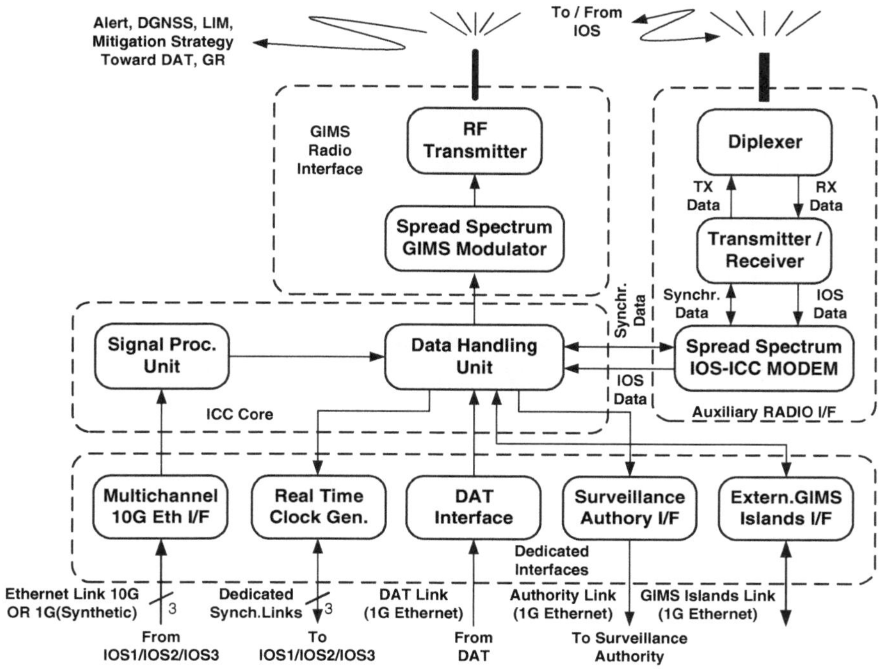

Fig. 16.4 ICC Block diagram

- The Data Handling Unit which is responsible of managing all the peripherals of the ICC including the Signal Processing Unit itself and the Auxiliary RADIO Interface. Furthermore it also implements inside the controller of the spread spectrum GIMS Modulator used to transmit information toward the users;

Dedicated Interfaces

It includes several interfaces to the equipment belonging to the GIMS or connected to the GIMS: IOS, DAT, Surveillance Authority and other GIMS Islands:

- The Multi-channel 10G Ethernet Interface: this is the interface used to receive data from the IOSs. In this block diagram three IOSs are foreseen in the GIMS architecture.
- The Real Time Clock Generator: this is the interface used to synchronise the IOSs through a suitable data protocol; a bidirectional link is foreseen for each IOS in order to implement the more generic approach for the time synchronisation;
- The DAT Interface: this is the interface necessary to communicate with the DAT equipment in order to receive the differential corrections to be broadcasted toward the users;

- The Surveillance Authority Interface: this is the interface necessary to connect the Authority to the GIMS; in the solution proposed by this block diagram only a cable connection is foreseen to this purpose;
- The External GIMS Islands Interface: for the communication among different GIMS at 1G Ethernet is here foreseen (only pre-alert or synthetic information must be exchanged between two or more GIMS);

GIMS RADIO Interface

It includes the radio interface between the IOS and GR to alert the users about the presence of the interferences and to provide them a suitable mitigation strategy. The main blocks are a Spread Spectrum GIMS Modulator and its RF Transmitter with Antenna. The GIMS Modulator will transmit the information concerning both the LAA (DGNSS and LIM) and the GIMS (Alert, Mitigation Strategy).

Auxiliary RADIO Interface

This section is the RADIO Interface which corresponds to the RADIO Interface already described for the IOS. It can be designed in order to be completely identical to the one used in the IOS. It is used to completely replace the cable interconnection (both for IOS Data and Time Synchronisation) in case of unavailability of the cabled network.

GR with Mitigation Add-On

Figure 16.5 shows the functional block diagram of a stand alone GR (highlighted part of the figure) and the functional block diagram of the DAT equipment (including LIM) with an embedded GR.

The final user will be equipped with the only GR (with mitigation capabilities) whereas the DAT and LIM architecture will have a more complex structure including this GR plus the dedicated equipments necessary to compute the differential corrections and to monitor the integrity of the Galileo signal.

Inside the Stand Alone GR (with mitigation facilities) four main functional blocks can be identified:

- An RF Receiver (down conversion and amplification) with the receiving antenna;
- A spread spectrum GIMS Demodulator to receive the Mitigation Strategy and the DGNSS data from the ICC;
- A Central Management Unit to manage the several units;
- The GR with mitigation algorithms implemented inside; this GR in case of SDIM algorithms must be equipped with a phased array antenna;

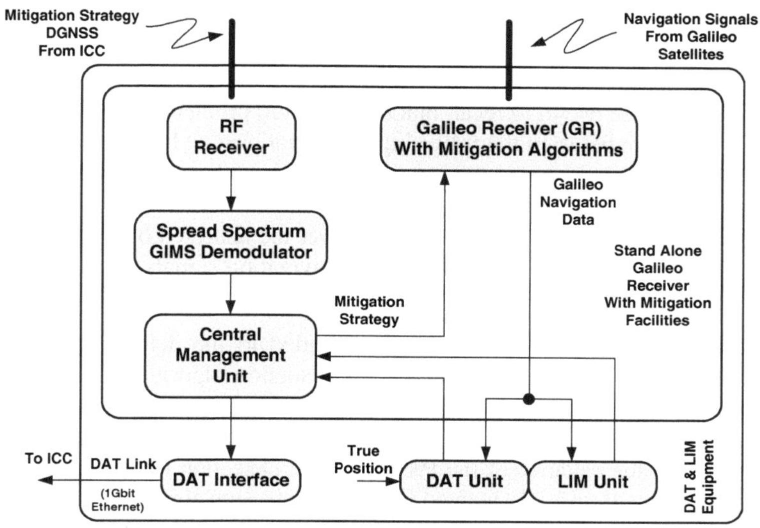

Fig. 16.5 GR/DAT/LIM with mitigation facilities

The DAT & LIM equipment is completed adding the specific functional blocks necessary for these purposes.

GIMS: Embedded Signal Processing

The previous system manages interferences with three main signal processing functions such as:

- Detection and automatic classification
- X, Y, Z localization of the sources
- Mitigation of the jamming interference

Detection and Automatic Classification

Several methods are currently available to detect signals in noise. Since many interferers can be modelled as cyclostationary signals (e.g. linear digital modulations like MPSK and QAM, pulsed signals, etc.…), their detection can be achieved by exploiting the existing theory developed by Gardner (1991). A method based on the energy of the signal is proposed in Urkowitz (1967); in such a method the energy can be alternatively substituted with a higher order moment of the signal (Nikias and Mendel 1993; Sattar and Salomonsson 2000). A trade off among these different approaches, based on their implementation complexity and on their effectiveness, led to the detection method proposed in this section. It is mainly based on Urkowitz (1967) to exploit its low complexity, suitably modified in order to obtain also some characteristics of the interferer typically provided by the cyclostationary analysis like the signal bandwidth. Before passing to describe the detector, it might be useful to recall that the detection problem can be formulated as a decision problem between the following two situations: (i) H_0: x(k) = n(k) and (ii) H_1: x(k) = n(k) + j(k); where x(k) is the complex, base-band, discrete signal at the detector input, n(k) is the noise component of the signal and j(k) is the jamming signal. The useful signal is supposed spread over the analysed bandwidth and fully immersed in the noise (this is the case of the navigation signals). The situations H_0 and H_1 correspond to two different phases of the detection process:

- *Calibration Phase*: only the noise is provided to the detector to analyse its statistical properties at the output of the detection algorithm and to extract a threshold on the basis of a given Probability of False Alarm (PFA);
- *Detection Phase*: the sum of interfering signal and noise is provided to the detector and the calculus of the Probability of Detection (PD) is carried out on the basis of the threshold obtained in the previous step applied to the Probability Density Function (PDF) of the signal x = j + n;

The detecting process can be divided into two cascaded stages:

- *Pre-Processing Stage*: here the input base-band signal is split into Frequency CHannels (FCH) frequency bins through using a polyphase filter which performs a coherent integration, then, for each bin, the power samples are non-coherently integrated over the time by a factor Sub-Integration Length (SIL) up to achieve and store Number of Bursts (NOB) energy samples. This stage works in the same way for both the calibration and detection phases;
- *Calibration/Detection Stage*: (i) during the calibration phase the inverse Complementary Cumulative Distribution Function (CCDF) of the stored power samples is used together to the PFA to extract the threshold to be used in the detection phase; (ii) during the detection phase the CCDF is used, together to the previously computed threshold, to extract the PD which is compared with a threshold to achieve a hard decision for each frequency bin;

The PD threshold has an initial value which can be programmed by the user. It is dynamically decreased by the algorithm until it gets one of the following two stop conditions:

- At least one of the FCH channels overcomes the threshold;
- The threshold reaches its minimum value (here set to PFA + 0.1);

The total number LEN of samples used by the detection process can be then defined as: LEN = NOB · SIL · FCH; it corresponds to an observation time T_{obs} = LEN/Fs where Fs is the adopted sampling frequency. Through simulating several scenarios, the following optimal parameters have been chosen: FCH = 32, SIL = 1024, NOB = 128, LEN = 2^{22}.

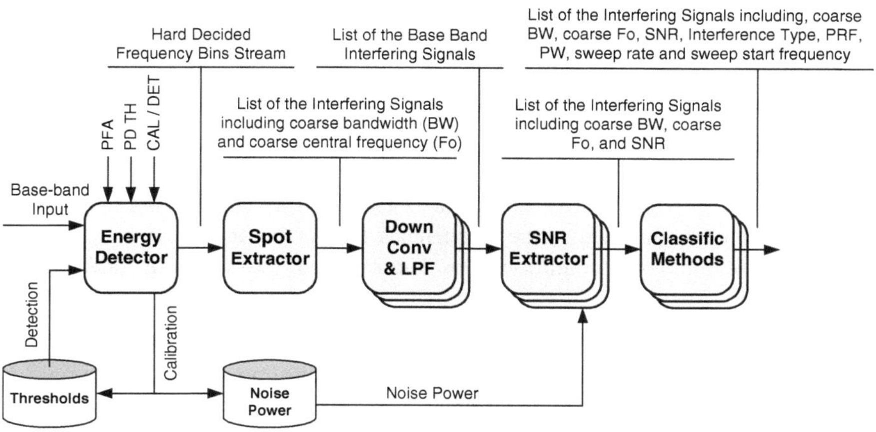

Fig. 16.6 Detection and classification architecture

Figure 16.6 shows the overall detection/classification architecture.

The input complex base-band signal is processed through several cascaded algorithms in order to detect and classify it on the basis of some parameters like bandwidth, central frequency, SNR, and so on.

After the energy detector the signal appears as a bit stream of FCH frequency bins which are arranged in spots when they are adjacent. The so obtained spots correspond to several, spectrally separated, interfering signals which must be processed separately from this point on after filtering them on the basis of:

- *Coarse Bandwidth* (*BW*): BW = M · Fs/FCH, where M is the number of frequency bins in the spot;
- *Coarse Central Frequency* (*CF*): CF = "centre of gravity of the spot";
  At this point it is possible to calculate the power of each signal and its:
- *Signal* (*Interference*) *to Noise Ratio* (*SNR*): The next algorithms (pulsed signal classifier and chirp estimator) add further details to the record of each signal, useful to identify it.

The algorithm adopted to recognise pulsed signals is based on the TOA (Time Of Arrival) of the rising and falling edges of the pulses and it is a modified version of the algorithm described in Mardia (1989). The modifications include the extraction of the Pulse Width (PW) of the signal beyond the Pulse Repetition Frequency (PRF) and some implementation details. The processing requires at least 50 pulses implying a constraint on the minimum detectable PRF: $PRF_{min} = 50/T_{obs}$.

The approach adopted to estimate the frequency and the frequency rate of a chirp signal is based on the FFT. The T_{obs} time interval is divided into 16 sub slots at the beginning of which a FFT over a sub-interval, $2^{16} = 65536$ points long, is performed. The frequencies $\{f_1 \ldots f_{16}\}$ corresponding to the maximum modulus of each FFT are extracted and used to classify the signal according to the rules:

- If the values $\{f_1 \ldots f_{16}\}$ are monotonically increasing or decreasing the signal is classified as sweep with start frequency equal to f_1 (suitably modified to take into account the frequency rate from the first sample of the observed interval) and with frequency rate equal to the mean value of the differences $f_k - f_{k-1}$, $k = 2 \ldots 16$;
- If the variance of the values $\{f_1 \ldots f_{16}\}$ is less than a given threshold (here established in 5 FFT bins) the signal is classified as CW having a frequency equal to the mean value of the frequencies f_k, $k = 1 \ldots 16$;
- If the previous two conditions are not satisfied the signal is classified as unknown;

The minimum and maximum detectable frequency rates, FR_{min} and FR_{max}, are respectively $FR_{min} = FFT_{resol}/FFT_{offset}$ (FFT_{offset} is the time distance between two successive FFTs and FFT_{resol} is the FFT resolution) and $FR_{max} = Fs/T_{obs}$ (sweep spread over all the frequency bins).

Figure 16.7 reports the performances achieved with the detection and classification strategy described above. Although the algorithm works in presence of multiple interferences, for sake of clarity, three different single interferences are analysed: (i) CW; (ii) pulsed (with two different PRF and PW values) and (iii) sweeping (with two frequency rate values). An isolated marker is put on the diagram as soon as the interferer is detected (the value in the vertical axis is don't care), after that each line starts when the algorithms succeed in classifying it.

The SNR is here computed as ratio between the Interferer power and the Noise power over the total observed bandwidth of 51.2 MHz. The observed time interval

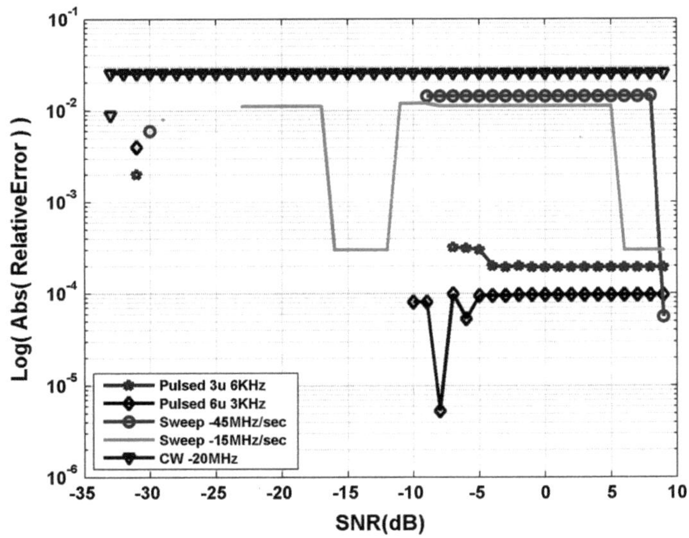

Fig. 16.7 Detection and classification performances

is here $T_{obs} = LEN/Fs = 2^{22}/51.2 \times 10^6 = 81.92$ ms. During this interval the sweep spans a bandwidth of $FR \cdot T_{obs}$.

The values assumed by each line in Fig. 16.7 are the accuracy achieved in recovery the central frequency for CW, the PRF for pulsed signals and the frequency rate for sweeps. Better accuracies are associated with the pulsed signals. The accuracies are however under the 2.5 % of the value to be estimated.

These Detection and Classification techniques have been extensively tested in a dedicated field campaign (Morgand and Ferreol 2008).

X, Y, Z Location of the Sources

AOA TDOA FDOA Techniques

Three main techniques are possible to locate radio electrical sources such as TDOA, FDOA and AOA.

The TDOA technique allows to estimate the delay of arrival between two stations. Three stations and three estimations are useful to estimate the X, Y, Z position. The main drawbacks of the TDOA technique are:

- The needs to transmit a temporal part of the signal received by the IOS to the ICC in order to process the time delay computation. This data transfer will generate a high data rate in the air medium between the two stations.
- The inefficiency in case of CW or narrow band signals of
  - the extraction of the time of arrival on a station.
  - to separate the multi-paths (direct and reflected).

Thus the TDOA technique has been rejected according to the inefficiency to process the CW or narrowband signals.

The FDOA technique based on the Doppler effect needs a mobile system to measure a frequency shift. The two following conditions are required:

- A source or a receiver on a vehicle with a constant speed
- The speed of the mobile measurement station must be high enough to obtain concrete measurements

Moreover, it is always difficult to maintain a constant speed on a rectilinear moving on the site.

Thus the FDOA technique has been rejected.

The AoA technique needs an array of elementary antennas in order to obtain on each station the DoA of the plane wave generated by the sources. Two stations set with a DoA measurement system will allow to find the position by the crossing of the two estimated directions.

This technique is suitable to the interference localization and robust against:

- The nature of the jammer such as CW, narrow or wide bandwidth
- multi path situations
- multi simultaneous source

The data useful to transfer between stations is limited to the angular values of each source. Thus a low rate link is sufficient between stations.

AOA Techniques

High resolution techniques have been employed in the MAGIC Project for the determination of Direction of Arrival of the sources. These techniques give better results in term of accuracy than for instance standard method like Capon Filter. In addition to that, in absence of coherent multi-path, high resolution techniques are unbiased.

The most famous algorithm is MUSIC. The method developed in this project is based on this algorithm and have been extended to coherent multi-paths.

MUSIC Algorithm

This algorithm is based on the eigen properties of the covariance matrix $\mathbf{R}_x = E\left[\mathbf{x}(t)\mathbf{x}(t)^H\right]$ of the observation vector $\mathbf{x}(t)$ where $E[.]$ the mathematical mean and ^{H} the transpose and conjugate operator. The expression of $\mathbf{R}_x$ is

$$\mathbf{R}_x = \mathbf{A}\ \mathbf{R}_s\ \mathbf{A}^H + \sigma^2\ \mathbf{I}_N \quad \text{where } \mathbf{R}_s = E[\mathbf{s}(t)\ \mathbf{s}(t)^H] \quad \text{and} [\mathbf{n}(t)\ \mathbf{n}(t)^H] = \sigma^2\ \mathbf{I}_N \tag{16.1}$$

And where

$$\mathbf{s}(t) = \begin{bmatrix} s_1(t-\tau_1) \\ \vdots \\ s_M(t-\tau_M) \\ b_1(t) \\ \vdots \\ b_P(t) \end{bmatrix} \quad \text{and } \mathbf{A} = \left[\mathbf{a}(\theta_{1d}) \quad \cdots \quad \mathbf{a}(\theta_{Md}) \quad \mathbf{a}(\theta_{1r}) \quad \cdots \quad \mathbf{a}(\theta_{Pr})\right] \tag{16.2}$$

$E[.]$ the mathematical mean and $\mathbf{x}(t)^H$ the transpose and conjugate of $\mathbf{x}(t)$.
The Fig 16.8 scheme shows that:

- The noise projector Π_b is deduced from an estimate $R_x(T_0)$ of the covariance matrix.
- The MUSIC criterion $c(\theta)$ is calculated from the steering vector $a(\theta)$ stored on a calibration file (Fig. 16.8).

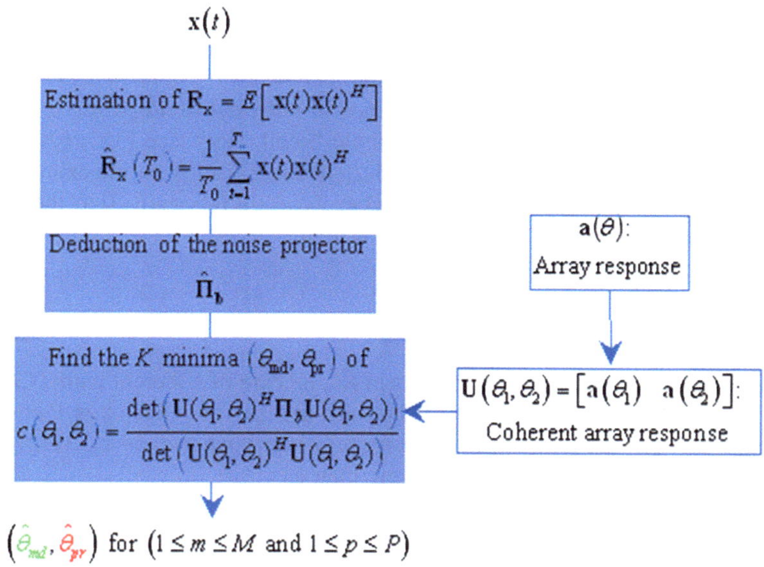

Fig. 16.8 MUSIC algorithms steps

The MUSIC criterion for two paths of directions $\theta_{11} = 100°$ and $\theta_{12} = 200°$ is presented in the following Figure. The array of $N = 5$ sensors is circular of radius 0.5λ. In this Fig. 16.9 the cases of incoherent and coherent multi-paths are simulated. The MUSIC criterion $J_{MUSIC}(\theta)$ is normalized between 0 and 1 and must be null in the directions $\theta = \theta_{11} = 100°$ and $\theta = \theta_{12} = 200°$.

According to Fig. 16.9, the K = 2 minima of MUSIC criterion $J_{Music}(\theta)$ are null when the multi-paths are not coherent and over 0.4 in the coherent case. Contrary to the not coherent case, the estimation of $\theta_{11} = 100°$ and $\theta_{12} = 200°$ is biased in the coherent case. In addition in the coherent case the dynamic between the secondary

Fig. 16.9 MUSIC criterions in case of incoherent or coherent multi-paths. The radius of the circular array with $N = 5$ sensors is 0.5λ

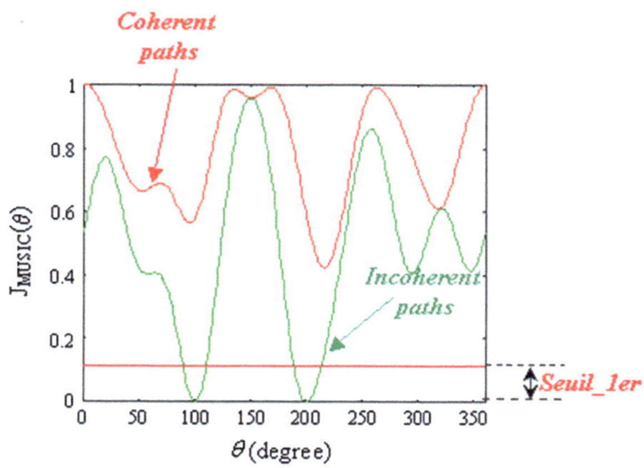

minima and the minima associated to the direction of paths, is less important than in the incoherent case. This last remark shows that in case of coherent multi-paths there is an important problem of ambiguity.

In that way, the detection of the presence of coherent multi-paths can be done with the used of a pseudo-spectrum threshold "seuil_1ier" (see Fig. 16.9).

When the minimum $J_{Music}(\theta m)$ verify $J_{Music}(\theta m) <$ seuil_1ier, the azimuth θm is the direction of a not coherent path and when $J_{Music}(\theta m) >$ seuil_1ier, the azimuth θm cannot be associated to a direction of arrival.

SC Algorithm

The SC (Self-Calibration) algorithm processes the case of coherent paths. Coherent paths are presents when $r_{mp} \rightarrow 0$. When the number of reflector is P = 1, it is straight to see that this condition of coherence is verify when $\tau_m \approx \tau_{mp}$ (Fig. 16.10).

The observation vector x(t) can be rewritten as

$$\begin{aligned} \mathbf{x}(t) &= \left(\mathbf{a}(\theta_{md}) + \rho_{mp}\mathbf{a}(\theta_{pr})\right)s_m(t-\tau_m) \\ &\quad + \sum_{m' \neq m} \mathbf{a}(\theta_{m'd})\, s_{m'}(t-\tau_{m'}) + \sum_{i=1}^{P} \mathbf{a}(\theta_{ir})\, u_i(t) + \mathbf{n}(t) \end{aligned} \tag{16.3}$$

The SC algorithm is:

- based on the following coherent steering vector modeling
- adapted MUSIC to the coherent steering vector modeling of (16.4).

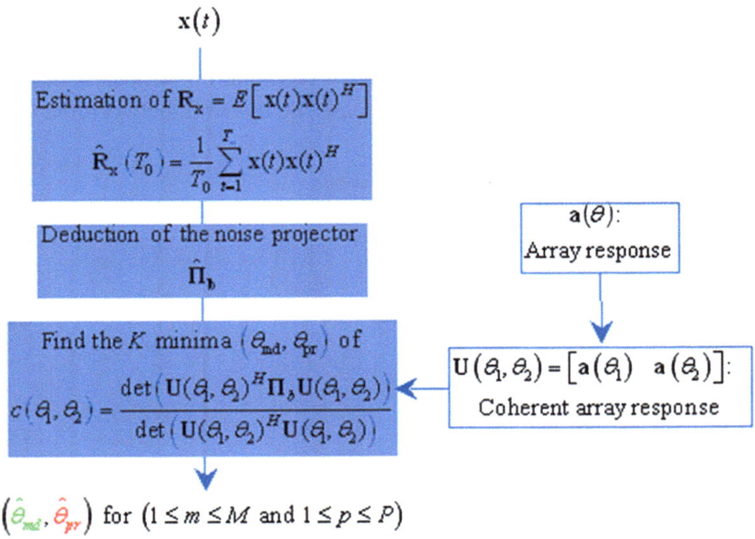

Fig. 16.10 SC algorithms steps

$$\mathbf{a}(\boldsymbol{\theta}, \boldsymbol{\rho}, P) = \sum_{p=1}^{P} \rho_p \, \mathbf{a}(\theta_p) = \mathbf{U}(\boldsymbol{\theta}) \, \boldsymbol{\rho}$$
$$\text{with } \mathbf{U}(\boldsymbol{\theta}) = [\mathbf{a}(\theta_1) \cdots \mathbf{a}(\theta_P)] \quad \text{and } \boldsymbol{\rho} = [\rho_1 \cdots \rho_P]^T \tag{16.4}$$

The calibration table associated to $U(\theta_1,\theta_2)$ is previously stored in a calibration file for $0 \le (\theta_1,\theta_2) \le 360°$ from the initial calibration file associated to a (θ) for $0 \le \theta \le 360°$.

Elementary AOA Module

In practice the MUSIC and SC algorithm are both used by applying the

- SC algorithm in presence of coherent multi-paths
- MUSIC algorithm without coherent paths

The choice of the AOA algorithm is done with the threshold "*seuil_1ier*" on the MUSIC pseudo-spectrum as presented in Fig. 16.9

The following Figure shows the Elementary AOA modules apply on the observation signal $\mathbf{x}(t)$ for $t_0 \le t \le t_0 + T_0$.

According to Fig. 16.11, the i-th application of the elementary AOA module gives Mi angle of arrivals $\theta_{m,A}$ (t_i) $(1 \le m \le M)$. As the directions $\theta_{m,A}$ (t_i) and $\theta_{m,A}$ (t_j) are not necessary associated to the same transmitter and M_i that can be different to M_j the purpose of the AOA synthesis is the determination of the

- main directions $\bar{\theta}_{m,A}$ for $1 \le m \le M$.
- standard deviation $\sigma_{\theta_{m},A}$ associated to $\bar{\theta}_{m,A}$.
- number M of main directions.

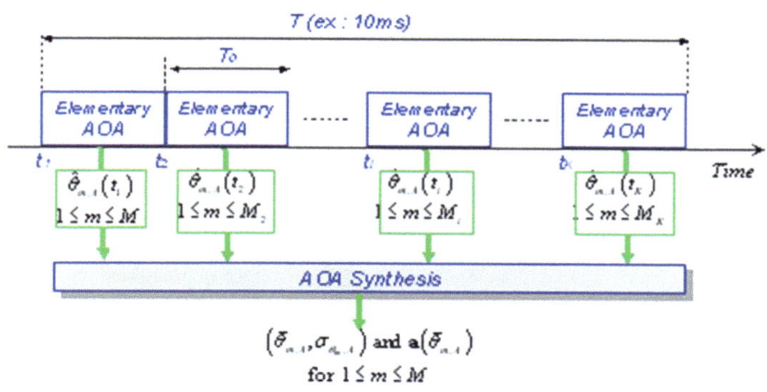

Fig. 16.11 Application of the elementary AOA estimation *K* times inside the time duration *T*

AOA-AOA Location

The AOA-AOA technique is divided in two steps

- The angular estimation θ_{mA}, θ_{mB} and θ_{mC} of each source ($1 \le m \le M$) in stations **A**, **B** and **C**.
- The localization process estimates the source location $\mathbf{E}_m$. The method evaluates the intersection $\mathbf{M} = \mathbf{E}_m$ between the three lines of direction θ_{mA}, θ_{mB} and θ_{mC}.

$$\begin{aligned} \theta_{mA} &= angle(\mathbf{AM}, \mathbf{AB}) \\ \theta_{mB} &= angle(\mathbf{BM}, \mathbf{AB}) \\ \theta_{mC} &= angle(\mathbf{CM}, \mathbf{AB}) \end{aligned} \tag{16.5}$$

The AOA-AOA module with three stations is illustrated in Fig. 16.12. The location block module is applied after an AOA synthesis on each AOA station. This technique doesn't not require a synchronous acquisition on each station. This is the reason why the AOA-AOA location is done with only one AOA station in MAGIC: The signal at location points A, B and C is stored when the MAGIC AOA station is located at position A, B and C.

AOA-AOA Module with Stations A, B and C

The AOA-AOA module is applied on a signal time duration T with N sensors and is described in Fig. 16.13.

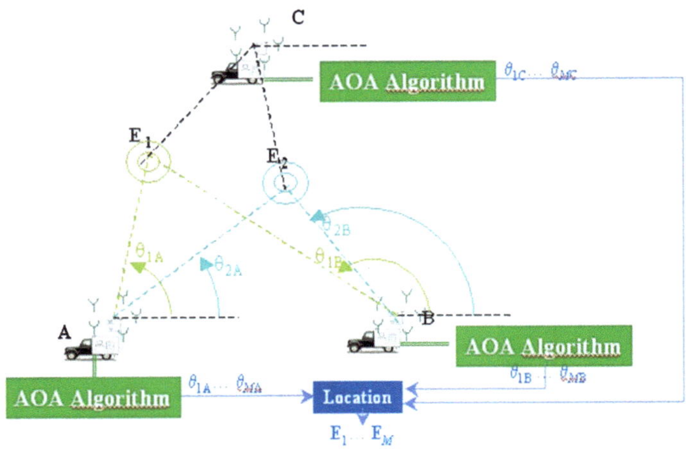

Fig. 16.12 AOA-AOA techniques in presence of M transmitters E_m ($1 \le m \le M$) with stations in A, B and C

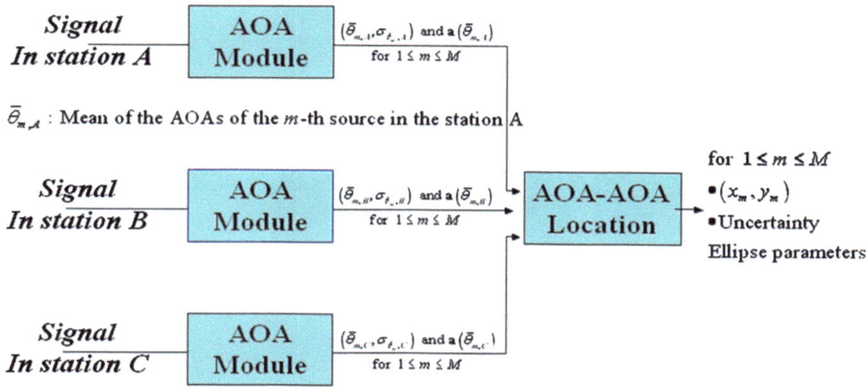

Fig. 16.13 AOA-AOA module

Location of DME

Localization of several DME have been done at about 100 km in the south west of Paris. The trial presented below took place at *Châteaudun*. The equipment located is a DME that transmit a double pulse at 1,195 MHz (Table 16.1).

The particularity of this trial is the position of each station of measurement. The graphic below shows that they spread in 5 km of distance and they formed more a little curve in south of DME than a classical triangle (Fig. 16.14).

This trial allows testing the precision of the MAGIC system in a new configuration. Moreover, distances between the interferer and the measurement station are much bigger than those in previous test because they are at least 3 km.

In spite of a configuration not optimal to get the best localization, the average error of angle of arrival is a little bite more than one degree and the error of

Table 16.1 Location of DME example

	Measurement point			DME	
	Station 1	Station 2	Station 3	Real position	Estimated position
X (m)	0	−2503.74	−5048.42	−2835.29	−2949.85
Y (m)	0	−1777.55	−1910.57	1,370	1473.35
Latitude (DMS)	48°2′ 36.59″	48°1′39.11″	48°1′34.81″	48°03′20.9″	
Longitude (DMS)	1°24′ 48.58″	1°22′47.51″	1°20′44.46″	1°22′31.4″	
Real angle of arrival (°)	154.21	96.01	55.99		
Estimated angle of arrival (°)	153.16	98.28	57.88		
Error of angle (°)	1.05	2.27	1.89		

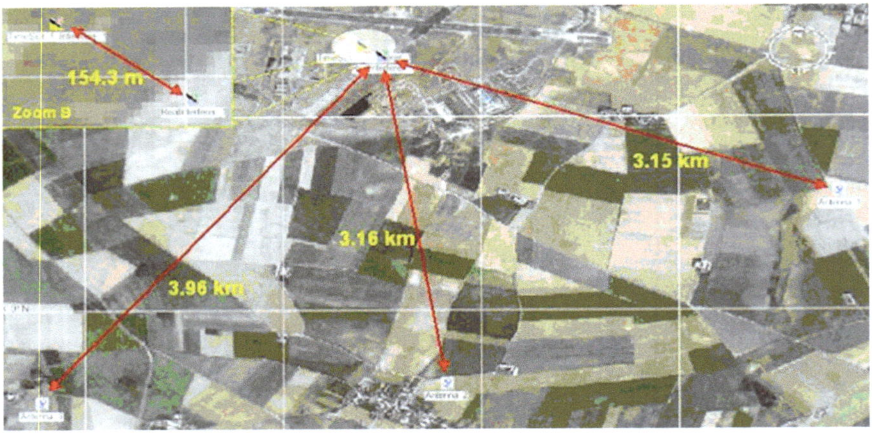

Fig. 16.14 Localization of DME at Châteaudun with quick plan

positioning between the real position and the estimated position is around 150 meters (zoom B), which represent an average of 5 % of error.

Isolation Synthesis Results on All Interferers

Table 16.2 shows the location error in meters and percentage on the distance D between the source and the interferers

Table 16.2 Distance errors on the interferer positions

	Error of localization (m)	Minimum distance between station and interferer (km)	Maximum distance between station and interferer (km)
Aerodrome	5.65	0.42	0.57
		Error: 1.3 %	Error: 1 %
Noues's lake	35 (interferer)	0.4	0.69
		Error: 8.7 %	Error: 5 %
	16 (reflector)	0.19	0.37
		Error: 8.4 %	Error: 4.3 %
Le Verdon's lake	38 (1st interferer)	0.6	1.52
		Error: 6.3 %	Error: 2.5 %
	23 (2nd interferer)	0.53	1.79
		Error: 4.3 %	Error: 1.2 %
Dammartin-en-Goële	123.75	2.13	2.27
		Error: 5.8 %	Error: 5.4 %
Châteaudun	154.3	3.15	3.96
		Error: 4.9 %	Error: 3.9 %

In average the error percentage is 5 % on the distance D and a maximum of 8.4 % is got on the lowest distance.

Conclusion on Isolation Techniques

The isolation technique implemented in the test-bed is the AOA-AOA: This technique needs multiple AOA stations. The study shows the need of use more than two stations in **A** and **B** in order to avoid false location in multi-transmitters context and improve the accuracy of the location.

- The isolation techniques are based on AOA algorithms that take into account.
- The coherent path case with self-calibration algorithm.
- The multi-transmitters case with MUSIC.
- The application of Self-Calibration algorithm or MUSIC algorithm is adapted to the signal context.
- The isolation studies realized in the MAGIC project shows that some new advanced research need to be done on the same way. These new proposed isolation works are:
- The use of new techniques for coherent paths. The purpose of these new techniques is to improve the power calculation of the self-calibration algorithm used in the test bed. These techniques are based on an interpolation of the array calibration in linear antenna in order to apply low cost techniques such as spatial smoothing.
- The use of location techniques that improve the maximum number of transmitters to be located through a global AOA-algorithm that will process jointly all the signals of all the stations.
- The maximum number of transmitters to be located can also be improved by the used of high order statistics or selective algorithm that use for instance the cyclo-stationarity of the input signals.

Mitigation of the Jamming Interferences

Interference signals in satellite based systems are often responsible of performances degradation. Several mitigation techniques can be adopted to reduce these disturbs avoiding, at the same time, to alter the wanted signal as much as possible. Among the methods described in literature the trade off carried out in the framework of our studies pinpointed two techniques: (i) the frequency domain mitigation (FDIM) approach proposed in Chen et al. (2002) and (ii) the space domain mitigation (SDIM) method given in Steyskal (1982). They are separately described below.

The FDIM technique consists in taking the Fourier transform of the interfered signal and in applying an adaptive mask in order to notch out the disturbing frequency bins. The processing applies then an inverse Fourier transform to bring back the signal to the time domain.

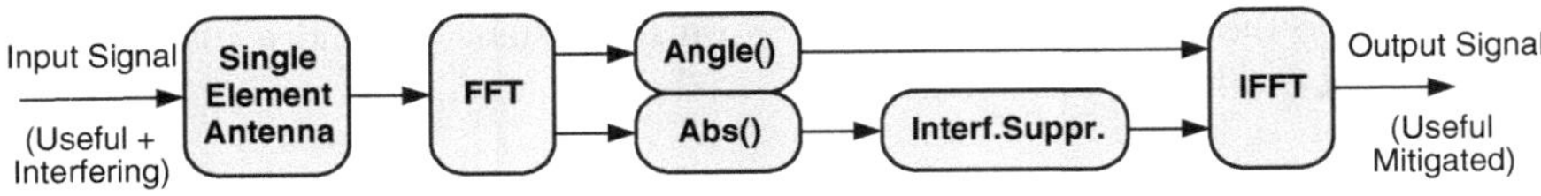

Fig. 16.15 FDIM model

Since the energy of the interfering signal is usually concentrated over a narrow band, compared to the band of the navigation spread spectrum signal, its magnitude in the frequency domain presents some peaks. The position and magnitude of these peaks can be detected via an envelope detector in order to determine a threshold and to apply a magnitude reduction according to some criterion. The threshold is often close to the magnitude response of the spread signal and it is assumed that the frequency bins exceeding such threshold contain interference. After reducing the magnitude values of these cells, the most energy of the interferer is cut out. Figure 16.15 shows the model of the FDIM technique.

There are three methods for processing the bins that exceed the fixed threshold and their neighbours:

- *Excision*: a drastic excision of the bins overcoming the threshold;
- *Hold to threshold*: only the part exceeding the threshold is considered interfering: the affected bins are held to threshold value;
- *Linear decrease*: the bins exceeding the threshold are multiplied by a coefficient;

The three suppression methods have different results and performance which also depends on the type of interfering signal.

The algorithm works in two phases: (i) calibration phase, during which a minimum threshold TH_{min} is determined as the maximum value of the noise PSD and (ii) adaptive phase, when the true threshold TH is computed as: $TH = TH_{min} + \eta \cdot TH_{adp}$ being TH_{adp} the adaptive contribution, achieved according to the mean value of the bins energy belonging to a given number of recent FFTs, and η a coefficient to be optimized by simulation whose values are typically between 0 and 1.

The FDIM technique does not require any knowledge of the interference type therefore can be applied to several interfering signals. As it was expected the algorithm best performs in the case of very narrow band interference, like CW, because the affected bins are totally masked and the effect on the information signal is almost negligible.

The performances of this technique have been analysed checking their behaviour in presence of several types of interfering signals. Figure 16.16 shows one of these cases where, in the considered frequency band of 51.2 MHz, one pulsed signal (PRF = 5 kHz, PW = 10 μs, SNR = 10 dB) and one sweeping signal (FR = 20 MHz/s, SNR = 5 dB) were observed for 20.48 ms time interval.

As it is possible to note the pulsed interference (on the left of Fig. 16.16) has been reduced by about 27 dB whereas the sweeping interference has been lowered by about 10 dB.

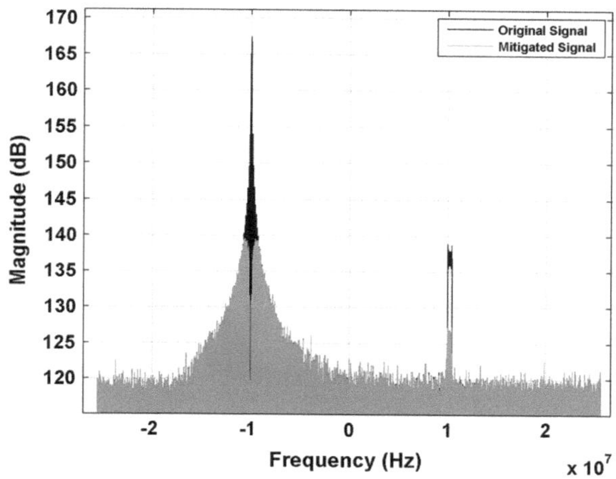

Fig. 16.16 FDIM mitigation spectrum

The SDIM technique exploits the presence of a phased array antenna to perform both Beam Forming and Null Steering. The Beam Forming is used to direct the main lobe of the antenna pattern toward the direction of the wanted signal whereas the Null Steering places nulls in the directions of the interfering signals. The Beam Forming is achieving by properly delaying the phases of the array elements affecting only the imaginary part of the weighting coefficients. Depending on the number of elements present in the array, some degrees of freedom are still available after the Beam Forming and they can be exploited for Null Steering. The Null Steering pattern $p_c(\mathbf{v})$ in a generic direction $\mathbf{v}$ is obtained by linearly combining multiple patterns, each one determined in order to null a given direction $\mathbf{v}_m$. Finally the desired antenna pattern is achieved by adapting $p_c(\mathbf{v})$ to the quiescent pattern $p_q(\mathbf{v})$ (i.e. the pattern obtained after the Beam Forming procedure), trying to not alter it in the direction of the wanted signal, by minimizing the mean squared error between them. A solution to this minimization problem is proposed in Steyskal (1982). Figure 16.17 shows the model of the SDIM technique.

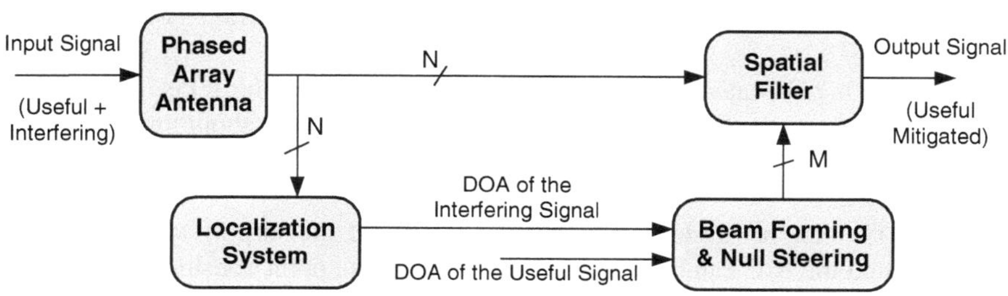

Fig. 16.17 SDIM model

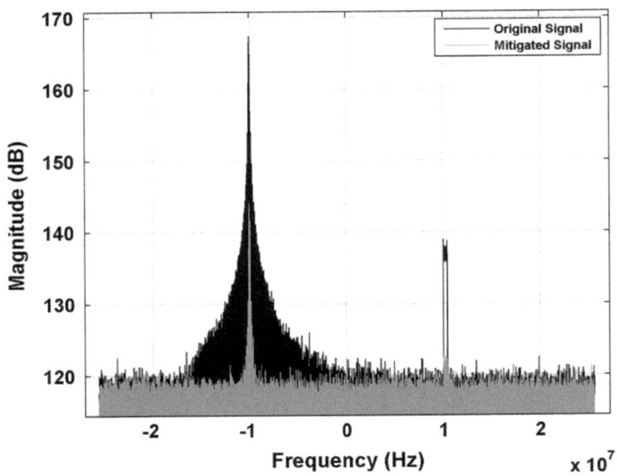

Fig. 16.18 SDIM mitigation spectrum

The position of a null in the antenna pattern virtually sets to zero the power received from that precise direction so, if the Direction Of Arrival (DOA) estimation of the interfering signals is precise enough, the attenuation of the interference may be very strong.

Also the performances of this technique have been analysed checking their behaviour in presence of several types of interfering signals. Figure 16.18 shows the response of the SDIM approach to the same scenario already analysed for the FDIM technique.

Both these techniques have been tested during a dedicated field campaign (Morgand and Ferreol 2008).

Possible GIMS Deployment

This phase is site-dependent because it involves activities strongly influenced by the selected site, due to the need to coordinate the activities with the local authorities and to incorporate the GIMS elements in the available infrastructure.

An example of the deployment in the Frankfurt Airport is shown in Fig. 16.19.

The position of the 3 IOSs has been chosen so that the coverage is wide as much as possible.

The ICC is collocated in an intermediate place between the IOSs.

There aren't particular outcomes from the MAGIC Project about the deployment phase but some practical considerations about the localization and the detection can be underlined.

The IOSs can be strategically positioned around the airport to have a good coverage and the ICC can be located in an intermediate point considering that it has to communicate with all the three IOSs. The deployment phase also has to integrate

Fig. 16.19 Frankfurt Airport: GIMS deployment

the GIMS architecture in the LAA exploiting the existing and re-usable infrastructure:

- The IOS may be collocated near the DAT of the LAA to share some resources;
- The ICC can be collocated in the control center supposed already existing.

Thus it can be deduced that the position of the GIMS units (IOSs and ICC) depends on the site characteristics and on the available infrastructure.

During this phase the management activity is very important to organize several actions:

- Site survey and constraints identification;
- Licenses requests;
- Coordination with the surveillance authorities;
- Acquisition of the necessary authorizations and licenses on site.

Conclusion on the GIMS System

According to the MAGIC project outcomes concerning the studied detection, classification, mitigation and isolation techniques, a target GIMS architecture has been defined and proposed.

This architecture mainly deals with the definition of the IOS, ICC and communication links between them and the external world.

The achieved IOS structure is based on three sections:

- The IOS Core which receive the interfering signals and processes them up to extract the synthetic parameters concerning the detection, classification and isolation outputs; it also include the communication interfaces toward the ICC;
- The Auxiliary Radio Interface which replace the cable based interfaces toward the ICC in case of necessity;
- The Synchronization Interface only necessary in case of TDOA-AOA approach;

The structure obtained for the ICC is based on the following sections:

- The ICC core which process the synthetic parameters coming from the IOS and manage the several peripherals required to communicate to and from external equipments (e.g. IOS, GR, DAT, external world);
- The GIMS Radio Interface necessary to communicate toward the GR and the DAT;
- The Dedicated Interfaces Section which includes all the required interfaces toward external equipments;
- The Auxiliary Radio Interface to be used to connect with the IOSs in case of unavailability of the cable interconnection;

The interconnections between the GIMS elements have been defined in order to reduce the possible interferences coming from the external environment because the first task of an interference detection system is to work in presence of interfering signals. This suggests using interconnections which are robust against the interferences:

- Cable Connections (this is the more robust approach being impossible to disturb it by definition);
- Spread Spectrum Radio Connections (the techniques based on the spread spectrum signals transmit typically a power under the noise floor spread in a relatively large bandwidth so it is robust against external interferences and doesn't create further disturbs);

The first ones are preferable where they can be used (e.g. airports where it is easy to deploy a cable network) while the second approach can be exploited where a cable network cannot be deployed (e.g. harbors).

Finally a plan on the GIMS development and deployment has been defined on the basis of the GIMS architecture briefly described above taking as reference its installation in an airport. This plan carries out an analysis of the:

- Human resources: personnel required to face the several phases necessary to carry through the design, the manufacturing, the testing, the installation of the GIMS;
- Equipment and materials: list of each component of the GIMS including the materials necessary to deploy them on a site (e.g. an airport);
- Tests to be undertaken: list of the tests necessary to check the correctness of the several equipment involved in each phase (design, manufacturing, etc.…);
- Overall GIMS Cost: summary of the costs associated to each aspect of the GIMS (e.g. human resources, materials, etc.…);
- Time Schedule: a typical example concerning the overall flow necessary to design, manufacture, test and install the GIMS on an airport has been provided;

These previous studies carried out cover all the aspects required to achieve a target GIMS from the definition of their elements up to the installation on the target site.

References

Chen X, Guo W, Zheng Y (2002) Frequency domain interference suppression in a DSSS system. In: IEEE 2002 international conference on communications, circuits and systems and west sino expositions, 2002, pp 247–251

Gardner WA (1991) Exploitation of spectral redundancy in cyclostationary signals. IEEE Signal Process Mag 8:14–36

Mardia HK (1989) New techniques for deinterleaving of repetitive sequences. IEE Proc 136 (4):149–154

Morgand P, Ferreol A, Rossini E (2008) Detection, isolation and mitigation of Galileo interferers. In: ENC-GNSS 2008 conference, Toulouse, France, 22–25 April 2008

Nikias CL, Mendel JM (1993) Signal processing with higher-order spectra. IEEE Signal Process Mag 10:10–37

Sattar F, Salomonsson G (2000) On detection using filter banks and higher order statistics. IEEE Trans Aerosp Electron Syst 36(4):1179–1189

Steyskal H (1982) Synthesis of antenna patterns with prescribed nulls. IEEE Trans Antenna Propag 30(2):273–279

Urkowitz H (1967) Energy detection of unknown deterministic signals. Proc IEEE 55(4):523–531

Chapter 17
Conclusions and Outlook

Jari Nurmi, Elena Simona Lohan, Stephan Sand and Heikki Hurskainen

Abstract Wireless positioning has become an important topic nowadays and it is likely to receive more and more attention in the near future, too. Already we have four complete or emerging satellite navigation systems and many local navigation satellite systems, operating either in a stand-alone mode or offering augmentation capabilities to the existing GPS system.

Wireless positioning has become an important topic nowadays and it is likely to receive more and more attention in the near future, too. Already we have four complete or emerging satellite navigation systems and many local navigation satellite systems, operating either in a stand-alone mode or offering augmentation capabilities to the existing GPS system.

While global-scale outdoor positioning technology is already mature, through the GNSS capabilities, indoor positioning has still many unsolved challenges.

Our book mainly addressed the outdoor positioning through the European Galileo system, covering all stages from the transmitter side and waveform modulations till the receiver design and PVT solution. Also hybridization aspects in cellular devices have been addressed in Chap. 8, as the building bridge towards seamless indoor to outdoor positioning.

A significant part of our book has been dedicated to Galileo receivers design, with both hardware and software paradigms, and by pointing out and tackling all different challenges met at various receiver blocks, e.g., front-end, acquisition, tracking, and navigation.

Less directly related to Galileo, but talking about another European initiative on GNSS is our Chap. 14, dealing with EGNOS, as a pre-cursor of Galileo. The Safety of Life service, which currently has been discontinued in Galileo for reasons of unsolved challenges regarding the integrity warranty, has also been discussed in

J. Nurmi · E.S. Lohan (✉) · H. Hurskainen
Department of Electronics and Communications Systems,
Tampere University of Technology, Tampere, Finland
e-mail: elena-simona.lohan@tut.fi

S. Sand
German Aerospace Center DLR, Oberpfaffenhofen, Germany

J. Nurmi et al. (eds.), *GALILEO Positioning Technology*,
Signals and Communication Technology 182, DOI 10.1007/978-94-007-1830-2_17

Chap. 15. SoL is currently re-profiled into an Integrity Monitoring service and research dedicated to receiver-level integrity monitoring is becoming more and more important.

Galileo system design is a huge topic and Galileo concept has been now around for more than 15 years. While our book addresses in detail the basic bricks of the system and several advances, such as advanced multipath mitigation and hybridization, there are still many open Galileo-related challenges.

For example, some of the main challenges in the future GNSS systems, and in particular in Galileo, remain: the interference mitigation, especially when dealing with intentional attacks, such as spoofing and meaconing, and the privacy and security of the location information. Indeed, in many existing and emerging Location Based Services, such as location-based billing, fleet tracking and transportation applications, achieving secure positioning solutions is of utmost importance. Privacy and security aspects in GNSS are expected to gain attention in the years to come.

Also, the usage of GNSS to remote sensing, weather forecasting and out-of-Earth applications, such as satellite tracking in LEO systems, or even Moon navigation are some other open directions in GNSS community.

People's daily activity and behavior are likely to be heavily influenced in the future by the availability, accuracy, integrity, privacy, and security aspects of the positioning solutions they will be offered, and it is a certain fact that GNSS will still play a significant role in all future wireless positioning systems. Also cognitive positioning concept is emerging and it is in a way complementary to the wider-spread cognitive communication concept. Cognitive positioning paradigm refers to an adaptive positioning system which is aware of the environment (and possibly of the target Location Based Service and of the users' needs) and uses this information for an enhanced localization solution. Cognitive positioning also refers to the capacity of selecting and optimally combining the relevant signals for positioning, in a world where more and more signals which can be used for positioning will be available (such as cellular signals, WPAN and WLAN signals, and basically any signal of opportunity).

Erratum to: GIMS: Galileo Interference Management System

Philippe Morgand, Eugenio Rossini and Anne Ferreol

Erratum to:

Chapter 16 in: J. Nurmi et al. (eds.), *GALILEO Positioning Technology*, DOI 10.1007/978-94-007-1830-2_16.

In the published version of Chapter 16, the author Eugenio Rossini, from the Space Engineering Spa, Rome, Italy, was omitted.

The online version of the original chapter can be found under
DOI 10.1007/978-94-007-1830-2_16

P. Morgand (✉) · A. Ferreol
Thales Communications, Gennevilliers, France
e-mail: philippe.morgand@fr.thalesgroup.com

E. Rossini
Space Engineering Spa, Rome, Italy

J. Nurmi et al. (eds.), *GALILEO Positioning Technology*,
Signals and Communication Technology 182, DOI 10.1007/978-94-007-1830-2_18

MIX
Papier aus verantwortungsvollen Quellen
Paper from responsible sources
FSC® C105338

If you have any concerns about our products,
you can contact us on
ProductSafety@springernature.com

In case Publisher is established outside the EU,
the EU authorized representative is:
Springer Nature Customer Service Center GmbH
Europaplatz 3, 69115 Heidelberg, Germany

Printed by Libri Plureos GmbH
in Hamburg, Germany